Solar System Data

Body	Mass (kg)	Mean Radius (m)	Period (s)	Distance from the Sun (m)
Mercury	3.18×10^{23}	2.43×10^6	7.60×10^6	5.79×10^{10}
Venus	4.88×10^{24}	6.06×10^6	1.94×10^7	1.08×10^{11}
Earth	5.98×10^{24}	6.37×10^6	3.156×10^7	1.496×10^{11}
Mars	6.42×10^{23}	3.37×10^6	5.94×10^7	2.28×10^{11}
Jupiter	1.90×10^{27}	6.99×10^7	3.74×10^8	7.78×10^{11}
Saturn	5.68×10^{26}	5.85×10^7	9.35×10^8	1.43×10^{12}
Uranus	8.68×10^{25}	2.33×10^7	2.64×10^9	2.87×10^{12}
Neptune	1.03×10^{26}	2.21×10^7	5.22×10^9	4.50×10^{12}
Pluto	$\approx 1.4 \times 10^{22}$	$\approx 1.5 \times 10^6$	7.82×10^9	5.91×10^{12}
Moon	7.36×10^{22}	1.74×10^6	—	—
Sun	1.991×10^{30}	6.96×10^8	—	—

Physical Data Often Used[a]

Average Earth–Moon distance	3.84×10^8 m
Average Earth–Sun distance	1.496×10^{11} m
Average radius of the Earth	6.37×10^6 m
Density of air (0°C and 1 atm)	1.29 kg/m^3
Density of water (20°C and 1 atm)	1.00×10^3 kg/m^3
Free-fall acceleration	9.80 m/s^2
Mass of the Earth	5.98×10^{24} kg
Mass of the Moon	7.36×10^{22} kg
Mass of the Sun	1.99×10^{30} kg
Standard atmospheric pressure	1.013×10^5 Pa

[a] These are the values of the constants as used in the text.

Some Prefixes for Powers of Ten

Power	Prefix	Abbreviation	Power	Prefix	Abbreviation
10^{-24}	yocto	y	10^1	deka	da
10^{-21}	zepto	z	10^2	hecto	h
10^{-18}	atto	a	10^3	kilo	k
10^{-15}	femto	f	10^6	mega	M
10^{-12}	pico	p	10^9	giga	G
10^{-9}	nano	n	10^{12}	tera	T
10^{-6}	micro	μ	10^{15}	peta	P
10^{-3}	milli	m	10^{18}	exa	E
10^{-2}	centi	c	10^{21}	zetta	Z
10^{-1}	deci	d	10^{24}	yotta	Y

PHYSICS

For Scientists and Engineers

Fifth Edition

Raymond A. Serway

Robert J. Beichner
North Carolina State University

John W. Jewett, Jr., Contributing Author
California State Polytechnic University—Pomona

SAUNDERS COLLEGE PUBLISHING

A Division of Harcourt College Publishers

FORT WORTH   PHILADELPHIA   SAN DIEGO   NEW YORK   ORLANDO   AUSTIN   SAN ANTONIO   TORONTO   MONTREAL   LONDON   SYDNEY   TOKYO

Publisher: Emily Barrosse
Publisher: John Vondeling
Marketing Manager: Pauline Mula
Developmental Editor: Susan Dust Pashos
Project Editor: Frank Messina
Production Manager: Charlene Catlett Squibb
Manager of Art and Design: Carol Clarkson Bleistine
Text and Cover Designer: Ruth A. Hoover

Cover Image and Credit: Victoria Falls, Zimbabwe, at sunset. *(© Schafer & Hill/Tony Stone Images)*
Frontmatter Images and Credits: Title page: ripples on water. *(© Yagi Studio/Superstock, Inc.)*; water droplets on flower *(© Richard H. Johnson/FPG International Corp.)*; p. vii: Speed skaters *(Bill Bachman/Photo Researchers, Inc.)*; p. viii: Sky surfer *(Jump Run Productions/The Image Bank)* and strings in a piano *(Charles D. Winters)*; p. ix: jogger *(Jim Cummins/FPG International)*; p. x: Charged metallic sphere *(E. R. Degginger/H. Armstrong Roberts)* and resistors on circuit board *(Superstock)*; p. xi: Fiber optic cable strands *(Dennis O'Clair/Tony Stone Images)*; p. xiii: Long-jumper *(Chuck Muhlstock/FPG International)*; p.xiv: Penny-farthing bicycle race *(© Steve Lovegrove/Tasmanian Photo Library)*; p. xv: "Corkscrew" roller coaster *(Robin Smith/Tony Stone Images)*; p.xvi: Cyclists pedaling uphill *(David Madison/Tony Stone Images)*; p. xvii: Twin Falls on the island of Kauai, Hawaii *(Bruce Byers/FPG)*; p. xviii: Bowling ball striking a pin *(Ben Rose/The Image Bank)*; p. xix: Sprinters at staggered starting positions *(© Gerard Vandystadt/Photo Researchers, Inc.)*; p. xx: U.S. Air Force F-117A stealth fighter in flight *(Courtesy of U.S. Air Force/Langley Air Force Base)*; p. xxi: Cheering crowd *(Gregg Adams/Tony Stone Images)*; p. xxiv: Welder *(© The Telegraph Colour Library/FPG)*; p. xxv: Basketball player dunking ball *(Ron Chapple/FPG International)*; p. xxvi: Athlete throwing discus *(Bruce Ayres/Tony Stone Images)*; p. xxvii: Chum salmon *(Daniel J. Cox/Tony Stone Images)*; p. xxix: Bottle-nosed dolphin *(Stuart Westmorland/Tony Stone Images)*.

PHYSICS FOR SCIENTISTS AND ENGINEERS, Fifth Edition
      Volume 1 (Chapters 1–15) ISBN 0-03-026944-X
      Volume 2 (Chapters 16–22) ISBN 0-03-026946-6
      Volume 3 (Chapters 23–34) ISBN 0-03-026947-4
      Volume 4 (Chapters 35–39) ISBN 0-03-026948-2

Library of Congress Catalog Card Number: 99-61820

Address for domestic orders:
Saunders College Publishing, 6277 Sea Harbor Drive, Orlando, FL 32887-6777
1-800-782-4479

Address for international orders:
International Customer Service, Harcourt, Inc.
6277 Sea Harbor Drive, Orlando FL 32887-6777
(407) 345-3800
Fax (407) 345-4060
e-mail hbintl@harcourt.com

Address for editorial correspondence:
Saunders College Publishing, Public Ledger Building, Suite 1250
150 S. Independence Mall West, Philadelphia, PA 19106-3412

Web Site Address
http://www.harcourtcollege.com

Printed in the United States of America
12345678 032 10 987654

Harcourt College Publishers

Where Learning Comes to Life

TECHNOLOGY

Technology is changing the learning experience, by increasing the power of your textbook and other learning materials; by allowing you to access more information, more quickly; and by bringing a wider array of choices in your course and content information sources.

Harcourt College Publishers has developed the most comprehensive Web sites, e-books, and electronic learning materials on the market to help you use technology to achieve your goals.

PARTNERS IN LEARNING

Harcourt partners with other companies to make technology work for you and to supply the learning resources you want and need. More importantly, Harcourt and its partners provide avenues to help you reduce your research time of numerous information sources.

Harcourt College Publishers and its partners offer increased opportunities to enhance your learning resources and address your learning style. With quick access to chapter-specific Web sites and e-books . . . from interactive study materials to quizzing, testing, and career advice . . . Harcourt and its partners bring learning to life.

Harcourt's partnership with Digital:Convergence™ brings :CRQ™ technology and the :CueCat™ reader to you and allows Harcourt to provide you with a complete and dynamic list of resources designed to help you achieve your learning goals. Just swipe the cue to view a list of Harcourt's partners and Harcourt's print and electronic learning solutions.

C 62 00 00 00 00 00 25 20

http://www.harcourtcollege.com/partners/

Contents Overview

Table of Contents

part V   Light and Optics   1105

part VI   Modern Physics   1245

Preface

*I*n writing this fifth edition of *Physics for Scientists and Engineers*, we have made a major effort to improve the clarity of presentation and to include new pedagogical features that help support the learning and teaching processes. Drawing on positive feedback from users of the fourth edition and reviewers' suggestions, we have made refinements in order to better meet the needs of students and teachers. We have also streamlined the supplements package, which now includes a CD-ROM containing student tutorials and interactive problem-solving software, as well as offerings on the World Wide Web.

This textbook is intended for a course in introductory physics for students majoring in science or engineering. The entire contents of the text could be covered in a three-semester course, but it is possible to use the material in shorter sequences with the omission of selected chapters and sections. The mathematical background of the student taking this course should ideally include one semester of calculus. If that is not possible, the student should be enrolled in a concurrent course in introductory calculus.

OBJECTIVES

This introductory physics textbook has two main objectives: to provide the student with a clear and logical presentation of the basic concepts and principles of physics, and to strengthen an understanding of the concepts and principles through a broad range of interesting applications to the real world. To meet these objectives, we have placed emphasis on sound physical arguments and problem-solving methodology. At the same time, we have attempted to motivate the student through practical examples that demonstrate the role of physics in other disciplines, including engineering, chemistry, and medicine.

CHANGES IN THE FIFTH EDITION

A large number of changes and improvements have been made in preparing the fifth edition of this text. Some of the new features are based on our experiences and on current trends in science education. Other changes have been incorporated in response to comments and suggestions offered by users of the fourth edition and by reviewers of the manuscript. The following represent the major changes in the fifth edition:

Improved Illustrations

- **Time-sequenced events** are represented by circled letters in selected mechanics illustrations. For example, Figure 2.1b (see page 25) shows such letters at the appropriate places on a position–time graph. This construction helps students "translate" the observed motion into its graphical representation.

- **Motion diagrams** are used early in the text to illustrate the difference between velocity and acceleration, concepts easily confused by beginning students. (For example, see Figure 2.9 on page 34, Figure 4.5 on page 81, and Figure 4.8 on page 84.) Students will benefit greatly from sketching their own motion diagrams, as they are asked to do in the Quick Quizzes found in Chapter 4.

- **Greater realism** is achieved with the superimposition of photographs and line art in selected figures (see pages 96 and 97). Also, the three-dimensional appearance of "blocks" in figures accompanying examples and problems in mechanics has been improved (see pages 142 and 143).

More Realistic Worked Examples   Readers familiar with the fourth edition may recall that Example 12.5 involved raising a cylinder onto a step of height h. In this idealized example, we calculated the minimum force **F** necessary to raise the cylinder, as well as the reaction force exerted by the step on the cylinder. In the fifth edition, we are pleased to present the revised Example 12.5 (see page 370), in which we calculate the force that a person must apply to a wheelchair's main wheel to roll it up over an uncut sidewalk curb. Although the revised Example 12.5 involves essentially the same calculation as its fourth-edition predecessor (with some changes in notation), we think that the increased realism makes the example more interesting and provides new motivation for studying physics.

Puzzlers   Every chapter begins with an interesting photograph and a caption that includes a Puzzler. Each Puzzler poses a thought-provoking question that is intended to motivate students' curiosity and enhance their interest in the chapter's subject matter. Part or all of the answer to each Puzzler is contained within the chapter text and is indicated by the 🧩 icon.

Chapter Outlines   The opening page of each chapter now includes an outline of the chapter's major headings. This outline gives students and instructors a preview of the chapter's content.

QuickLabs   This new feature encourages students to perform simple experiments on their own, thereby engaging them actively in the learning process. Most QuickLab experiments can be performed with low-cost items such as string, rubber bands, tape, a ruler, drinking straws, and balloons. In most cases, students are asked to observe the outcome of an experiment and to explain their results in terms of what they have learned in the chapter. When appropriate, students are asked to record data and to graph their results.

Quick Quizzes   Several Quick Quiz questions are included in each chapter to provide students with opportunities to test their understanding of the physical concepts presented. Many questions are written in multiple-choice format and require students to make decisions and defend them on the basis of sound reasoning. Some of them have been written to help students overcome common misconceptions. (Instructors should look to the Instructor's Notes in the margins of the Instructor's Annotated Edition for tips regarding certain Quick Quizzes.) Answers to all Quick Quiz questions are found at the end of each chapter.

Marginal Comments and Icons   To provide students with further guidance, common misconceptions and pitfalls are pointed out in comments in the margin of the text. Often, references to the *Saunders Core Concepts in Physics CD-ROM* and useful World Wide Web site addresses are given in these comments to encourage students to expand their understanding of physical concepts. The 💿 icon in the text margin refers students to the specific module and screen number(s) of the *Saunders Core*

Concepts in Physics CD-ROM that deals with the topic under discussion. A text illustration, example, Quick Quiz, or problem marked with the 🖳 icon indicates that it is accompanied by an Interactive Physics™ simulation that can be found on the *Student Tools CD-ROM*. See the Student Ancillaries section (page xvii) for descriptions of these two electronic learning packages.

Applications Some chapters include Applications, which are about the same length as or slightly longer than worked examples. The Applications demonstrate to students how the physical principles covered in a chapter apply to practical problems of everyday life or engineering. For instance, Applications discuss antilock brakes within the context of static and kinetic friction (see Chapter 5); analyze the tension and compressional forces on the structural components of a truss bridge (see Chapter 12); explore the power delivered in automobile and diesel engine cycles (see Chapter 22); and discuss the construction and circuit wiring of holiday lightbulb strings (see Chapter 28).

Problems A substantial revision of the end-of-chapter problems was made in an effort to improve their clarity and quality. Approximately 20 percent of the problems (about 650) are new, and most of these new problems are at the intermediate level (as identified by blue problem numbers). Many of the new problems require students to make order-of-magnitude calculations. All problems have been carefully edited and reworded when necessary. Solutions to approximately 20 percent of the end-of-chapter problems are included in the *Student Solutions Manual and Study Guide*. These problems are identified in the text by boxes around their numbers. A smaller subset of solutions are posted on the World Wide Web (**http://www.saunderscollege.com/physics/**) and are accessible to students and instructors using *Physics for Scientists and Engineers*. These problems are identified in the text by the **WEB** icon. See the next section for a complete description of other features of the problem set.

Line-by-Line Revision The entire text has been carefully edited to improve clarity of presentation and precision of language. We believe that the result is a book that is both accurate and enjoyable to read.

Typographical and Notation Changes The Text Features section (see page xvi) mentions the use of **boldface** type and screens for emphasizing important statements and definitions. Boldfaced passages in the text of the fifth edition replace the less legible passages appearing in italics in the fourth edition. Similarly, the symbols for vectors stand out very clearly from the surrounding text owing to the strong boldface type used in the fifth edition. As a step toward making equations more transparent and therefore more easily understood, the use of the subscripts "i" and "f" for initial and final values replaces the fourth edition's older notation, which makes use of subscript 0 (usually pronounced "naught") for an initial value and no subscript for a final value. In equations describing motion or direction, variables carry the subscripts x, y, or z whenever added clarity is needed.

Content Changes Examination of the full Table of Contents might lead one to the impression that the content and organization of the textbook are essentially the same as those of the fourth edition. However, a number of subtle yet significant improvements in content have been made. Following are some examples:

- Section 16.8 contains a more complete and careful derivation of the power or rate of energy transfer for sinusoidal waves on strings. A similar development occurs in Section 17.3, which deals with the intensity of periodic sound waves.

- Section 18.2 contains an improved discussion of the envelope function of a standing wave.

- Chapter 20 contains an updated discussion of the distinction between heat and internal energy. Both heat and work are described and clarified as means of changing the energy of a system.

- Chapter 22 contains a new description of microstates and macrostates of a system, beginning with Section 22.6 on entropy and continuing through the end of the chapter.

- Section 24.3 contains a new list of guidelines for choosing a gaussian surface, allowing the student to take advantage of the symmetry of a charge distribution when determining the electric field.

- Chapter 25 contains new two- and three-dimensional graphs of the electric potential near a point charge and an electric dipole.

- In Chapter 27 and in following chapters, we use "Ohm's law" to refer only to the direct proportionality between current density and electric field seen in some (but not all) materials. See Section 27.2 and the corresponding Instructor's Note for a full explanation.

- Section 29.3 now makes explicit comparison between the potential energy of an electric dipole in an electric field and that of a magnetic dipole in a magnetic field. The section also contains new examples on satellite attitude control and the d'Arsonval galvanometer.

- Chapter 33 contains new information on rectifier circuits, including diodes. The material on rectifiers and filter circuits is now included in Optional Section 33.9, which follows the section on transformers and power transmission.

- In Chapter 35, reflection and refraction are now covered in separate sections, and discussion of Huygens's principle now precedes the section on dispersion and prisms. New Figure 35.8 illustrates retroreflection, which has many practical applications.

- Section 38.2 contains a new subsection considering two-slit diffraction patterns, in which the effects of diffraction and interference are combined.

- Within Section 39.4, new subsections cover space–time graphs and the relativistic Doppler effect. References to the concept of "rest mass" have been deleted.

Many sections in these and other chapters have been streamlined, deleted, or combined with other sections to allow for a more balanced presentation. Looking ahead to the extended version of this text, the former Chapter 44 on superconductivity in the fourth edition of *Physics for Scientists and Engineers with Modern Physics* has been deleted, and an abridged section on this topic has been added to Chapter 43. Some of the sections deleted from the fourth edition may be found on the textbook's Web sites for both instructors and students.

Instructor's Notes   For the first time, tips and comments are offered to instructors in blue marginal Instructor's Notes, which appear only in the Instructor's Annotated Edition. These annotations expand on common student misconceptions; call attention to certain worked examples, QuickLabs, and Quick Quizzes; or cite key physics education research literature that bears on the topic at hand. In some chapters, Instructor's Notes appear as footnotes in the end-of-chapter problem sets; these notes point out related groups of problems found in other chapters of the textbook. The Instructor's Annotated Edition includes Chapters 1 to 39.

CONTENT

The material in this book covers fundamental topics in classical physics and provides an introduction to modern physics. The book is divided into six parts. Part 1 (Chapters 1 to 15) deals with the fundamentals of Newtonian mechanics and the physics of fluids, Part 2 (Chapters 16 to 18) covers wave motion and sound, Part 3 (Chapters 19 to 22) addresses heat and thermodynamics, Part 4 (Chapters 23 to 34) treats electricity and magnetism, Part 5 (Chapters 35 to 38) covers light and optics, and Part 6 (Chapters 39 to 46) deals with relativity and modern physics. Each part opener includes an overview of the subject matter covered in that part, as well as some historical perspectives.

TEXT FEATURES

Most instructors would agree that the textbook selected for a course should be the student's primary guide for understanding and learning the subject matter. Furthermore, the textbook should be easily accessible and should be styled and written to facilitate instruction and learning. With these points in mind, we have included many pedagogical features in the textbook that are intended to enhance its usefulness to both students and instructors. These features are as follows:

Previews   Most chapters begin with a brief preview that includes a discussion of the chapters' objectives and content.

Important Statements and Equations   Most important statements and definitions are set in boldface type or are highlighted with a tan background screen for added emphasis and ease of review. Similarly, important equations are highlighted with a tan background screen to facilitate location.

Problem-Solving Hints   We have included general strategies for solving the types of problems featured both in the examples and in the end-of-chapter problems. This feature helps students to identify necessary steps in problem-solving and to eliminate any uncertainty they might have. Problem-Solving Hints are highlighted with a light blue-gray screen for emphasis and ease of location.

Marginal Notes   Comments and notes appearing in the margin can be used to locate important statements, equations, and concepts in the text.

Illustrations   The three-dimensional appearance of many illustrations has been improved in this fifth edition.

Mathematical Level   We have introduced calculus gradually, keeping in mind that students often take introductory courses in calculus and physics concurrently. Most steps are shown when basic equations are developed, and reference is often made to mathematical appendices at the end of the textbook. Vector products are introduced later in the text, where they are needed in physical applications. The dot product is introduced in Chapter 7, which addresses work and energy; the cross product is introduced in Chapter 11, which deals with rotational dynamics.

Worked Examples   A large number of worked examples of varying difficulty are presented to promote students' understanding of concepts. In many cases, the examples serve as models for solving the end-of-chapter problems. The examples are set off in boxes, and the answers to examples with numerical solutions are highlighted with a light blue-gray screen.

Worked Example Exercises   Many of the worked examples are followed immediately by exercises with answers. These exercises are intended to promote interactivity between the student and the textbook and to immediately reinforce the student's understanding of concepts and problem-solving techniques. The exercises represent extensions of the worked examples.

Conceptual Examples   As in the fourth edition, we have made a concerted effort to emphasize critical thinking and the teaching of physical concepts. We have accomplished this by including Conceptual Examples (for instance, see page 41). These examples provide students with a means of reviewing and applying the concepts presented in a section. Some Conceptual Examples demonstrate the connection between concepts presented in a chapter and other disciplines. The Conceptual Examples can serve as models for students when they are asked to respond to end-of-chapter questions, which are largely conceptual in nature.

Questions   Questions requiring verbal responses are provided at the end of each chapter. Over 1,000 questions are included in this edition. Some questions provide students with a means of testing their mastery of the concepts presented in the chapter. Others could serve as a basis for initiating classroom discussions. Answers to selected questions are included in the *Student Solutions Manual and Study Guide.*

Significant Figures   Significant figures in both worked examples and end-of-chapter problems have been handled with care. Most numerical examples and problems are worked out to either two or three significant figures, depending on the accuracy of the data provided.

Problems   An extensive set of problems is included at the end of each chapter; in all, over 3,000 problems are given throughout the text. Answers to odd-numbered problems are provided at the end of the book in a section whose pages have colored edges for ease of location. For the convenience of both the student and the instructor, about two thirds of the problems are keyed to specific sections of the chapter. The remaining problems, labeled "Additional Problems," are not keyed to specific sections.

Usually, the problems within a given section are presented so that the straightforward problems (those with black problem numbers) appear first; these straightforward problems are followed by problems of increasing difficulty. For ease of identification, the numbers of intermediate-level problems are printed in blue, and those of a small number of challenging problems are printed in magenta.

Review Problems   Many chapters include review problems that require the student to draw on numerous concepts covered in the chapter, as well as on those discussed in previous chapters. These problems could be used by students in preparing for tests and by instructors for special assignments and classroom discussions.

Paired Problems   Some end-of-chapter numerical problems are paired with the same problems in symbolic form. Two paired problems are identified by a common tan background screen.

Computer- and Calculator-Based Problems   Most chapters include one or more problems whose solution requires the use of a computer or graphing calculator. These problems are identified by the 💻 icon. Modeling of physical phenomena enables students to obtain graphical representations of variables and to perform numerical analyses.

Units   The international system of units (SI) is used throughout the text. The British engineering system of units (conventional system) is used only to a limited extent in the chapters on mechanics, heat, and thermodynamics.

Summaries   Each chapter contains a summary that reviews the important concepts and equations discussed in that chapter.

Appendices and Endpapers   Several appendices are provided at the end of the text-book. Most of the appendix material represents a review of mathematical concepts and techniques used in the text, including scientific notation, algebra, geometry, trigonometry, differential calculus, and integral calculus. Reference to these appendices is made throughout the text. Most mathematical review sections in the appendices include worked examples and exercises with answers. In addition to the mathematical reviews, the appendices contain tables of physical data, conversion factors, atomic masses, and the SI units of physical quantities, as well as a periodic table of the elements. Other useful information, including fundamental constants and physical data, planetary data, a list of standard prefixes, mathematical symbols, the Greek alphabet, and standard abbreviations of units of measure, appears on the endpapers.

ANCILLARIES

The ancillary package has been updated substantially and streamlined in response to suggestions from users of the fourth edition. The most essential changes in the student package are a *Student Solutions Manual and Study Guide* with a tighter focus on problem-solving, the *Student Tools CD-ROM,* and the *Saunders Core Concepts in Physics CD-ROM* developed by Archipelago Productions. Instructors will find increased support for their teaching efforts with new electronic materials.

Student Ancillaries

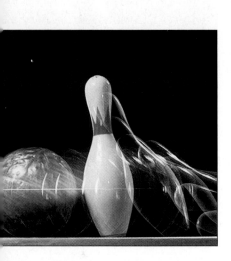

Student Solutions Manual and Study Guide by John R. Gordon, Ralph McGrew, and Raymond A. Serway, with contributions by Duane Deardorff. This two-volume manual features detailed solutions to 20 percent of the end-of-chapter problems from the textbook. Problems in the textbook whose complete solutions are found in the manual are identified by boxes around their numbers. The solutions to many problems follow the **GOAL** protocol described in the textbook (see page 47). The manual also features a list of important equations and concepts, as well as answers to selected end-of-chapter questions.

Pocket Guide by V. Gordon Lind. This 5″ × 7″ paperback is a section-by-section capsule of the textbook and serves as a handy guide for looking up important concepts, formulas, and problem-solving hints.

Student Tools CD-ROM   This CD-ROM contains tools that are designed to enhance the learning of physical concepts and train students to become better problem-solvers. It includes a textbook version of the highly acclaimed Interactive Physics™ software by MSC Working Knowledge and more than 100 Interactive Physics™ simulations keyed to appropriate figures, worked examples, Quick Quizzes, and selected end-of-chapter problems (as identified by the 📖 icon).

Saunders Core Concepts in Physics CD-ROM This CD-ROM package developed by Archipelago Productions applies the power of multimedia to the introductory physics course, offering full-motion animation and video, engaging interactive graphics, clear and concise text, and guiding narration. *Saunders Core Concepts in Physics CD-ROM* focuses on those concepts students usually find most difficult in the course, drawing from topics in mechanics, thermodynamics, electric fields, magnetic fields, and optics. The animations and graphics are presented to aid the student in developing accurate conceptual models of difficult topics—topics often too complex to be explained in words or chalkboard illustrations. The CD-ROM also presents step-by-step explorations of problem-solving strategies and provides animations of problems in order to promote conceptual understanding and sharpen problem-solving skills. Topics in the textbook that are further explored on the CD-ROM are identified by marginal 💿 icons that give the appropriate module and screen number(s). Students should look to the CD-ROM to aid in their understanding of these topics.

Student Web Site Students will have access to an abundance of material at **http://www.saunderscollege.com/physics/.** The Web Site features special topic essays by guest authors, practice problems with answers, and optional topics that accompany selected chapters of the textbook. Also included are selected solutions from the *Student Solutions Manual and Study Guide,* a sampling of the *Pocket Guide,* and a glossary that includes more than 300 physics terms. Students also can take practice quizzes in our Practice Exercises and Testing area.

Physics Laboratory Manual, Second Edition by David Loyd. Updated and redesigned, this manual supplements the learning of basic physical principles while introducing laboratory procedures and equipment. Each chapter includes a pre-laboratory assignment, objectives, an equipment list, the theory behind the experiment, step-by-step experimental procedures, and questions. A laboratory report form is provided for each experiment so that students can record data and make calculations. Students are encouraged to apply statistical analysis to their data so that they can develop their ability to judge the validity of their results.

So You Want to Take Physics: A Preparatory Course with Calculus by Rodney Cole. This introductory-level book is useful to those students who need additional preparation before or during a calculus-based course in physics. The friendly, straightforward style makes it easier to understand how mathematics is used in the context of physics.

Life Science Applications for Physics by Jerry Faughn. This book provides examples, readings, and problems from the biological sciences as they relate to physics. Topics include "Friction in Human Joints," "Physics of the Human Circulatory System," "Physics of the Nervous System," and "Ultrasound and Its Applications." This supplement is useful in those courses taken by a significant number of pre-med students.

Instructor's Ancillaries

Instructor's Manual with Solutions by Ralph McGrew, Jeff Saul, and Charles Teague, with contributions by Duane Deardorff and Rhett Allain. This manual contains chapter summaries, answers to even-numbered problems, and complete worked solutions to all the problems in the textbook. The solutions to problems new to the fifth edition are marked for easy identification by the instructor. New to this edition of the manual are suggestions on how to teach difficult topics and how to help students overcome common misconceptions. These suggestions are based on recent research in physics education.

Instructor's Web Site   The instructor's area at **http://www.saunderscollege.com/physics/** includes a listing of overhead transparencies; a guide to relevant experiments in David Loyd's *Physics Laboratory Manual, Second Edition*; a correlation guide between sections in *Physics for Scientists and Engineers* and modules in the *Saunders Core Concepts in Physics CD-ROM*; supplemental problems with answers; optional topics to accompany selected chapters of the textbook; and a problems correlation guide.

Instructor's Resource CD-ROM   This CD-ROM accompanying the fifth edition of *Physics for Scientists and Engineers* has been created to provide instructors with an exciting new tool for classroom presentation. The CD-ROM contains a collection of graphics files of line art from the textbook. These files can be opened directly, can be imported into a variety of presentation packages, or can be used in the presentation package included on the CD-ROM. The labels for each piece of art have been enlarged and boldfaced to facilitate classroom viewing. The CD-ROM contains electronic files of the *Instructor's Manual, Test Bank,* and *Supplemental Problems with Answers.*

CAPA: A Computer-Assisted Personalized Approach   CAPA is a network system for learning, teaching, assessment, and administration. It provides students with personalized problem sets, quizzes, and examinations consisting of qualitative conceptual problems and quantitative problems, including problems from *Physics for Scientists and Engineers.* CAPA was developed through a collaborative effort of the Physics–Astronomy, Computer Science, and Chemistry Departments at Michigan State University. Students are given instant feedback and relevant hints via the Internet and may correct errors without penalty before an assignment's due date. The system records each student's participation and performance on assignments, quizzes, and examinations; and records are available on-line to both the individual student and to his or her instructor. For more information, visit the CAPA Web site at **http://www.pa.msu.edu/educ/CAPA/**

WebAssign: A Web-Based Homework System   WebAssign is a Web-based homework delivery, collection, grading, and recording service developed at North Carolina State University. Instructors who sign up for WebAssign can assign homework to their students, using questions and problems taken directly from *Physics for Scientists and Engineers.* WebAssign gives students immediate feedback on their homework that helps them to master information and skills, leading to greater competence and better grades. WebAssign can free instructors from the drudgery of grading homework and recording scores, allowing them to devote more time to meeting with students and preparing classroom presentations. Details about and a demonstration of WebAssign are available at **http://webassign.net/info**. For more information about ordering this service, contact WebAssign at **webassign@ncsu.edu**

Homework Service   With this service, instructors can reduce their grading workload by assigning thought-provoking homework problems using the World Wide Web. Instructors browse problem banks that include problems from *Physics for Scientists and Engineers,* select those they wish to assign to their students, and then let the Homework Service take over the delivery and grading. This system was developed and is maintained by Fred Moore at the University of Texas (**moore@physics.utexas.edu**). Students download their unique problems, submit their answers, and obtain immediate feedback; if students' answers are incorrect, they can resubmit them. This rapid grading feature facilitates effective learning. After the due date of their assignments, students can obtain the solutions to their problems. Minimal on-line connect time is required. The Homework Service uses algorithm-based problems: This means that each student solves sets of problems that are different from those given to other students. Details about and a demonstration of this service are available at **http://hw10.ph.utexas.edu/instInst.html**

Printed Test Bank by Edward Adelson. The *Printed Test Bank* contains approximately 2,300 multiple-choice questions. It is provided for the instructor who does not have access to a computer. About 20% of the old test items have been replaced with new, concept-based, thought-provoking questions.

Computerized Test Bank Available in Windows™ and Macintosh® formats, the *Computerized Test Bank* contains more than 2,300 multiple-choice questions, representing every chapter of the text. The *Computerized Test Bank* enables the instructor to create many unique tests by allowing the editing of questions and the addition of new questions. The software program solves all problems and prints each answer on a separate grading key. All questions have been reviewed for accuracy.

Overhead Transparency Acetates This collection of transparencies consists of 300 full-color figures from the text and features large print for easy viewing in the classroom.

Instructor's Manual for Physics Laboratory Manual by David Loyd. Each chapter contains a discussion of the experiment, teaching hints, answers to selected questions, and a post-laboratory quiz with short-answer and essay questions. It also includes a list of the suppliers of scientific equipment and a summary of the equipment needed for each of the laboratory experiments in the manual.

Saunders College Publishing, a division of Harcourt College Publishers, may provide complementary instructional aids and supplements or supplement packages to those adopters qualified under our adoption policy. Please contact your sales representative for more information. If as an adopter or potential user you receive supplements you do not need, please return them to your sales representative or send them to

Attn: Returns Department
Troy Warehouse
465 South Lincoln Drive
Troy, MO 63379

TEACHING OPTIONS

The topics in this textbook are presented in the following sequence: classical mechanics, mechanical waves, and heat and thermodynamics, followed by electricity and magnetism, electromagnetic waves, optics, and relativity. This presentation represents a more traditional sequence, with the subject of mechanical waves being presented before the topics of electricity and magnetism. Some instructors may prefer to cover this material after completing electricity and magnetism (i.e., after Chapter 34). The chapter on relativity was placed near the end of the text because this topic often is treated as an introduction to the era of "modern physics." If time permits, instructors may choose to cover Chapter 39 in the first semester after completing Chapter 14, as it concludes the material on Newtonian mechanics.

   Instructors teaching a two-semester sequence can delete sections and chapters without any loss of continuity. We have labeled these as "Optional" in the Table of Contents and in the appropriate sections of the text. For student enrichment, instructors can assign some of these sections or chapters as extra reading.

ACKNOWLEDGMENTS

The fifth edition of this textbook was prepared with the guidance and assistance of many professors who reviewed part or all of the manuscript, the pre-revision text, or both. We wish to acknowledge the following scholars and express our sincere appreciation for their suggestions, criticisms, and encouragement:

Edward Adelson, *Ohio State University*
Roger Bengtson, *University of Texas at Austin*
Joseph Biegen, *Broome Community College*
Ronald J. Bieniek, *University of Missouri at Rolla*
Ronald Brown, *California Polytechnic State University—San Luis Obispo*
Michael E. Browne, *University of Idaho*
Tim Burns, *Leeward Community College*
Randall Caton, *Christopher Newport University*
Sekhar Chivukula, *Boston University*
Alfonso Díaz-Jiménez, *ADJOIN Research Center*
N. John DiNardo, *Drexel University*
F. Eugene Dunnum, *University of Florida*
William Ellis, *Cornell University*
F. Paul Esposito, *University of Cincinnati*
Paul Fahey, *University of Scranton*

Arnold Feldman, *University of Hawaii at Manoa*
Alexander Firestone, *Iowa State University*
Robert Forsythe, *Broome Community College*
Philip Fraundorf, *University of Missouri at St. Louis*
John Gerty, *Broome Community College*
John B. Gruber, *San Jose State University*
Frank Hayes, *McMaster University (Canada)*
John Hubisz, *North Carolina State University*
Joey Huston, *Michigan State University*
Calvin S. Kalman, *Concordia University*
Natalie Kerr, M.D., *University of Tennessee, Memphis*
Peter Killen, *University of Queensland (Australia)*
Earl Koller, *Stevens Institute of Technology*

David LaGraffe, *U.S. Military Academy*
Ying-Cheng Lai, *University of Kansas*
Donald Larson, *Drexel University*
Robert Lieberman, *Cornell University*
Ralph McGrew, *Broome Community College*
David Mills, *Monash University (Australia)*
Clement J. Moses, *Utica College*
Peter Parker, *Yale University*
John Parsons, *Columbia University*
Arnold Perlmutter, *University of Miami*
Henry Schriemer, *Queen's University (Canada)*
Paul Snow, *University of Bath (U.K.)*
Edward W. Thomas, *Georgia Institute of Technology*
Charles C. Vuille, *Embry-Riddle Aeronautical University*
Xiaojun Wang, *Georgia Southern University*
Gail Welsh, *Salisbury State University*

This book was carefully checked for accuracy by James H. Smith *(University of Illinois at Urbana-Champaign)*, Gregory Snow *(University of Nebraska—Lincoln)*, Edward Gibson *(California State University—Sacramento)*, Ronald Jodoin *(Rochester Institute of Technology)*, Arnold Perlmutter *(University of Miami)*, Michael Paesler *(North Carolina State University)*, and Clement J. Moses *(Utica College)*.

We thank the following people for their suggestions and assistance during the preparation of earlier editions of this textbook:

George Alexandrakis, *University of Miami*
Elmer E. Anderson, *University of Alabama*
Wallace Arthur, *Fairleigh Dickinson University*
Duane Aston, *California State University—Sacramento*
Stephen Baker, *Rice University*
Richard Barnes, *Iowa State University*
Stanley Bashkin, *University of Arizona*
Robert Bauman, *University of Alabama*
Marvin Blecher, *Virginia Polytechnic Institute and State University*
Jeffrey J. Braun, *University of Evansville*
Kenneth Brownstein, *University of Maine*
William A. Butler, *Eastern Illinois University*
Louis H. Cadwell, *Providence College*

Ron Canterna, *University of Wyoming*
Bo Casserberg, *University of Minnesota*
Soumya Chakravarti, *California Polytechnic State University*
C. H. Chan, *The University of Alabama at Huntsville*
Edward Chang, *University of Massachusetts at Amherst*
Don Chodrow, *James Madison University*
Clifton Bob Clark, *University of North Carolina at Greensboro*
Walter C. Connolly, *Appalachian State University*
Hans Courant, *University of Minnesota*
Lance E. De Long, *University of Kentucky*
James L. DuBard, *Binghamton-Southern College*

F. Paul Esposito, *University of Cincinnati*
Jerry S. Faughn, *Eastern Kentucky University*
Paul Feldker, *Florissant Valley Community College*
Joe L. Ferguson, *Mississippi State University*
R. H. Garstang, *University of Colorado at Boulder*
James B. Gerhart, *University of Washington*
John R. Gordon, *James Madison University*
Clark D. Hamilton, *National Bureau of Standards*
Mark Heald, *Swarthmore College*
Herb Helbig, *Rome Air Development Center*
Howard Herzog, *Broome Community College*
Paul Holoday, *Henry Ford Community College*

Jerome W. Hosken, *City College of San Francisco*

Larry Hmurcik, *University of Bridgeport*

William Ingham, *James Madison University*

Mario Iona, *University of Denver*

Karen L. Johnston, *North Carolina State University*

Brij M. Khorana, *Rose-Hulman Institute of Technology*

Larry Kirkpatrick, *Montana State University*

Carl Kocher, *Oregon State University*

Robert E. Kribel, *Jacksonville State University*

Barry Kunz, *Michigan Technological University*

Douglas A. Kurtze, *Clarkson University*

Fred Lipschultz, *University of Connecticut*

Francis A. Liuima, *Boston College*

Robert Long, *Worcester Polytechnic Institute*

Roger Ludin, *California Polytechnic State University*

Nolen G. Massey, *University of Texas at Arlington*

Charles E. McFarland, *University of Missouri at Rolla*

Ralph V. McGrew, *Broome Community College*

James Monroe, *The Pennsylvania State University, Beaver Campus*

Bruce Morgan, *United States Naval Academy*

Clement J. Moses, *Utica College*

Curt Moyer, *Clarkson University*

David Murdock, *Tennessee Technological University*

A. Wilson Nolle, *University of Texas at Austin*

Thomas L. O'Kuma, *San Jacinto College North*

Fred A. Otter, *University of Connecticut*

George Parker, *North Carolina State University*

William F. Parks, *University of Missouri at Rolla*

Philip B. Peters, *Virginia Military Institute*

Eric Peterson, *Highland Community College*

Richard Reimann, *Boise State University*

Joseph W. Rudmin, *James Madison University*

Jill Rugare, *DeVry Institute of Technology*

Charles Scherr, *University of Texas at Austin*

Eric Sheldon, *University of Massachusetts—Lowell*

John Shelton, *College of Lake County*

Stan Shepard, *The Pennsylvania State University*

James H. Smith, *University of Illinois at Urbana-Champaign*

Richard R. Sommerfield, *Foothill College*

Kervork Spartalian, *University of Vermont*

Robert W. Stewart, *University of Victoria*

James Stith, *American Institute of Physics*

Charles D. Teague, *Eastern Kentucky University*

Edward W. Thomas, *Georgia Institute of Technology*

Carl T. Tomizuka, *University of Arizona*

Herman Trivilino, *San Jacinto College North*

Som Tyagi, *Drexel University*

Steve Van Wyk, *Chapman College*

Joseph Veit, *Western Washington University*

T. S. Venkataraman, *Drexel University*

Noboru Wada, *Colorado School of Mines*

James Walker, *Washington State University*

Gary Williams, *University of California, Los Angeles*

George Williams, *University of Utah*

Edward Zimmerman, *University of Nebraska, Lincoln*

Earl Zwicker, *Illinois Institute of Technology*

We are grateful to Ralph McGrew for organizing the end-of-chapter problems, writing many new problems, and his suggestions for improving the content of the textbook. The new end-of-chapter problems were written by Rich Cohen, John DiNardo, Robert Forsythe, Ralph McGrew, and Ronald Bieniek, with suggestions by Liz McGrew, Alexandra Héder, and Richard McGrew. We thank Laurent Hodges for permission to use selected end-of-chapter problems. We are grateful to John R. Gordon, Ralph McGrew, and Duane Deardorff for writing the *Student Solutions Manual and Study Guide,* and we thank Michael Rudmin for its attractive layout. Ralph McGrew, Jeff Saul, and Charles Teague have prepared an excellent *Instructor's Manual,* and we thank them. We thank Gloria Langer, Linda Miller, and Jennifer Serway for their excellent work in preparing the *Instructor's Manual* and the supplemental materials that appear on our Web site.

Special thanks and recognition go to the professional staff at Saunders College Publishing—in particular, Susan Pashos, Sally Kusch, Carol Bleistine, Frank Messina, Suzanne Hakanen, Ruth Hoover, Alexandra Buczek, Pauline Mula, Walter Neary, and John Vondeling—for their fine work during the development and production of this textbook. We are most appreciative of the intelligent line editing by Irene Nunes, the final copy editing by Sue Nelson and Mary Patton, the excellent artwork produced by Rolin Graphics, and the dedicated photo research efforts of Dena Digilio Betz.

Finally, we are deeply indebted to our wives and children for their love, support, and long-term sacrifices.

Raymond A. Serway
Chapel Hill, North Carolina

Robert J. Beichner
Raleigh, North Carolina

John W. Jewett, Jr.
Pomona, California

To the Student

*I*t is appropriate to offer some words of advice that should be of benefit to you, the student. Before doing so, we assume that you have read the Preface, which describes the various features of the text that will help you through the course.

HOW TO STUDY

Very often instructors are asked, "How should I study physics and prepare for examinations?" There is no simple answer to this question, but we would like to offer some suggestions that are based on our own experiences in learning and teaching over the years.

First and foremost, maintain a positive attitude toward the subject matter, keeping in mind that physics is the most fundamental of all natural sciences. Other science courses that follow will use the same physical principles, so it is important that you understand and are able to apply the various concepts and theories discussed in the text.

CONCEPTS AND PRINCIPLES

It is essential that you understand the basic concepts and principles before attempting to solve assigned problems. You can best accomplish this goal by carefully reading the textbook before you attend your lecture on the covered material. When reading the text, you should jot down those points that are not clear to you. We've purposely left wide margins in the text to give you space for doing this. Also be sure to make a diligent attempt at answering the questions in the Quick Quizzes as you come to them in your reading. We have worked hard to prepare questions that help you judge for yourself how well you understand the material. The QuickLabs provide an occasional break from your reading and will help you to experience some of the new concepts you are trying to learn. During class, take careful notes and ask questions about those ideas that are unclear to you. Keep in mind that few people are able to absorb the full meaning of scientific material after only one reading. Several readings of the text and your notes may be necessary. Your lectures and laboratory work supplement reading of the textbook and should clarify some of the more difficult material. You should minimize your memorization of material. Successful memorization of passages from the text, equations, and derivations does not necessarily indicate that you understand the material. Your understanding of the material will be enhanced through a combination of efficient study habits, discussions with other students and with instructors, and your ability to solve the problems presented in the textbook. Ask questions whenever you feel clarification of a concept is necessary.

STUDY SCHEDULE

It is important that you set up a regular study schedule, preferably one that is daily. Make sure that you to read the syllabus for the course and adhere to the schedule set by your instructor. The lectures will be much more meaningful if you read the corresponding textual material before attending them. As a general rule, you should devote about two hours of study time for every hour you are in class. If you are having trouble with the course, seek the advice of the instructor or other students who have taken the course. You may find it necessary to seek further instruction from experienced students. Very often, instructors offer review sessions in addition to regular class periods. It is important that you avoid the practice of delaying study until a day or two before an exam. More often than not, this approach has disastrous results. Rather than undertake an all-night study session, briefly review the basic concepts and equations and get a good night's rest. If you feel you need additional help in understanding the concepts, in preparing for exams, or in problem-solving, we suggest that you acquire a copy of the *Student Solutions Manual and Study Guide* that accompanies this textbook; this manual should be available at your college bookstore.

USE THE FEATURES

You should make full use of the various features of the text discussed in the preface. For example, marginal notes are useful for locating and describing important equations and concepts, and **boldfaced** type indicates important statements and definitions. Many useful tables are contained in the Appendices, but most are incorporated in the text where they are most often referenced. Appendix B is a convenient review of mathematical techniques.

Answers to odd-numbered problems are given at the end of the textbook, answers to Quick Quizzes are located at the end of each chapter, and answers to selected end-of-chapter questions are provided in the *Student Solutions Manual and Study Guide*. The exercises (with answers) that follow some worked examples represent extensions of those examples; in most of these exercises, you are expected to perform a simple calculation (see Example 4.7 on page 90). Their purpose is to test your problem-solving skills as you read through the text. Problem-Solving Hints are included in selected chapters throughout the text and give you additional information about how you should solve problems. The Table of Contents provides an overview of the entire text, while the Index enables you to locate specific material quickly. Footnotes sometimes are used to supplement the text or to cite other references on the subject discussed.

After reading a chapter, you should be able to define any new quantities introduced in that chapter and to discuss the principles and assumptions that were used to arrive at certain key relations. The chapter summaries and the review sections of the *Student Solutions Manual and Study Guide* should help you in this regard. In some cases, it may be necessary for you to refer to the index of the text to locate certain topics. You should be able to correctly associate with each physical quantity the symbol used to represent that quantity and the unit in which the quantity is specified. Furthermore, you should be able to express each important relation in a concise and accurate prose statement.

PROBLEM-SOLVING

R. P. Feynman, Nobel laureate in physics, once said, "You do not know anything until you have practiced." In keeping with this statement, we strongly advise that you develop the skills necessary to solve a wide range of problems. Your ability to solve problems will be one of the main tests of your knowledge of physics, and therefore you should try to solve as many problems as possible. It is good practice to try to find alternate solutions to the same problem. For example, you can solve problems in mechanics using Newton's laws, but very often an alternative method that draws on energy considerations is more direct. You should not deceive yourself into thinking that you understand a problem merely because you have seen it solved in class. You must be able to solve the problem and similar problems on your own.

The approach to solving problems should be carefully planned. A systematic plan is especially important when a problem involves several concepts. First, read the problem several times until you are confident you understand what is being asked. Look for any key words that will help you interpret the problem and perhaps allow you to make certain assumptions. Your ability to interpret a question properly is an integral part of problem-solving. Second, you should acquire the habit of writing down the information given in a problem and those quantities that need to be found; for example, you might construct a table listing both the quantities given and the quantities to be found. This procedure is sometimes used in the worked examples of the textbook. Finally, after you have decided on the method you feel is appropriate for a given problem, proceed with your solution. General problem-solving strategies of this type are included in the text and are highlighted with a light blue-gray screen. We have also developed the **GOAL** protocol (see page 47) to help guide you through complex problems. If you follow the steps of this procedure (**G**ather information, **O**rganize your approach, carry out your **A**nalysis, and finally **L**earn from your work), you will not only find it easier to come up with a solution, but you will also gain more from your efforts.

Often, students fail to recognize the limitations of certain formulas or physical laws in a particular situation. It is very important that you understand and remember the assumptions that underlie a particular theory or formalism. For example, certain equations in kinematics apply only to a particle moving with constant acceleration. These equations are not valid for describing motion whose acceleration is not constant, such as the motion of an object connected to a spring or the motion of an object through a fluid.

General Problem-Solving Strategy

Most courses in general physics require the student to learn the skills of problem-solving, and exams are largely composed of problems that test such skills. This brief section describes some useful ideas that will enable you to increase your accuracy in solving problems, enhance your understanding of physical concepts, eliminate initial panic or lack of direction in approaching a problem, and organize your work. One way to help accomplish these goals is to adopt a problem-solving strategy. Many chapters in this text include Problem-Solving Hints that should help you through the "rough spots."

In developing problem-solving strategies, five basic steps are commonly followed:

• Draw a suitable diagram with appropriate labels and coordinate axes (if needed).

- As you examine what is being asked in the problem, identify the basic physical principle (or principles) that are involved, listing the knowns and the unknowns.

- Select a basic relationship or derive an equation that can be used to find the unknown, and then solve the equation for the unknown symbolically.

- Substitute the given values along with the appropriate units into the equation.

- Obtain a numerical value for the unknown. The problem is verified and receives a check mark if the following questions can be properly answered: Do the units match? Is the answer reasonable? Is the plus or minus sign proper or meaningful?

One of the purposes of this strategy is to promote accuracy. Properly drawn diagrams can eliminate many sign errors. Diagrams also help to isolate the physical principles of the problem. Obtaining symbolic solutions and carefully labeling knowns and unknowns will help eliminate other careless errors. The use of symbolic solutions should help you think in terms of the physics of the problem. A check of units at the end of the problem can indicate a possible algebraic error. The physical layout and organization of your problem will make the final product more understandable and easier to follow. Once you have developed an organized system for examining problems and extracting relevant information, you will become a more confident problem-solver.

EXPERIMENTS

Physics is a science based on experimental observations. In view of this fact, we recommend that you try to supplement your reading of the text by performing various types of "hands-on" experiments, either at home or in the laboratory. Most chapters include one or two QuickLabs that describe simple experiments you can do on your own. These can be used to test ideas and models discussed in class or in the textbook. For example, the common Slinky™ toy is an excellent tool for studying traveling waves; a ball swinging on the end of a long string can be used to investigate pendulum motion; various masses attached to the end of a vertical spring or rubber band can be used to determine their elastic nature; an old pair of Polaroid sunglasses, some discarded lenses, and a magnifying glass are the components of various experiments in optics; and the approximate measure of the acceleration of gravity can be determined simply by measuring with a stopwatch the time it takes for a ball to drop from a known height. The list of such experiments is endless. When physical models are not available, be imaginative and try to develop models of your own.

NEW MEDIA

We strongly encourage you to use one or more of the following multimedia products that accompany this textbook. It is far easier to understand physics if you see it in action, and these new materials will enable you to become a part of that action.

Student Tools CD-ROM   The dual-platform (Windows™- and Macintosh®-compatible) *Student Tools CD-ROM* is available with each new copy of the textbook. This CD-ROM contains a textbook version of the Interactive Physics™ program by MSC Working Knowledge. Interactive Physics™ simulations are keyed to the following figures, worked examples, Quick Quizzes, and end-of-chapter problems (identified in the text with the 🖥 icon).

Saunders Core Concepts in Physics CD-ROM　In addition, you may purchase the *Saunders Core Concepts in Physics CD-ROM* developed by Archipelago Productions. This CD-ROM provides a complete multimedia presentation of selected topics in mechanics, thermodynamics, electromagnetism, and optics. It contains more than 350 movies—both animated and live video—that bring to life laboratory demonstrations, "real-world" examples, graphic models, and step-by-step explanations of essential mathematics. Those CD-ROM modules that supplement the material in *Physics for Scientists and Engineers* are identified in the margin of the text by the 💿 icon.

AN INVITATION TO PHYSICS

It is our sincere hope that you too will find physics an exciting and enjoyable experience and that you will profit from this experience, regardless of your chosen profession. Welcome to the exciting world of physics!

The scientist does not study nature because it is useful; he studies it because he delights in it, and he delights in it because it is beautiful. If nature were not beautiful, it would not be worth knowing, and if nature were not worth knowing, life would not be worth living.

—Henri Poincaré

About the Authors

Raymond A. Serway received his doctorate at Illinois Institute of Technology and is Professor Emeritus at James Madison University. In 1990, he received the Madison Scholar award at James Madison University, where he taught for 17 years. Dr. Serway began his teaching career at Clarkson University, where he conducted research and taught from 1967 to 1980. He was the recipient of the Distinguished Teaching Award at Clarkson University in 1977 and of the Alumni Achievement Award from Utica College in 1985. As Guest Scientist at the IBM Research Laboratory in Zurich, Switzerland, he worked with K. Alex Müller, 1987 Nobel Prize recipient. Dr. Serway also was a visiting scientist at Argonne National Laboratory, where he collaborated with his mentor and friend, Sam Marshall. In addition to earlier editions of this textbook, Dr. Serway is the author of *Principles of Physics, Second Edition,* and co-author of *College Physics, Fifth Edition,* and *Modern Physics, Second Edition*; he also is the author of the high-school textbook *Physics,* published by Holt, Rinehart, & Winston. In addition, Dr. Serway has published more than 40 research papers in the field of condensed matter physics and has given more than 60 presentations at professional meetings. Dr. Serway and his wife Elizabeth enjoy traveling, golfing, and spending quality time with their four children and four grandchildren.

Robert J. Beichner received his doctorate at the State University of New York at Buffalo. Currently, he is Associate Professor of Physics at North Carolina State University, where he directs the Physics Education Research and Development Group. He has more than 20 years of teaching experience at the community-college, four year–college, and university levels. His research interests are centered on improving physics instruction: In his work, he has published studies of video-based laboratories, collaborative learning, technology-supplemented learning environments, and the assessment of student understanding of various physics topics. Dr. Beichner has held several leadership roles in the field of physics education research and has given numerous talks and colloquia on his work. In addition to being an author of this textbook, he is the co-author of two CD-ROMs, several commercially available software packages, and two books for preservice elementary school teachers. Dr. Beichner is an avid sea kayaker and enjoys spending time with his wife Mary and their two daughters, Sarah and Julie.

John W. Jewett, Jr. earned his doctorate at Ohio State University, specializing in optical and magnetic properties of condensed matter. He is currently Professor of Physics at California State Polytechnic University—Pomona. Throughout his teaching career, Dr. Jewett has been active in promoting science education. In addition to receiving four National Science Foundation grants, he helped found and direct the Southern California Area Modern Physics Institute (SCAMPI). He also is the director of Science IMPACT (Institute for Modern Pedagogy and Creative Teaching), which works with teachers and schools to develop effective science curricula. Both organizations operate in the United States and abroad. Dr. Jewett's honors include four Meritorious Performance and Professional Promise awards, selection as Outstanding Professor at California State Polytechnic University for 1991–1992, and the Excellence in Undergraduate Physics Teaching Award from the American Association of Physics Teachers (AAPT) in 1998. He has given many presentations both domestically and abroad, including multiple presentations at national meetings of the AAPT. He will co-author the third edition of *Principles of Physics* with Dr. Serway. Dr. Jewett enjoys playing piano, traveling, and collecting antique quack medical devices, as well as spending time with his wife Lisa and their children.

Pedagogical Color Chart

Part 1 (Chapters 1–15) : Mechanics

Displacement and position vectors

Linear (**v**) and angular (**ω**) velocity vectors

Velocity component vectors

Force vectors (**F**)

Force component vectors

Acceleration vectors (**a**)

Acceleration component vectors

Linear (**p**) and angular (**L**) momentum vectors

Torque vectors (**τ**)

Linear or rotational motion directions

Springs

Pulleys

Part 4 (Chapters 23–34) : Electricity and Magnetism

Electric fields

Magnetic fields

Positive charges

Negative charges

Resistors

Batteries and other dc power supplies

Switches

Capacitors

Inductors (coils)

Voltmeters

Ammeters

Galvanometers

ac Generators

Ground symbol

Part 5 (Chapters 35–38) : Light and Optics

Light rays

Lenses and prisms

Mirrors

Objects

Images

part 2

Mechanical Waves

As we learned in Chapter 13, most elastic objects oscillate when a force is applied to them and then removed. That is, once such an object is distorted, its shape tends to be restored to some equilibrium configuration. Even the atoms in a solid oscillate about some equilibrium position, as if they were connected to their neighbors by imaginary springs.

Wave motion—the subject we study next—is closely related to the phenomenon of oscillation. Sound waves, earthquake waves, waves on stretched strings, and water waves are all produced by some source of oscillation. As a sound wave travels through the air, the air molecules oscillate back and forth; as a water wave travels across a pond, the water molecules oscillate up and down and backward and forward. In general, as waves travel through any medium, the particles of the medium move in repetitive cycles. Therefore, the motion of the particles bears a strong resemblance to the periodic motion of an oscillating pendulum or a mass attached to a spring.

To explain many other phenomena in nature, we must understand the concepts of oscillations and waves. For instance, although sky-scrapers and bridges appear to be rigid, they actually oscillate, a fact that the architects and engineers who design and build them must take into account. To understand how radio and television work, we must understand the origin and nature of electromagnetic waves and how they propagate through space. Finally, much of what scientists have learned about atomic structure has come from information carried by waves. Therefore, we must first study oscillations and waves if we are to understand the concepts and theories of atomic physics.

◀ Copyright Warren Bolster/Tony Stone Images.

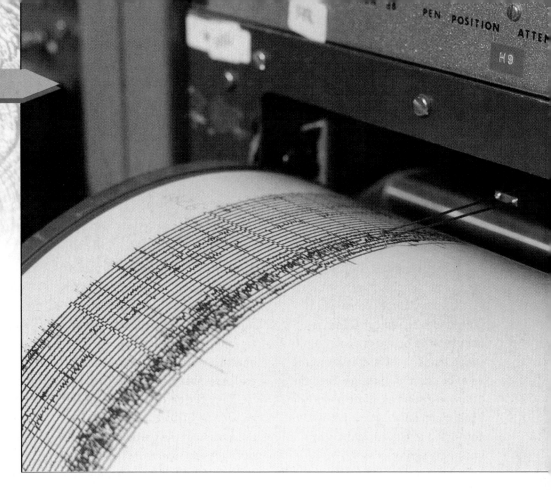

✳ PUZZLER

A simple seismograph can be constructed with a spring-suspended pen that draws a line on a slowly unrolling strip of paper. The paper is mounted on a structure attached to the ground. During an earthquake, the pen remains nearly stationary while the paper shakes beneath it. How can a few jagged lines on a piece of paper allow scientists at a seismograph station to determine the distance to the origin of an earthquake? *(Ken M. Johns/Photo Researchers, Inc.)*

c h a p t e r

16

Wave Motion

Chapter Outline

16.1 Basic Variables of Wave Motion

16.2 Direction of Particle Displacement

16.3 One-Dimensional Traveling Waves

16.4 Superposition and Interference

16.5 The Speed of Waves on Strings

16.6 Reflection and Transmission

16.7 Sinusoidal Waves

16.8 Rate of Energy Transfer by Sinusoidal Waves on Strings

16.9 *(Optional)* The Linear Wave Equation

M ost of us experienced waves as children when we dropped a pebble into a pond. At the point where the pebble hits the water's surface, waves are created. These waves move outward from the creation point in expanding circles until they reach the shore. If you were to examine carefully the motion of a leaf floating on the disturbed water, you would see that the leaf moves up, down, and sideways about its original position but does not undergo any net displacement away from or toward the point where the pebble hit the water. The water molecules just beneath the leaf, as well as all the other water molecules on the pond's surface, behave in the same way. That is, the water *wave* moves from the point of origin to the shore, but the water is not carried with it.

An excerpt from a book by Einstein and Infeld gives the following remarks concerning wave phenomena:[1]

> A bit of gossip starting in Washington reaches New York [by word of mouth] very quickly, even though not a single individual who takes part in spreading it travels between these two cities. There are two quite different motions involved, that of the rumor, Washington to New York, and that of the persons who spread the rumor. The wind, passing over a field of grain, sets up a wave which spreads out across the whole field. Here again we must distinguish between the motion of the wave and the motion of the separate plants, which undergo only small oscillations... The particles constituting the medium perform only small vibrations, but the whole motion is that of a progressive wave. The essentially new thing here is that for the first time we consider the motion of something which is not matter, but energy propagated through matter.

The world is full of waves, the two main types being *mechanical* waves and *electromagnetic* waves. We have already mentioned examples of mechanical waves: sound waves, water waves, and "grain waves." In each case, some physical medium is being disturbed—in our three particular examples, air molecules, water molecules, and stalks of grain. Electromagnetic waves do not require a medium to propagate; some examples of electromagnetic waves are visible light, radio waves, television signals, and x-rays. Here, in Part 2 of this book, we study only mechanical waves.

The wave concept is abstract. When we observe what we call a water wave, what we see is a rearrangement of the water's surface. Without the water, there would be no wave. A wave traveling on a string would not exist without the string. Sound waves could not travel through air if there were no air molecules. With mechanical waves, what we interpret as a wave corresponds to the propagation of a disturbance through a medium.

Interference patterns produced by outward-spreading waves from many drops of liquid falling into a body of water. *(Martin Dohrn/ Science Photo Library/Photo Researchers, Inc.)*

[1] A. Einstein and L. Infeld, *The Evolution of Physics*, New York, Simon & Schuster, 1961. Excerpt from "What Is a Wave?"

The mechanical waves discussed in this chapter require (1) some source of disturbance, (2) a medium that can be disturbed, and (3) some physical connection through which adjacent portions of the medium can influence each other. We shall find that all waves carry energy. The amount of energy transmitted through a medium and the mechanism responsible for that transport of energy differ from case to case. For instance, the power of ocean waves during a storm is much greater than the power of sound waves generated by a single human voice.

16.1 ▸ BASIC VARIABLES OF WAVE MOTION

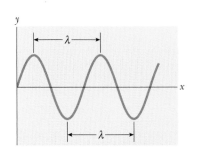

Figure 16.1 The wavelength λ of a wave is the distance between adjacent crests, adjacent troughs, or any other comparable adjacent identical points.

Imagine you are floating on a raft in a large lake. You slowly bob up and down as waves move past you. As you look out over the lake, you may be able to see the individual waves approaching. The point at which the displacement of the water from its normal level is highest is called the **crest** of the wave. The distance from one crest to the next is called the **wavelength** λ (Greek letter lambda). More generally, the wavelength is **the minimum distance between any two identical points (such as the crests) on adjacent waves,** as shown in Figure 16.1.

If you count the number of seconds between the arrivals of two adjacent waves, you are measuring the **period** T of the waves. In general, the period is **the time required for two identical points (such as the crests) of adjacent waves to pass by a point.**

The same information is more often given by the inverse of the period, which is called the **frequency** f. In general, the frequency of a periodic wave is **the number of crests (or troughs, or any other point on the wave) that pass a given point in a unit time interval.** The maximum displacement of a particle of the medium is called the **amplitude** A of the wave. For our water wave, this represents the highest distance of a water molecule above the undisturbed surface of the water as the wave passes by.

Waves travel with a specific speed, and this speed depends on the properties of the medium being disturbed. For instance, sound waves travel through room-temperature air with a speed of about 343 m/s (781 mi/h), whereas they travel through most solids with a speed greater than 343 m/s.

16.2 ▸ DIRECTION OF PARTICLE DISPLACEMENT

One way to demonstrate wave motion is to flick one end of a long rope that is under tension and has its opposite end fixed, as shown in Figure 16.2. In this manner, a single wave bump (called a *wave pulse*) is formed and travels along the rope with a definite speed. This type of disturbance is called a **traveling wave,** and Figure 16.2 represents four consecutive "snapshots" of the creation and propagation of the traveling wave. The rope is the medium through which the wave travels. Such a single pulse, in contrast to a train of pulses, has no frequency, no period, and no wavelength. However, the pulse does have definite amplitude and definite speed. As we shall see later, the properties of this particular medium that determine the speed of the wave are the tension in the rope and its mass per unit length. The shape of the wave pulse changes very little as it travels along the rope.[2]

As the wave pulse travels, each small segment of the rope, as it is disturbed, moves in a direction perpendicular to the wave motion. Figure 16.3 illustrates this

---

[2] Strictly speaking, the pulse changes shape and gradually spreads out during the motion. This effect is called *dispersion* and is common to many mechanical waves, as well as to electromagnetic waves. We do not consider dispersion in this chapter.

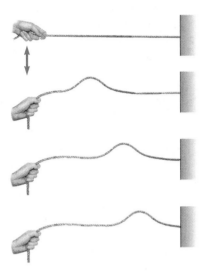

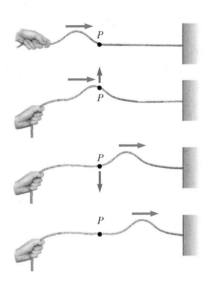

Figure 16.2 A wave pulse traveling down a stretched rope. The shape of the pulse is approximately unchanged as it travels along the rope.

Figure 16.3 A pulse traveling on a stretched rope is a transverse wave. The direction of motion of any element *P* of the rope (blue arrows) is perpendicular to the direction of wave motion (red arrows).

point for one particular segment, labeled *P*. Note that no part of the rope ever moves in the direction of the wave.

> A traveling wave that causes the particles of the disturbed medium to move perpendicular to the wave motion is called a **transverse wave.**

Transverse wave

Compare this with another type of wave—one moving down a long, stretched spring, as shown in Figure 16.4. The left end of the spring is pushed briefly to the right and then pulled briefly to the left. This movement creates a sudden compression of a region of the coils. The compressed region travels along the spring (to the right in Figure 16.4). The compressed region is followed by a region where the coils are extended. Notice that the direction of the displacement of the coils is *parallel* to the direction of propagation of the compressed region.

> A traveling wave that causes the particles of the medium to move parallel to the direction of wave motion is called a **longitudinal wave.**

Longitudinal wave

Sound waves, which we shall discuss in Chapter 17, are another example of longitudinal waves. The disturbance in a sound wave is a series of high-pressure and low-pressure regions that travel through air or any other material medium.

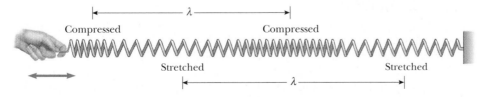

Figure 16.4 A longitudinal wave along a stretched spring. The displacement of the coils is in the direction of the wave motion. Each compressed region is followed by a stretched region.

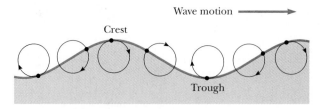

Figure 16.5 The motion of water molecules on the surface of deep water in which a wave is propagating is a combination of transverse and longitudinal displacements, with the result that molecules at the surface move in nearly circular paths. Each molecule is displaced both horizontally and vertically from its equilibrium position.

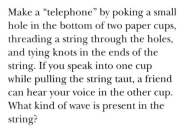

QuickLab

Make a "telephone" by poking a small hole in the bottom of two paper cups, threading a string through the holes, and tying knots in the ends of the string. If you speak into one cup while pulling the string taut, a friend can hear your voice in the other cup. What kind of wave is present in the string?

Some waves in nature exhibit a combination of transverse and longitudinal displacements. Surface water waves are a good example. When a water wave travels on the surface of deep water, water molecules at the surface move in nearly circular paths, as shown in Figure 16.5. Note that the disturbance has both transverse and longitudinal components. The transverse displacement is seen in Figure 16.5 as the variations in vertical position of the water molecules. The longitudinal displacement can be explained as follows: As the wave passes over the water's surface, water molecules at the crests move in the direction of propagation of the wave, whereas molecules at the troughs move in the direction opposite the propagation. Because the molecule at the labeled crest in Figure 16.5 will be at a trough after half a period, its movement in the direction of the propagation of the wave will be canceled by its movement in the opposite direction. This holds for every other water molecule disturbed by the wave. Thus, there is no net displacement of any water molecule during one complete cycle. Although the *molecules* experience no net displacement, the *wave* propagates along the surface of the water.

The three-dimensional waves that travel out from the point under the Earth's surface at which an earthquake occurs are of both types—transverse and longitudinal. The longitudinal waves are the faster of the two, traveling at speeds in the range of 7 to 8 km/s near the surface. These are called **P waves,** with "P" standing for *primary* because they travel faster than the transverse waves and arrive at a seismograph first. The slower transverse waves, called **S waves** (with "S" standing for *secondary*), travel through the Earth at 4 to 5 km/s near the surface. By recording the time interval between the arrival of these two sets of waves at a seismograph, the distance from the seismograph to the point of origin of the waves can be determined. A single such measurement establishes an imaginary sphere centered on the seismograph, with the radius of the sphere determined by the difference in arrival times of the P and S waves. The origin of the waves is located somewhere on that sphere. The imaginary spheres from three or more monitoring stations located far apart from each other intersect at one region of the Earth, and this region is where the earthquake occurred.

Quick Quiz 16.1

(a) In a long line of people waiting to buy tickets, the first person leaves and a pulse of motion occurs as people step forward to fill the gap. As each person steps forward, the gap moves through the line. Is the propagation of this gap transverse or longitudinal? (b) Consider the "wave" at a baseball game: people stand up and shout as the wave arrives at their location, and the resultant pulse moves around the stadium. Is this wave transverse or longitudinal?

16.3 ONE-DIMENSIONAL TRAVELING WAVES

Consider a wave pulse traveling to the right with constant speed v on a long, taut string, as shown in Figure 16.6. The pulse moves along the x axis (the axis of the string), and the transverse (vertical) displacement of the string (the medium) is measured along the y axis. Figure 16.6a represents the shape and position of the pulse at time $t = 0$. At this time, the shape of the pulse, whatever it may be, can be represented as $y = f(x)$. That is, y, which is the vertical position of any point on the string, is some definite function of x. The displacement y, sometimes called the *wave function*, depends on both x and t. For this reason, it is often written $y(x, t)$, which is read "y as a function of x and t." Consider a particular point P on the string, identified by a specific value of its x coordinate. Before the pulse arrives at P, the y coordinate of this point is zero. As the wave passes P, the y coordinate of this point increases, reaches a maximum, and then decreases to zero. Therefore, **the wave function y represents the y coordinate of any point P of the medium at any time t.**

Because its speed is v, the wave pulse travels to the right a distance vt in a time t (see Fig. 16.6b). If the shape of the pulse does not change with time, we can represent the wave function y for all times after $t = 0$. Measured in a stationary reference frame having its origin at O, the wave function is

$$y = f(x - vt) \tag{16.1}$$

Wave traveling to the right

If the wave pulse travels to the left, the string displacement is

$$y = f(x + vt) \tag{16.2}$$

Wave traveling to the left

For any given time t, the wave function y as a function of x defines a curve representing the shape of the pulse at this time. This curve is equivalent to a "snapshot" of the wave at this time. For a pulse that moves without changing shape, the speed of the pulse is the same as that of any feature along the pulse, such as the crest shown in Figure 16.6. To find the speed of the pulse, we can calculate how far the crest moves in a short time and then divide this distance by the time interval. To follow the motion of the crest, we must substitute some particular value, say x_0, in Equation 16.1 for $x - vt$. Regardless of how x and t change individually, we must require that $x - vt = x_0$ in order to stay with the crest. This expression therefore represents the equation of motion of the crest. At $t = 0$, the crest is at $x = x_0$; at a

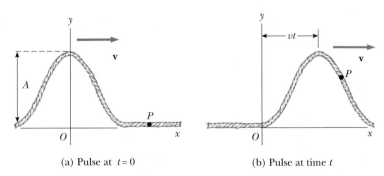

(a) Pulse at $t = 0$ (b) Pulse at time t

Figure 16.6 A one-dimensional wave pulse traveling to the right with a speed v. (a) At $t = 0$, the shape of the pulse is given by $y = f(x)$. (b) At some later time t, the shape remains unchanged and the vertical displacement of any point P of the medium is given by $y = f(x - vt)$.

time dt later, the crest is at $x = x_0 + v\,dt$. Therefore, in a time dt, the crest has moved a distance $dx = (x_0 + v\,dt) - x_0 = v\,dt$. Hence, the wave speed is

$$v = \frac{dx}{dt} \qquad\qquad \textbf{(16.3)}$$

EXAMPLE 16.1   A Pulse Moving to the Right

A wave pulse moving to the right along the x axis is represented by the wave function

$$y(x, t) = \frac{2}{(x - 3.0t)^2 + 1}$$

where x and y are measured in centimeters and t is measured in seconds. Plot the wave function at $t = 0$, $t = 1.0$ s, and $t = 2.0$ s.

Solution   First, note that this function is of the form $y = f(x - vt)$. By inspection, we see that the wave speed is $v = 3.0$ cm/s. Furthermore, the wave amplitude (the maximum value of y) is given by $A = 2.0$ cm. (We find the maximum value of the function representing y by letting $x - 3.0t = 0$.) The wave function expressions are

$$y(x, 0) = \frac{2}{x^2 + 1} \qquad \text{at } t = 0$$

$$y(x, 1.0) = \frac{2}{(x - 3.0)^2 + 1} \qquad \text{at } t = 1.0 \text{ s}$$

$$y(x, 2.0) = \frac{2}{(x - 6.0)^2 + 1} \qquad \text{at } t = 2.0 \text{ s}$$

We now use these expressions to plot the wave function versus x at these times. For example, let us evaluate $y(x, 0)$ at $x = 0.50$ cm:

$$y(0.50, 0) = \frac{2}{(0.50)^2 + 1} = 1.6 \text{ cm}$$

Likewise, at $x = 1.0$ cm, $y(1.0, 0) = 1.0$ cm, and at $x = 2.0$ cm, $y(2.0, 0) = 0.40$ cm. Continuing this procedure for other values of x yields the wave function shown in Figure 16.7a. In a similar manner, we obtain the graphs of $y(x, 1.0)$ and $y(x, 2.0)$, shown in Figure 16.7b and c, respectively. These snapshots show that the wave pulse moves to the right without changing its shape and that it has a constant speed of 3.0 cm/s.

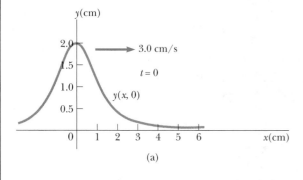

(a)

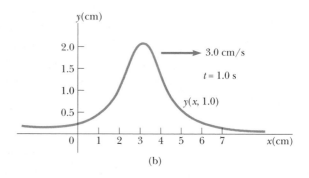

(b)

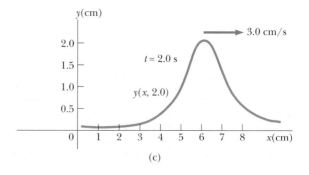

(c)

Figure 16.7   Graphs of the function $y(x, t) = 2/[(x - 3.0t)^2 + 1]$ at (a) $t = 0$, (b) $t = 1.0$ s, and (c) $t = 2.0$ s.

16.4 ▷ SUPERPOSITION AND INTERFERENCE

Many interesting wave phenomena in nature cannot be described by a single moving pulse. Instead, one must analyze complex waves in terms of a combination of many traveling waves. To analyze such wave combinations, one can make use of the **superposition principle:**

> If two or more traveling waves are moving through a medium, the resultant wave function at any point is the algebraic sum of the wave functions of the individual waves.

Linear waves obey the superposition principle

Waves that obey this principle are called *linear waves* and are generally characterized by small amplitudes. Waves that violate the superposition principle are called *nonlinear waves* and are often characterized by large amplitudes. In this book, we deal only with linear waves.

One consequence of the superposition principle is that **two traveling waves can pass through each other without being destroyed or even altered.** For instance, when two pebbles are thrown into a pond and hit the surface at different places, the expanding circular surface waves do not destroy each other but rather pass through each other. The complex pattern that is observed can be viewed as two independent sets of expanding circles. Likewise, when sound waves from two sources move through air, they pass through each other. The resulting sound that one hears at a given point is the resultant of the two disturbances.

Figure 16.8 is a pictorial representation of superposition. The wave function for the pulse moving to the right is y_1, and the wave function for the pulse moving

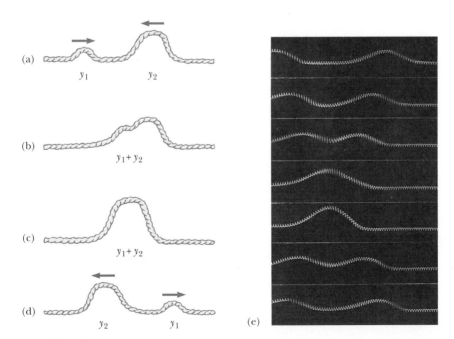

Figure 16.8   (a–d) Two wave pulses traveling on a stretched string in opposite directions pass through each other. When the pulses overlap, as shown in (b) and (c), the net displacement of the string equals the sum of the displacements produced by each pulse. Because each pulse displaces the string in the positive direction, we refer to the superposition of the two pulses as *constructive interference.* (e) Photograph of superposition of two equal, symmetric pulses traveling in opposite directions on a stretched spring. *(e, Education Development Center, Newton, MA)*

Interference of water waves produced in a ripple tank. The sources of the waves are two objects that oscillate perpendicular to the surface of the tank. *(Courtesy of Central Scientific Company)*

to the left is y_2. The pulses have the same speed but different shapes. Each pulse is assumed to be symmetric, and the displacement of the medium is in the positive y direction for both pulses. (Note, however, that the superposition principle applies even when the two pulses are not symmetric.) When the waves begin to overlap (Fig. 16.8b), the wave function for the resulting complex wave is given by $y_1 + y_2$.

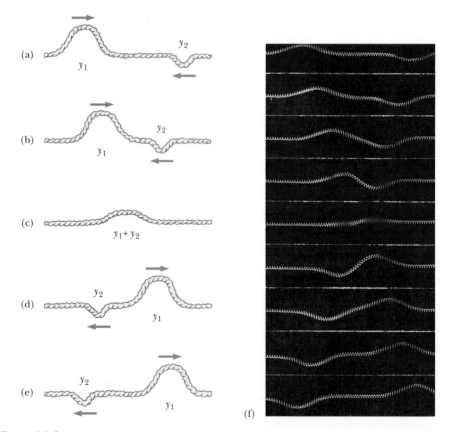

Figure 16.9   (a–e) Two wave pulses traveling in opposite directions and having displacements that are inverted relative to each other. When the two overlap in (c), their displacements partially cancel each other. (f) Photograph of superposition of two symmetric pulses traveling in opposite directions, where one pulse is inverted relative to the other. *(f, Education Development Center, Newton, MA)*

When the crests of the pulses coincide (Fig. 16.8c), the resulting wave given by $y_1 + y_2$ is symmetric. The two pulses finally separate and continue moving in their original directions (Fig. 16.8d). Note that the pulse shapes remain unchanged, as if the two pulses had never met!

The combination of separate waves in the same region of space to produce a resultant wave is called **interference.** For the two pulses shown in Figure 16.8, the displacement of the medium is in the positive y direction for both pulses, and the resultant wave (created when the pulses overlap) exhibits a displacement greater than that of either individual pulse. Because the displacements caused by the two pulses are in the same direction, we refer to their superposition as **constructive interference.**

Now consider two pulses traveling in opposite directions on a taut string where one pulse is inverted relative to the other, as illustrated in Figure 16.9. In this case, when the pulses begin to overlap, the resultant wave is given by $y_1 + y_2$, but the values of the function y_2 are negative. Again, the two pulses pass through each other; however, because the displacements caused by the two pulses are in opposite directions, we refer to their superposition as **destructive interference.**

Quick Quiz 16.2

Two pulses are traveling toward each other at 10 cm/s on a long string, as shown in Figure 16.10. Sketch the shape of the string at $t = 0.6$ s.

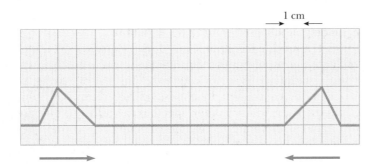

Figure 16.10   The pulses on this string are traveling at 10 cm/s.

16.5   THE SPEED OF WAVES ON STRINGS

In this section, we focus on determining the speed of a transverse pulse traveling on a taut string. Let us first conceptually argue the parameters that determine the speed. If a string under tension is pulled sideways and then released, the tension is responsible for accelerating a particular segment of the string back toward its equilibrium position. According to Newton's second law, the acceleration of the segment increases with increasing tension. If the segment returns to equilibrium more rapidly due to this increased acceleration, we would intuitively argue that the wave speed is greater. Thus, we expect the wave speed to increase with increasing tension.

Likewise, we can argue that the wave speed decreases if the mass per unit length of the string increases. This is because it is more difficult to accelerate a massive segment of the string than a light segment. If the tension in the string is T (not to be confused with the same symbol used for the period) and its mass per

The strings of this piano vary in both tension and mass per unit length. These differences in tension and density, in combination with the different lengths of the strings, allow the instrument to produce a wide range of sounds. *(Charles D. Winters)*

unit length is μ (Greek letter mu), then, as we shall show, the wave speed is

Speed of a wave on a stretched string

$$v = \sqrt{\frac{T}{\mu}}$$

(16.4)

First, let us verify that this expression is dimensionally correct. The dimensions of T are $\mathrm{ML/T^2}$, and the dimensions of μ are $\mathrm{M/L}$. Therefore, the dimensions of T/μ are $\mathrm{L^2/T^2}$; hence, the dimensions of $\sqrt{T/\mu}$ are $\mathrm{L/T}$—indeed, the dimensions of speed. No other combination of T and μ is dimensionally correct if we assume that they are the only variables relevant to the situation.

Now let us use a mechanical analysis to derive Equation 16.4. On our string under tension, consider a pulse moving to the right with a uniform speed v measured relative to a stationary frame of reference. Instead of staying in this reference frame, it is more convenient to choose as our reference frame one that moves along with the pulse with the same speed as the pulse, so that the pulse is at rest within the frame. This change of reference frame is permitted because Newton's laws are valid in either a stationary frame or one that moves with constant velocity. In our new reference frame, a given segment of the string initially to the right of the pulse moves to the left, rises up and follows the shape of the pulse, and then continues to move to the left. Figure 16.11a shows such a segment at the instant it is located at the top of the pulse.

The small segment of the string of length Δs shown in Figure 16.11a, and magnified in Figure 16.11b, forms an approximate arc of a circle of radius R. In our moving frame of reference (which is moving to the right at a speed v along with the pulse), the shaded segment is moving to the left with a speed v. This segment has a centripetal acceleration equal to v^2/R, which is supplied by components of the tension **T** in the string. The force **T** acts on either side of the segment and tangent to the arc, as shown in Figure 16.11b. The horizontal components of **T** cancel, and each vertical component $T \sin \theta$ acts radially toward the center of the arc. Hence, the total radial force is $2T \sin \theta$. Because the segment is small, θ is small, and we can use the small-angle approximation $\sin \theta \approx \theta$. Therefore, the total radial force is

$$\sum F_r = 2T \sin \theta \approx 2T\theta$$

The segment has a mass $m = \mu \Delta s$. Because the segment forms part of a circle and subtends an angle 2θ at the center, $\Delta s = R(2\theta)$, and hence

$$m = \mu \Delta s = 2\mu R \theta$$

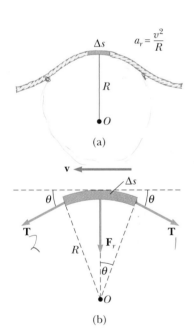

Figure 16.11 (a) To obtain the speed v of a wave on a stretched string, it is convenient to describe the motion of a small segment of the string in a moving frame of reference. (b) In the moving frame of reference, the small segment of length Δs moves to the left with speed v. The net force on the segment is in the radial direction because the horizontal components of the tension force cancel.

If we apply Newton's second law to this segment, the radial component of motion gives

$$\sum F_r = ma = \frac{mv^2}{R}$$

$$2T\theta = \frac{2\mu R\theta v^2}{R}$$

Solving for v gives Equation 16.4.

Notice that this derivation is based on the assumption that the pulse height is small relative to the length of the string. Using this assumption, we were able to use the approximation $\sin\theta \approx \theta$. Furthermore, the model assumes that the tension T is not affected by the presence of the pulse; thus, T is the same at all points on the string. Finally, this proof does *not* assume any particular shape for the pulse. Therefore, we conclude that a pulse of *any shape* travels along the string with speed $v = \sqrt{T/\mu}$ without any change in pulse shape.

EXAMPLE 16.2   The Speed of a Pulse on a Cord

A uniform cord has a mass of 0.300 kg and a length of 6.00 m (Fig. 16.12). The cord passes over a pulley and supports a 2.00-kg object. Find the speed of a pulse traveling along this cord.

Solution   The tension T in the cord is equal to the weight of the suspended 2.00-kg mass:

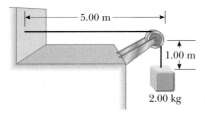

Figure 16.12   The tension T in the cord is maintained by the suspended object. The speed of any wave traveling along the cord is given by $v = \sqrt{T/\mu}$.

$$T = mg = (2.00\text{ kg})(9.80\text{ m/s}^2) = 19.6\text{ N}$$

(This calculation of the tension neglects the small mass of the cord. Strictly speaking, the cord can never be exactly horizontal, and therefore the tension is not uniform.) The mass per unit length μ of the cord is

$$\mu = \frac{m}{\ell} = \frac{0.300\text{ kg}}{6.00\text{ m}} = 0.0500\text{ kg/m}$$

Therefore, the wave speed is

$$v = \sqrt{\frac{T}{\mu}} = \sqrt{\frac{19.6\text{ N}}{0.050\,0\text{ kg/m}}} = \boxed{19.8\text{ m/s}}$$

Exercise   Find the time it takes the pulse to travel from the wall to the pulley.

Answer   0.253 s.

Quick Quiz 16.3

Suppose you create a pulse by moving the free end of a taut string up and down once with your hand. The string is attached at its other end to a distant wall. The pulse reaches the wall in a time t. Which of the following actions, taken by itself, decreases the time it takes the pulse to reach the wall? More than one choice may be correct.
(a)   Moving your hand more quickly, but still only up and down once by the same amount.
(b)   Moving your hand more slowly, but still only up and down once by the same amount.
(c)   Moving your hand a greater distance up and down in the same amount of time.
(d)   Moving your hand a lesser distance up and down in the same amount of time.
(e)   Using a heavier string of the same length and under the same tension.
(f)   Using a lighter string of the same length and under the same tension.
(g)   Using a string of the same linear mass density but under decreased tension.
(h)   Using a string of the same linear mass density but under increased tension.

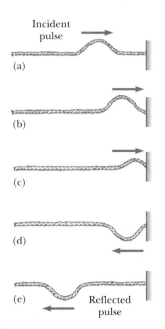

Figure 16.13   The reflection of a traveling wave pulse at the fixed end of a stretched string. The reflected pulse is inverted, but its shape is unchanged.

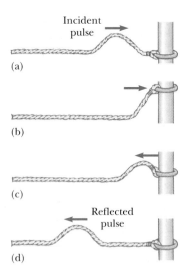

Figure 16.14   The reflection of a traveling wave pulse at the free end of a stretched string. The reflected pulse is not inverted.

16.6 ▶ REFLECTION AND TRANSMISSION

We have discussed traveling waves moving through a uniform medium. We now consider how a traveling wave is affected when it encounters a change in the medium. For example, consider a pulse traveling on a string that is rigidly attached to a support at one end (Fig. 16.13). When the pulse reaches the support, a severe change in the medium occurs—the string ends. The result of this change is that the wave undergoes **reflection**—that is, the pulse moves back along the string in the opposite direction.

Note that the reflected pulse is inverted. This inversion can be explained as follows: When the pulse reaches the fixed end of the string, the string produces an upward force on the support. By Newton's third law, the support must exert an equal and opposite (downward) reaction force on the string. This downward force causes the pulse to invert upon reflection.

Now consider another case: this time, the pulse arrives at the end of a string that is free to move vertically, as shown in Figure 16.14. The tension at the free end is maintained because the string is tied to a ring of negligible mass that is free to slide vertically on a smooth post. Again, the pulse is reflected, but this time it is not inverted. When it reaches the post, the pulse exerts a force on the free end of the string, causing the ring to accelerate upward. The ring overshoots the height of the incoming pulse, and then the downward component of the tension force pulls the ring back down. This movement of the ring produces a reflected pulse that is not inverted and that has the same amplitude as the incoming pulse.

Finally, we may have a situation in which the boundary is intermediate between these two extremes. In this case, part of the incident pulse is reflected and part undergoes **transmission**—that is, some of the pulse passes through the boundary. For instance, suppose a light string is attached to a heavier string, as shown in Figure 16.15. When a pulse traveling on the light string reaches the boundary between the two, part of the pulse is reflected and inverted and part is transmitted to the heavier string. The reflected pulse is inverted for the same reasons described earlier in the case of the string rigidly attached to a support.

Note that the reflected pulse has a smaller amplitude than the incident pulse. In Section 16.8, we shall learn that the energy carried by a wave is related to its amplitude. Thus, according to the principle of the conservation of energy, when the pulse breaks up into a reflected pulse and a transmitted pulse at the boundary, the sum of the energies of these two pulses must equal the energy of the incident pulse. Because the reflected pulse contains only part of the energy of the incident pulse, its amplitude must be smaller.

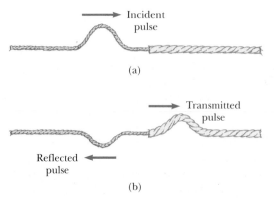

Figure 16.15   (a) A pulse traveling to the right on a light string attached to a heavier string. (b) Part of the incident pulse is reflected (and inverted), and part is transmitted to the heavier string.

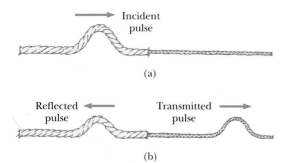

(a)

Incident pulse

Reflected pulse   Transmitted pulse

(b)

Figure 16.16 (a) A pulse traveling to the right on a heavy string attached to a lighter string. (b) The incident pulse is partially reflected and partially transmitted, and the reflected pulse is not inverted.

When a pulse traveling on a heavy string strikes the boundary between the heavy string and a lighter one, as shown in Figure 16.16, again part is reflected and part is transmitted. In this case, the reflected pulse is not inverted.

In either case, the relative heights of the reflected and transmitted pulses depend on the relative densities of the two strings. If the strings are identical, there is no discontinuity at the boundary and no reflection takes place.

According to Equation 16.4, the speed of a wave on a string increases as the mass per unit length of the string decreases. In other words, a pulse travels more slowly on a heavy string than on a light string if both are under the same tension. The following general rules apply to reflected waves: **When a wave pulse travels from medium A to medium B and $v_A > v_B$ (that is, when B is denser than A), the pulse is inverted upon reflection. When a wave pulse travels from medium A to medium B and $v_A < v_B$ (that is, when A is denser than B), the pulse is not inverted upon reflection.**

16.7 SINUSOIDAL WAVES

In this section, we introduce an important wave function whose shape is shown in Figure 16.17. The wave represented by this curve is called a **sinusoidal wave** because the curve is the same as that of the function $\sin \theta$ plotted against θ. The sinusoidal wave is the simplest example of a periodic continuous wave and can be used to build more complex waves, as we shall see in Section 18.8. The red curve represents a snapshot of a traveling sinusoidal wave at $t = 0$, and the blue curve represents a snapshot of the wave at some later time t. At $t = 0$, the function describing the positions of the particles of the medium through which the sinusoidal wave is traveling can be written

$$y = A \sin\left(\frac{2\pi}{\lambda} x\right) \tag{16.5}$$

where the constant A represents the wave amplitude and the constant λ is the wavelength. Thus, we see that the position of a particle of the medium is the same whenever x is increased by an integral multiple of λ. If the wave moves to the right with a speed v, then the wave function at some later time t is

$$y = A \sin\left[\frac{2\pi}{\lambda}(x - vt)\right] \tag{16.6}$$

That is, the traveling sinusoidal wave moves to the right a distance vt in the time t, as shown in Figure 16.17. Note that the wave function has the form $f(x - vt)$ and

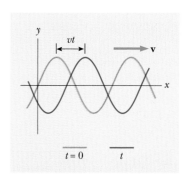

Figure 16.17 A one-dimensional sinusoidal wave traveling to the right with a speed v. The red curve represents a snapshot of the wave at $t = 0$, and the blue curve represents a snapshot at some later time t.

so represents a wave traveling to the right. If the wave were traveling to the left, the quantity $x - vt$ would be replaced by $x + vt$, as we learned when we developed Equations 16.1 and 16.2.

By definition, the wave travels a distance of one wavelength in one period T. Therefore, the wave speed, wavelength, and period are related by the expression

$$v = \frac{\lambda}{T} \qquad \textbf{(16.7)}$$

Substituting this expression for v into Equation 16.6, we find that

$$y = A \sin\left[2\pi\left(\frac{x}{\lambda} - \frac{t}{T}\right)\right] \qquad \textbf{(16.8)}$$

This form of the wave function clearly shows the *periodic* nature of y. At any given time t (a snapshot of the wave), y has the *same* value at the positions x, $x + \lambda$, $x + 2\lambda$, and so on. Furthermore, at any given position x, the value of y is the same at times t, $t + T$, $t + 2T$, and so on.

We can express the wave function in a convenient form by defining two other quantities, the **angular wave number** k and the **angular frequency** ω:

Angular wave number

$$k \equiv \frac{2\pi}{\lambda} \qquad \textbf{(16.9)}$$

Angular frequency

$$\omega \equiv \frac{2\pi}{T} \qquad \textbf{(16.10)}$$

Using these definitions, we see that Equation 16.8 can be written in the more compact form

Wave function for a sinusoidal wave

$$y = A \sin(kx - \omega t) \qquad \textbf{(16.11)}$$

The frequency of a sinusoidal wave is related to the period by the expression

Frequency

$$f = \frac{1}{T} \qquad \textbf{(16.12)}$$

The most common unit for frequency, as we learned in Chapter 13, is second^{-1}, or **hertz** (Hz). The corresponding unit for T is seconds.

Using Equations 16.9, 16.10, and 16.12, we can express the wave speed v originally given in Equation 16.7 in the alternative forms

$$v = \frac{\omega}{k} \qquad \textbf{(16.13)}$$

Speed of a sinusoidal wave

$$v = \lambda f \qquad \textbf{(16.14)}$$

The wave function given by Equation 16.11 assumes that the vertical displacement y is zero at $x = 0$ and $t = 0$. This need not be the case. If it is not, we generally express the wave function in the form

General expression for a sinusoidal wave

$$y = A \sin(kx - \omega t + \phi) \qquad \textbf{(16.15)}$$

where ϕ is the **phase constant,** just as we learned in our study of periodic motion in Chapter 13. This constant can be determined from the initial conditions.

EXAMPLE 16.3   A Traveling Sinusoidal Wave

A sinusoidal wave traveling in the positive x direction has an amplitude of 15.0 cm, a wavelength of 40.0 cm, and a frequency of 8.00 Hz. The vertical displacement of the medium at $t = 0$ and $x = 0$ is also 15.0 cm, as shown in Figure 16.18. (a) Find the angular wave number k, period T, angular frequency ω, and speed v of the wave.

Solution (a) Using Equations 16.9, 16.10, 16.12, and 16.14, we find the following:

$$k = \frac{2\pi}{\lambda} = \frac{2\pi \text{ rad}}{40.0 \text{ cm}} = \boxed{0.157 \text{ rad/cm}}$$

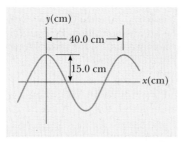

Figure 16.18 A sinusoidal wave of wavelength $\lambda = 40.0$ cm and amplitude $A = 15.0$ cm. The wave function can be written in the form $y = A\cos(kx - \omega t)$.

$$\omega = 2\pi f = 2\pi(8.00 \text{ s}^{-1}) = \boxed{50.3 \text{ rad/s}}$$

$$T = \frac{1}{f} = \frac{1}{8.00 \text{ s}^{-1}} = \boxed{0.125 \text{ s}}$$

$$v = \lambda f = (40.0 \text{ cm})(8.00 \text{ s}^{-1}) = \boxed{320 \text{ cm/s}}$$

(b) Determine the phase constant ϕ, and write a general expression for the wave function.

Solution Because $A = 15.0$ cm and because $y = 15.0$ cm at $x = 0$ and $t = 0$, substitution into Equation 16.15 gives

$$15.0 = (15.0)\sin\phi \quad \text{or} \quad \sin\phi = 1$$

We may take the principal value $\phi = \pi/2$ rad (or 90°). Hence, the wave function is of the form

$$y = A\sin\left(kx - \omega t + \frac{\pi}{2}\right) = A\cos(kx - \omega t)$$

By inspection, we can see that the wave function must have this form, noting that the cosine function has the same shape as the sine function displaced by 90°. Substituting the values for A, k, and ω into this expression, we obtain

$$y = (15.0 \text{ cm})\cos(0.157x - 50.3t)$$

Sinusoidal Waves on Strings

In Figure 16.2, we demonstrated how to create a pulse by jerking a taut string up and down once. To create a train of such pulses, normally referred to as a *wave train,* or just plain *wave,* we can replace the hand with an oscillating blade. If the wave consists of a train of identical cycles, whatever their shape, the relationships $f = 1/T$ and $v = f\lambda$ among speed, frequency, period, and wavelength hold true. We can make more definite statements about the wave function if the source of the waves vibrates in simple harmonic motion. Figure 16.19 represents snapshots of the wave created in this way at intervals of $T/4$. Note that because the end of the blade oscillates in simple harmonic motion, **each particle of the string, such as that at P, also oscillates vertically with simple harmonic motion.** This must be the case because each particle follows the simple harmonic motion of the blade. Therefore, every segment of the string can be treated as a simple harmonic oscillator vibrating with a frequency equal to the frequency of oscillation of the blade.[3] Note that although each segment oscillates in the y direction, the wave travels in the x direction with a speed v. Of course, this is the definition of a transverse wave.

---

[3] In this arrangement, we are assuming that a string segment always oscillates in a vertical line. The tension in the string would vary if a segment were allowed to move sideways. Such motion would make the analysis very complex.

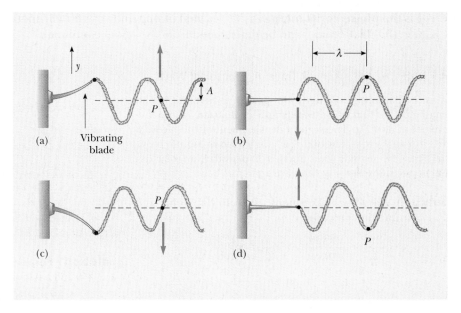

Figure 16.19   One method for producing a train of sinusoidal wave pulses on a string. The left end of the string is connected to a blade that is set into oscillation. Every segment of the string, such as the point P, oscillates with simple harmonic motion in the vertical direction.

If the wave at $t = 0$ is as described in Figure 16.19b, then the wave function can be written as

$$y = A \sin(kx - \omega t)$$

We can use this expression to describe the motion of any point on the string. The point P (or any other point on the string) moves only vertically, and so its x coordinate remains constant. Therefore, the **transverse speed v_y** (not to be confused with the wave speed v) and the **transverse acceleration a_y** are

$$v_y = \frac{dy}{dt}\bigg]_{x=\text{constant}} = \frac{\partial y}{\partial t} = -\omega A \cos(kx - \omega t) \qquad \textbf{(16.16)}$$

$$a_y = \frac{dv_y}{dt}\bigg]_{x=\text{constant}} = \frac{\partial v_y}{\partial t} = -\omega^2 A \sin(kx - \omega t) \qquad \textbf{(16.17)}$$

In these expressions, we must use partial derivatives (see Section 8.6) because y depends on both x and t. In the operation $\partial y/\partial t$, for example, we take a derivative with respect to t while holding x constant. The maximum values of the transverse speed and transverse acceleration are simply the absolute values of the coefficients of the cosine and sine functions:

$$v_{y,\,\text{max}} = \omega A \qquad \textbf{(16.18)}$$

$$a_{y,\,\text{max}} = \omega^2 A \qquad \textbf{(16.19)}$$

The transverse speed and transverse acceleration do not reach their maximum values simultaneously. The transverse speed reaches its maximum value (ωA) when $y = 0$, whereas the transverse acceleration reaches its maximum value ($\omega^2 A$) when $y = \pm A$. Finally, Equations 16.18 and 16.19 are identical in mathematical form to the corresponding equations for simple harmonic motion, Equations 13.10 and 13.11.

Quick Quiz 16.4

A sinusoidal wave is moving on a string. If you increase the frequency f of the wave, how do the transverse speed, wave speed, and wavelength change?

EXAMPLE 16.4   **A Sinusoidally Driven String**

The string shown in Figure 16.19 is driven at a frequency of 5.00 Hz. The amplitude of the motion is 12.0 cm, and the wave speed is 20.0 m/s. Determine the angular frequency ω and angular wave number k for this wave, and write an expression for the wave function.

Solution   Using Equations 16.10, 16.12, and 16.13, we find that

$$\omega = \frac{2\pi}{T} = 2\pi f = 2\pi(5.00 \text{ Hz}) = \boxed{31.4 \text{ rad/s}}$$

$$k = \frac{\omega}{v} = \frac{31.4 \text{ rad/s}}{20.0 \text{ m/s}} = \boxed{1.57 \text{ rad/m}}$$

Because $A = 12.0$ cm $= 0.120$ m, we have

$$y = A \sin(kx - \omega t) = (0.120 \text{ m}) \sin(1.57x - 31.4t)$$

Exercise   Calculate the maximum values for the transverse speed and transverse acceleration of any point on the string.

Answer   3.77 m/s; 118 m/s².

16.8   RATE OF ENERGY TRANSFER BY SINUSOIDAL WAVES ON STRINGS

As waves propagate through a medium, they transport energy. We can easily demonstrate this by hanging an object on a stretched string and then sending a pulse down the string, as shown in Figure 16.20. When the pulse meets the suspended object, the object is momentarily displaced, as illustrated in Figure 16.20b. In the process, energy is transferred to the object because work must be done for it to move upward. This section examines the rate at which energy is transported along a string. We shall assume a one-dimensional sinusoidal wave in the calculation of the energy transferred.

Consider a sinusoidal wave traveling on a string (Fig. 16.21). The source of the energy being transported by the wave is some external agent at the left end of the string; this agent does work in producing the oscillations. As the external agent performs work on the string, moving it up and down, energy enters the system of the string and propagates along its length. Let us focus our attention on a segment of the string of length Δx and mass Δm. Each such segment moves vertically with simple harmonic motion. Furthermore, all segments have the same angular frequency ω and the same amplitude A. As we found in Chapter 13, the elastic potential energy U associated with a particle in simple harmonic motion is $U = \frac{1}{2}ky^2$, where the simple harmonic motion is in the y direction. Using the relationship $\omega^2 = k/m$ developed in Equations 13.16 and 13.17, we can write this as

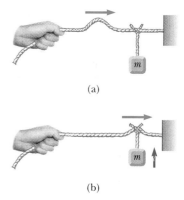

(a)

(b)

Figure 16.20   (a) A pulse traveling to the right on a stretched string on which an object has been suspended. (b) Energy is transmitted to the suspended object when the pulse arrives.

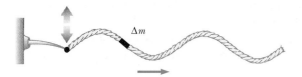

Figure 16.21   A sinusoidal wave traveling along the x axis on a stretched string. Every segment moves vertically, and every segment has the same total energy.

$U = \frac{1}{2}m\omega^2 y^2$. If we apply this equation to the segment of mass Δm, we see that the potential energy of this segment is

$$\Delta U = \frac{1}{2}(\Delta m)\omega^2 y^2$$

Because the mass per unit length of the string is $\mu = \Delta m/\Delta x$, we can express the potential energy of the segment as

$$\Delta U = \frac{1}{2}(\mu \Delta x)\omega^2 y^2$$

As the length of the segment shrinks to zero, $\Delta x \to dx$, and this expression becomes a differential relationship:

$$dU = \frac{1}{2}(\mu dx)\omega^2 y^2$$

We replace the general displacement y of the segment with the wave function for a sinusoidal wave:

$$dU = \frac{1}{2}\mu\omega^2 [A\sin(kx - \omega t)]^2\, dx = \frac{1}{2}\mu\omega^2 A^2 \sin^2(kx - \omega t)\, dx$$

If we take a snapshot of the wave at time $t = 0$, then the potential energy in a given segment is

$$dU = \frac{1}{2}\mu\omega^2 A^2 \sin^2 kx\, dx$$

To obtain the total potential energy in one wavelength, we integrate this expression over all the string segments in one wavelength:

$$U_\lambda = \int dU = \int_0^\lambda \frac{1}{2}\mu\omega^2 A^2 \sin^2 kx\, dx = \frac{1}{2}\mu\omega^2 A^2 \int_0^\lambda \sin^2 kx\, dx$$

$$= \frac{1}{2}\mu\omega^2 A^2 \left[\frac{1}{2}x - \frac{1}{4k}\sin 2kx\right]_0^\lambda = \frac{1}{2}\mu\omega^2 A^2 (\frac{1}{2}\lambda) = \frac{1}{4}\mu\omega^2 A^2 \lambda$$

Because it is in motion, each segment of the string also has kinetic energy. When we use this procedure to analyze the total kinetic energy in one wavelength of the string, we obtain the same result:

$$K_\lambda = \int dK = \frac{1}{4}\mu\omega^2 A^2 \lambda$$

The total energy in one wavelength of the wave is the sum of the potential and kinetic energies:

$$E_\lambda = U_\lambda + K_\lambda = \frac{1}{2}\mu\omega^2 A^2 \lambda \tag{16.20}$$

As the wave moves along the string, this amount of energy passes by a given point on the string during one period of the oscillation. Thus, the power, or rate of energy transfer, associated with the wave is

Power of a wave

$$\mathcal{P} = \frac{E_\lambda}{\Delta t} = \frac{\frac{1}{2}\mu\omega^2 A^2 \lambda}{T} = \frac{1}{2}\mu\omega^2 A^2\left(\frac{\lambda}{T}\right)$$

$$\mathcal{P} = \frac{1}{2}\mu\omega^2 A^2 v \tag{16.21}$$

This shows that the rate of energy transfer by a sinusoidal wave on a string is proportional to (a) the wave speed, (b) the square of the frequency, and (c) the square of the amplitude. In fact: **the rate of energy transfer in any sinusoidal wave is proportional to the square of the angular frequency and to the square of the amplitude.**

EXAMPLE 16.5 **Power Supplied to a Vibrating String**

A taut string for which $\mu = 5.00 \times 10^{-2}$ kg/m is under a tension of 80.0 N. How much power must be supplied to the string to generate sinusoidal waves at a frequency of 60.0 Hz and an amplitude of 6.00 cm?

Solution The wave speed on the string is, from Equation 16.4,

$$v = \sqrt{\frac{T}{\mu}} = \sqrt{\frac{80.0 \text{ N}}{5.00 \times 10^{-2} \text{ kg/m}}} = 40.0 \text{ m/s}$$

Because $f = 60.0$ Hz, the angular frequency ω of the sinus-

oidal waves on the string has the value

$$\omega = 2\pi f = 2\pi(60.0 \text{ Hz}) = 377 \text{ s}^{-1}$$

Using these values in Equation 16.21 for the power, with $A = 6.00 \times 10^{-2}$ m, we obtain

$$\mathcal{P} = \tfrac{1}{2}\mu\omega^2 A^2 v$$
$$= \tfrac{1}{2}(5.00 \times 10^{-2} \text{ kg/m})(377 \text{ s}^{-1})^2$$
$$\times (6.00 \times 10^{-2} \text{ m})^2(40.0 \text{ m/s})$$
$$= \boxed{512 \text{ W}}$$

Optional Section

16.9 > THE LINEAR WAVE EQUATION

In Section 16.3 we introduced the concept of the wave function to represent waves traveling on a string. All wave functions $y(x, t)$ represent solutions of an equation called the *linear wave equation*. This equation gives a complete description of the wave motion, and from it one can derive an expression for the wave speed. Furthermore, the linear wave equation is basic to many forms of wave motion. In this section, we derive this equation as applied to waves on strings.

Suppose a traveling wave is propagating along a string that is under a tension T. Let us consider one small string segment of length Δx (Fig. 16.22). The ends of the segment make small angles θ_A and θ_B with the x axis. The net force acting on the segment in the vertical direction is

$$\sum F_y = T\sin\theta_B - T\sin\theta_A = T(\sin\theta_B - \sin\theta_A)$$

Because the angles are small, we can use the small-angle approximation $\sin\theta \approx \tan\theta$ to express the net force as

$$\sum F_y \approx T(\tan\theta_B - \tan\theta_A)$$

However, the tangents of the angles at A and B are defined as the slopes of the string segment at these points. Because the slope of a curve is given by $\partial y/\partial x$, we have

$$\sum F_y \approx T\left[\left(\frac{\partial y}{\partial x}\right)_B - \left(\frac{\partial y}{\partial x}\right)_A\right] \qquad \textbf{(16.22)}$$

We now apply Newton's second law to the segment, with the mass of the segment given by $m = \mu\Delta x$:

$$\sum F_y = ma_y = \mu\Delta x\left(\frac{\partial^2 y}{\partial t^2}\right) \qquad \textbf{(16.23)}$$

Combining Equation 16.22 with Equation 16.23, we obtain

$$\mu\Delta x\left(\frac{\partial^2 y}{\partial t^2}\right) = T\left[\left(\frac{\partial y}{\partial x}\right)_B - \left(\frac{\partial y}{\partial x}\right)_A\right]$$

$$\frac{\mu}{T}\frac{\partial^2 y}{\partial t^2} = \frac{(\partial y/\partial x)_B - (\partial y/\partial x)_A}{\Delta x} \qquad \textbf{(16.24)}$$

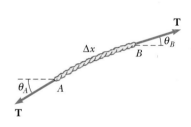

Figure 16.22 A segment of a string under tension T. The slopes at points A and B are given by $\tan\theta_A$ and $\tan\theta_B$, respectively.

The right side of this equation can be expressed in a different form if we note that the partial derivative of any function is defined as

$$\frac{\partial f}{\partial x} \equiv \lim_{\Delta x \to 0} \frac{f(x + \Delta x) - f(x)}{\Delta x}$$

If we associate $f(x + \Delta x)$ with $(\partial y/\partial x)_B$ and $f(x)$ with $(\partial y/\partial x)_A$, we see that, in the limit $\Delta x \to 0$, Equation 16.24 becomes

Linear wave equation

$$\frac{\mu}{T} \frac{\partial^2 y}{\partial t^2} = \frac{\partial^2 y}{\partial x^2} \qquad \text{(16.25)}$$

This is the linear wave equation as it applies to waves on a string.

We now show that the sinusoidal wave function (Eq. 16.11) represents a solution of the linear wave equation. If we take the sinusoidal wave function to be of the form $y(x, t) = A \sin(kx - \omega t)$, then the appropriate derivatives are

$$\frac{\partial^2 y}{\partial t^2} = -\omega^2 A \sin(kx - \omega t)$$

$$\frac{\partial^2 y}{\partial x^2} = -k^2 A \sin(kx - \omega t)$$

Substituting these expressions into Equation 16.25, we obtain

$$-\frac{\mu \omega^2}{T} \sin(kx - \omega t) = -k^2 \sin(kx - \omega t)$$

This equation must be true for all values of the variables x and t in order for the sinusoidal wave function to be a solution of the wave equation. Both sides of the equation depend on x and t through the same function $\sin(kx - \omega t)$. Because this function divides out, we do indeed have an identity, provided that

$$k^2 = \frac{\mu \omega^2}{T}$$

Using the relationship $v = \omega/k$ (Eq. 16.13) in this expression, we see that

$$v^2 = \frac{\omega^2}{k^2} = \frac{T}{\mu}$$

$$v = \sqrt{\frac{T}{\mu}}$$

which is Equation 16.4. This derivation represents another proof of the expression for the wave speed on a taut string.

The linear wave equation (Eq. 16.25) is often written in the form

Linear wave equation in general

$$\frac{\partial^2 y}{\partial x^2} = \frac{1}{v^2} \frac{\partial^2 y}{\partial t^2} \qquad \text{(16.26)}$$

This expression applies in general to various types of traveling waves. For waves on strings, y represents the vertical displacement of the string. For sound waves, y corresponds to displacement of air molecules from equilibrium or variations in either the pressure or the density of the gas through which the sound waves are propagating. In the case of electromagnetic waves, y corresponds to electric or magnetic field components.

We have shown that the sinusoidal wave function (Eq. 16.11) is one solution of the linear wave equation (Eq. 16.26). Although we do not prove it here, the linear wave equation is satisfied by *any* wave function having the form $y = f(x \pm vt)$. Furthermore, we have seen that the linear wave equation is a direct consequence of Newton's second law applied to any segment of the string.

SUMMARY

A **transverse wave** is one in which the particles of the medium move in a direction *perpendicular* to the direction of the wave velocity. An example is a wave on a taut string. A **longitudinal wave** is one in which the particles of the medium move in a direction *parallel* to the direction of the wave velocity. Sound waves in fluids are longitudinal. You should be able to identify examples of both types of waves.

Any one-dimensional wave traveling with a speed v in the x direction can be represented by a wave function of the form

$$y = f(x \pm vt) \qquad \textbf{(16.1, 16.2)}$$

where the positive sign applies to a wave traveling in the negative x direction and the negative sign applies to a wave traveling in the positive x direction. The shape of the wave at any instant in time (a snapshot of the wave) is obtained by holding t constant.

The **superposition principle** specifies that when two or more waves move through a medium, the resultant wave function equals the algebraic sum of the individual wave functions. When two waves combine in space, they interfere to produce a resultant wave. The **interference** may be **constructive** (when the individual displacements are in the same direction) or **destructive** (when the displacements are in opposite directions).

The **speed** of a wave traveling on a taut string of mass per unit length μ and tension T is

$$v = \sqrt{\frac{T}{\mu}} \qquad \textbf{(16.4)}$$

A wave is totally or partially reflected when it reaches the end of the medium in which it propagates or when it reaches a boundary where its speed changes discontinuously. If a wave pulse traveling on a string meets a fixed end, the pulse is reflected and inverted. If the pulse reaches a free end, it is reflected but not inverted.

The **wave function** for a one-dimensional sinusoidal wave traveling to the right can be expressed as

$$y = A \sin\left[\frac{2\pi}{\lambda}(x - vt)\right] = A \sin(kx - \omega t) \qquad \textbf{(16.6, 16.11)}$$

where A is the **amplitude,** λ is the **wavelength,** k is the **angular wave number,** and ω is the **angular frequency.** If T is the **period** and f the **frequency,** v, k and ω can be written

$$v = \frac{\lambda}{T} = \lambda f \qquad \textbf{(16.7, 16.14)}$$

$$k \equiv \frac{2\pi}{\lambda} \qquad \textbf{(16.9)}$$

$$\omega \equiv \frac{2\pi}{T} = 2\pi f \qquad \textbf{(16.10, 16.12)}$$

You should know how to find the equation describing the motion of particles in a wave from a given set of physical parameters.

The **power** transmitted by a sinusoidal wave on a stretched string is

$$\mathcal{P} = \tfrac{1}{2}\mu \omega^2 A^2 v \qquad \textbf{(16.21)}$$

QUESTIONS

1. Why is a wave pulse traveling on a string considered a transverse wave?
2. How would you set up a longitudinal wave in a stretched spring? Would it be possible to set up a transverse wave in a spring?
3. By what factor would you have to increase the tension in a taut string to double the wave speed?
4. When traveling on a taut string, does a wave pulse always invert upon reflection? Explain.
5. Can two pulses traveling in opposite directions on the same string reflect from each other? Explain.
6. Does the vertical speed of a segment of a horizontal, taut string, through which a wave is traveling, depend on the wave speed?
7. If you were to shake one end of a taut rope periodically three times each second, what would be the period of the sinusoidal waves set up in the rope?
8. A vibrating source generates a sinusoidal wave on a string under constant tension. If the power delivered to the string is doubled, by what factor does the amplitude change? Does the wave speed change under these circumstances?
9. Consider a wave traveling on a taut rope. What is the difference, if any, between the speed of the wave and the speed of a small segment of the rope?
10. If a long rope is hung from a ceiling and waves are sent up the rope from its lower end, they do not ascend with constant speed. Explain.

11. What happens to the wavelength of a wave on a string when the frequency is doubled? Assume that the tension in the string remains the same.
12. What happens to the speed of a wave on a taut string when the frequency is doubled? Assume that the tension in the string remains the same.
13. How do transverse waves differ from longitudinal waves?
14. When all the strings on a guitar are stretched to the same tension, will the speed of a wave along the more massive bass strings be faster or slower than the speed of a wave on the lighter strings?
15. If you stretch a rubber hose and pluck it, you can observe a pulse traveling up and down the hose. What happens to the speed of the pulse if you stretch the hose more tightly? What happens to the speed if you fill the hose with water?
16. In a longitudinal wave in a spring, the coils move back and forth in the direction of wave motion. Does the speed of the wave depend on the maximum speed of each coil?
17. When two waves interfere, can the amplitude of the resultant wave be greater than either of the two original waves? Under what conditions?
18. A solid can transport both longitudinal waves and transverse waves, but a fluid can transport only longitudinal waves. Why?

PROBLEMS

1, *2*, **3** = straightforward, intermediate, challenging   ☐ = full solution available in the *Student Solutions Manual and Study Guide*
WEB = solution posted at **http://www.saunderscollege.com/physics/**   🖥 = Computer useful in solving problem   📱 = Interactive Physics
☐ = paired numerical/symbolic problems

Section 16.1   Basic Variables of Wave Motion

Section 16.2   Direction of Particle Displacement

Section 16.3   One-Dimensional Traveling Waves

1. At $t = 0$, a transverse wave pulse in a wire is described by the function

$$y = \frac{6}{x^2 + 3}$$

where x and y are in meters. Write the function $y(x, t)$ that describes this wave if it is traveling in the positive x direction with a speed of 4.50 m/s.

2. Two wave pulses A and B are moving in opposite directions along a taut string with a speed of 2.00 cm/s. The amplitude of A is twice the amplitude of B. The pulses are shown in Figure P16.2 at $t = 0$. Sketch the shape of the string at $t = 1$, 1.5, 2, 2.5, and 3 s.

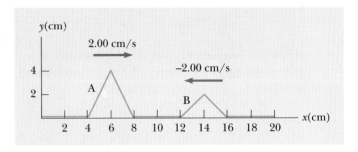

Figure P16.2

3. A wave moving along the x axis is described by

$$y(x, t) = 5.00e^{-(x+5.00t)^2}$$

where x is in meters and t is in seconds. Determine (a) the direction of the wave motion and (b) the speed of the wave.

4. Ocean waves with a crest-to-crest distance of 10.0 m can be described by the equation

$$y(x, t) = (0.800 \text{ m}) \sin[0.628(x - vt)]$$

where $v = 1.20$ m/s. (a) Sketch $y(x, t)$ at $t = 0$. (b) Sketch $y(x, t)$ at $t = 2.00$ s. Note how the entire wave form has shifted 2.40 m in the positive x direction in this time interval.

5. Two points, A and B, on the surface of the Earth are at the same longitude and 60.0° apart in latitude. Suppose that an earthquake at point A sends two waves toward point B. A transverse wave travels along the surface of the Earth at 4.50 km/s, and a longitudinal wave travels straight through the body of the Earth at 7.80 km/s. (a) Which wave arrives at point B first? (b) What is the time difference between the arrivals of the two waves at point B? Take the radius of the Earth to be 6 370 km.

6. A seismographic station receives S and P waves from an earthquake, 17.3 s apart. Suppose that the waves have traveled over the same path at speeds of 4.50 km/s and 7.80 km/s, respectively. Find the distance from the seismometer to the epicenter of the quake.

Section 16.4   Superposition and Interference

WEB 7. Two sinusoidal waves in a string are defined by the functions

$$y_1 = (2.00 \text{ cm}) \sin(20.0x - 32.0t)$$

and

$$y_2 = (2.00 \text{ cm}) \sin(25.0x - 40.0t)$$

where y and x are in centimeters and t is in seconds. (a) What is the phase difference between these two waves at the point $x = 5.00$ cm at $t = 2.00$ s? (b) What is the positive x value closest to the origin for which the two phases differ by $\pm \pi$ at $t = 2.00$ s? (This is where the sum of the two waves is zero.)

8. Two waves in one string are described by the wave functions

$$y_1 = 3.0 \cos(4.0x - 1.6t)$$

and

$$y_2 = 4.0 \sin(5.0x - 2.0t)$$

where y and x are in centimeters and t is in seconds. Find the superposition of the waves $y_1 + y_2$ at the points (a) $x = 1.00$, $t = 1.00$; (b) $x = 1.00$, $t = 0.500$; (c) $x = 0.500$, $t = 0$. (Remember that the arguments of the trigonometric functions are in radians.)

9. Two pulses traveling on the same string are described by the functions

$$y_1 = \frac{5}{(3x - 4t)^2 + 2}$$

and

$$y_2 = \frac{-5}{(3x + 4t - 6)^2 + 2}$$

(a) In which direction does each pulse travel? (b) At what time do the two cancel? (c) At what point do the two waves always cancel?

Section 16.5   The Speed of Waves on Strings

10. A phone cord is 4.00 m long. The cord has a mass of 0.200 kg. A transverse wave pulse is produced by plucking one end of the taut cord. The pulse makes four trips down and back along the cord in 0.800 s. What is the tension in the cord?

11. Transverse waves with a speed of 50.0 m/s are to be produced in a taut string. A 5.00-m length of string with a total mass of 0.060 0 kg is used. What is the required tension?

12. A piano string having a mass per unit length 5.00 × 10^{-3} kg/m is under a tension of 1 350 N. Find the speed with which a wave travels on this string.

13. An astronaut on the Moon wishes to measure the local value of g by timing pulses traveling down a wire that has a large mass suspended from it. Assume that the wire has a mass of 4.00 g and a length of 1.60 m, and that a 3.00-kg mass is suspended from it. A pulse requires 36.1 ms to traverse the length of the wire. Calculate g_{Moon} from these data. (You may neglect the mass of the wire when calculating the tension in it.)

14. Transverse pulses travel with a speed of 200 m/s along a taut copper wire whose diameter is 1.50 mm. What is the tension in the wire? (The density of copper is 8.92 g/cm³.)

15. Transverse waves travel with a speed of 20.0 m/s in a string under a tension of 6.00 N. What tension is required to produce a wave speed of 30.0 m/s in the same string?

16. A simple pendulum consists of a ball of mass M hanging from a uniform string of mass m and length L, with $m \ll M$. If the period of oscillation for the pendulum is T, determine the speed of a transverse wave in the string when the pendulum hangs at rest.

17. The elastic limit of a piece of steel wire is 2.70 × 10^9 Pa. What is the maximum speed at which transverse wave pulses can propagate along this wire before this stress is exceeded? (The density of steel is 7.86 × 10^3 kg/m³.)

18. **Review Problem.** A light string with a mass per unit length of 8.00 g/m has its ends tied to two walls separated by a distance equal to three-fourths the length of the string (Fig. P16.18). An object of mass m is sus-

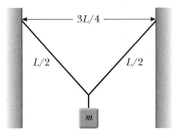

Figure P16.18

pended from the center of the string, putting a tension in the string. (a) Find an expression for the transverse wave speed in the string as a function of the hanging mass. (b) How much mass should be suspended from the string to produce a wave speed of 60.0 m/s?

19. **Review Problem.**   A light string with a mass of 10.0 g and a length $L = 3.00$ m has its ends tied to two walls that are separated by the distance $D = 2.00$ m. Two objects, each with a mass $M = 2.00$ kg, are suspended from the string, as shown in Figure P16.19. If a wave pulse is sent from point A, how long does it take for it to travel to point B?

20. **Review Problem.**   A light string of mass m and length L has its ends tied to two walls that are separated by the distance D. Two objects, each of mass M, are suspended from the string, as shown in Figure P16.19. If a wave pulse is sent from point A, how long does it take to travel to point B?

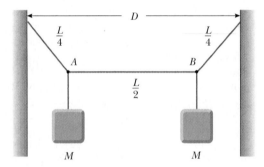

Figure P16.19   Problems 19 and 20.

WEB **21.** A 30.0-m steel wire and a 20.0-m copper wire, both with 1.00-mm diameters, are connected end to end and are stretched to a tension of 150 N. How long does it take a transverse wave to travel the entire length of the two wires?

Section 16.6   Reflection and Transmission

22. A series of pulses, each of amplitude 0.150 m, are sent down a string that is attached to a post at one end. The pulses are reflected at the post and travel back along the string without loss of amplitude. What is the displacement at a point on the string where two pulses are crossing (a) if the string is rigidly attached to the post? (b) if the end at which reflection occurs is free to slide up and down?

Section 16.7   Sinusoidal Waves

23. (a) Plot y versus t at $x = 0$ for a sinusoidal wave of the form $y = (15.0 \text{ cm}) \cos(0.157x - 50.3t)$, where x and y are in centimeters and t is in seconds. (b) Determine

the period of vibration from this plot and compare your result with the value found in Example 16.3.

24. For a certain transverse wave, the distance between two successive crests is 1.20 m, and eight crests pass a given point along the direction of travel every 12.0 s. Calculate the wave speed.

25. A sinusoidal wave is traveling along a rope. The oscillator that generates the wave completes 40.0 vibrations in 30.0 s. Also, a given maximum travels 425 cm along the rope in 10.0 s. What is the wavelength?

26. Consider the sinusoidal wave of Example 16.3, with the wave function

$$y = (15.0 \text{ cm}) \cos(0.157x - 50.3t)$$

At a certain instant, let point A be at the origin and point B be the first point along the x axis where the wave is 60.0° out of phase with point A. What is the coordinate of point B?

27. When a particular wire is vibrating with a frequency of 4.00 Hz, a transverse wave of wavelength 60.0 cm is produced. Determine the speed of wave pulses along the wire.

28. A sinusoidal wave traveling in the $-x$ direction (to the left) has an amplitude of 20.0 cm, a wavelength of 35.0 cm, and a frequency of 12.0 Hz. The displacement of the wave at $t = 0$, $x = 0$ is $y = -3.00$ cm; at this same point, a particle of the medium has a positive velocity. (a) Sketch the wave at $t = 0$. (b) Find the angular wave number, period, angular frequency, and wave speed of the wave. (c) Write an expression for the wave function $y(x, t)$.

29. A sinusoidal wave train is described by the equation

$$y = (0.25 \text{ m}) \sin(0.30x - 40t)$$

where x and y are in meters and t is in seconds. Determine for this wave the (a) amplitude, (b) angular frequency, (c) angular wave number, (d) wavelength, (e) wave speed, and (f) direction of motion.

30. A transverse wave on a string is described by the expression

$$y = (0.120 \text{ m}) \sin(\pi x/8 + 4\pi t)$$

(a) Determine the transverse speed and acceleration of the string at $t = 0.200$ s for the point on the string located at $x = 1.60$ m. (b) What are the wavelength, period, and speed of propagation of this wave?

WEB **31.** (a) Write the expression for y as a function of x and t for a sinusoidal wave traveling along a rope in the *negative x* direction with the following characteristics: $A = 8.00$ cm, $\lambda = 80.0$ cm, $f = 3.00$ Hz, and $y(0, t) = 0$ at $t = 0$. (b) Write the expression for y as a function of x and t for the wave in part (a), assuming that $y(x, 0) = 0$ at the point $x = 10.0$ cm.

32. A transverse sinusoidal wave on a string has a period $T = 25.0$ ms and travels in the negative x direction with a speed of 30.0 m/s. At $t = 0$, a particle on the string at

$x = 0$ has a displacement of 2.00 cm and travels downward with a speed of 2.00 m/s. (a) What is the amplitude of the wave? (b) What is the initial phase angle? (c) What is the maximum transverse speed of the string? (d) Write the wave function for the wave.

33. A sinusoidal wave of wavelength 2.00 m and amplitude 0.100 m travels on a string with a speed of 1.00 m/s to the right. Initially, the left end of the string is at the origin. Find (a) the frequency and angular frequency, (b) the angular wave number, and (c) the wave function for this wave. Determine the equation of motion for (d) the left end of the string and (e) the point on the string at $x = 1.50$ m to the right of the left end. (f) What is the maximum speed of any point on the string?

34. A sinusoidal wave on a string is described by the equation

$$y = (0.51 \text{ cm}) \sin(kx - \omega t)$$

where $k = 3.10$ rad/cm and $\omega = 9.30$ rad/s. How far does a wave crest move in 10.0 s? Does it move in the positive or negative x direction?

35. A wave is described by $y = (2.00 \text{ cm}) \sin(kx - \omega t)$, where $k = 2.11$ rad/m, $\omega = 3.62$ rad/s, x is in meters, and t is in seconds. Determine the amplitude, wavelength, frequency, and speed of the wave.

36. A transverse traveling wave on a taut wire has an amplitude of 0.200 mm and a frequency of 500 Hz. It travels with a speed of 196 m/s. (a) Write an equation in SI units of the form $y = A \sin(kx - \omega t)$ for this wave. (b) The mass per unit length of this wire is 4.10 g/m. Find the tension in the wire.

37. A wave on a string is described by the wave function

$$y = (0.100 \text{ m}) \sin(0.50x - 20t)$$

(a) Show that a particle in the string at $x = 2.00$ m executes simple harmonic motion. (b) Determine the frequency of oscillation of this particular point.

Section 16.8 Rate of Energy Transfer by Sinusoidal Waves on Strings

38. A taut rope has a mass of 0.180 kg and a length of 3.60 m. What power must be supplied to the rope to generate sinusoidal waves having an amplitude of 0.100 m and a wavelength of 0.500 m and traveling with a speed of 30.0 m/s?

39. A two-dimensional water wave spreads in circular wave fronts. Show that the amplitude A at a distance r from the initial disturbance is proportional to $1/\sqrt{r}$. (*Hint:* Consider the energy carried by one outward-moving ripple.)

40. Transverse waves are being generated on a rope under constant tension. By what factor is the required power increased or decreased if (a) the length of the rope is doubled and the angular frequency remains constant, (b) the amplitude is doubled and the angular fre-

quency is halved, (c) both the wavelength and the amplitude are doubled, and (d) both the length of the rope and the wavelength are halved?

WEB 41. Sinusoidal waves 5.00 cm in amplitude are to be transmitted along a string that has a linear mass density of 4.00×10^{-2} kg/m. If the source can deliver a maximum power of 300 W and the string is under a tension of 100 N, what is the highest vibrational frequency at which the source can operate?

42. It is found that a 6.00-m segment of a long string contains four complete waves and has a mass of 180 g. The string is vibrating sinusoidally with a frequency of 50.0 Hz and a peak-to-valley displacement of 15.0 cm. (The "peak-to-valley" distance is the vertical distance from the farthest positive displacement to the farthest negative displacement.) (a) Write the function that describes this wave traveling in the positive x direction. (b) Determine the power being supplied to the string.

43. A sinusoidal wave on a string is described by the equation

$$y = (0.15 \text{ m}) \sin(0.80x - 50t)$$

where x and y are in meters and t is in seconds. If the mass per unit length of this string is 12.0 g/m, determine (a) the speed of the wave, (b) the wavelength, (c) the frequency, and (d) the power transmitted to the wave.

44. A horizontal string can transmit a maximum power of $\mathcal{P}$ (without breaking) if a wave with amplitude A and angular frequency ω is traveling along it. To increase this maximum power, a student folds the string and uses the "double string" as a transmitter. Determine the maximum power that can be transmitted along the "double string," supposing that the tension is constant.

(Optional)
Section 16.9 The Linear Wave Equation

45. (a) Evaluate A in the scalar equality $(7 + 3)4 = A$. (b) Evaluate A, B, and C in the vector equality $7.00\mathbf{i} + 3.00\mathbf{k} = A\mathbf{i} + B\mathbf{j} + C\mathbf{k}$. Explain how you arrive at your answers. (c) The functional equality or identity

$$A + B \cos(Cx + Dt + E) = (7.00 \text{ mm}) \cos(3x + 4t + 2)$$

is true for all values of the variables x and t, which are measured in meters and in seconds, respectively. Evaluate the constants A, B, C, D, and E. Explain how you arrive at your answers.

46. Show that the wave function $y = e^{b(x - vt)}$ is a solution of the wave equation (Eq. 16.26), where b is a constant.

47. Show that the wave function $y = \ln[b(x - vt)]$ is a solution to Equation 16.26, where b is a constant.

48. (a) Show that the function $y(x, t) = x^2 + v^2t^2$ is a solution to the wave equation. (b) Show that the function above can be written as $f(x + vt) + g(x - vt)$, and determine the functional forms for f and g. (c) Repeat parts (a) and (b) for the function $y(x, t) = \sin(x) \cos(vt)$.

ADDITIONAL PROBLEMS

49. The "wave" is a particular type of wave pulse that can sometimes be seen propagating through a large crowd gathered at a sporting arena to watch a soccer or American football match (Fig. P16.49). The particles of the medium are the spectators, with zero displacement corresponding to their being in the seated position and maximum displacement corresponding to their being in the standing position and raising their arms. When a large fraction of the spectators participate in the wave motion, a somewhat stable pulse shape can develop. The wave speed depends on people's reaction time, which is typically on the order of 0.1 s. Estimate the order of magnitude, in minutes, of the time required for such a wave pulse to make one circuit around a large sports stadium. State the quantities you measure or estimate and their values.

Figure P16.49 (*Gregg Adams/Tony Stone Images*)

50. A traveling wave propagates according to the expression $y = (4.0 \text{ cm}) \sin(2.0x - 3.0t)$, where x is in centimeters and t is in seconds. Determine (a) the amplitude, (b) the wavelength, (c) the frequency, (d) the period, and (e) the direction of travel of the wave.

WEB **51.** The wave function for a traveling wave on a taut string is (in SI units)

$$y(x, t) = (0.350 \text{ m}) \sin(10\pi t - 3\pi x + \pi/4)$$

(a) What are the speed and direction of travel of the wave? (b) What is the vertical displacement of the string at $t = 0$, $x = 0.100$ m? (c) What are the wavelength and frequency of the wave? (d) What is the maximum magnitude of the transverse speed of the string?

52. Motion picture film is projected at 24.0 frames per second. Each frame is a photograph 19.0 mm in height. At what constant speed does the film pass into the projector?

53. **Review Problem.** A block of mass M, supported by a string, rests on an incline making an angle θ with the horizontal (Fig. P16.53). The string's length is L, and its mass is $m \ll M$. Derive an expression for the time it takes a transverse wave to travel from one end of the string to the other.

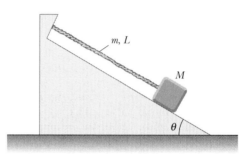

Figure P16.53

54. (a) Determine the speed of transverse waves on a string under a tension of 80.0 N if the string has a length of 2.00 m and a mass of 5.00 g. (b) Calculate the power required to generate these waves if they have a wavelength of 16.0 cm and an amplitude of 4.00 cm.

55. **Review Problem.** A 2.00-kg block hangs from a rubber cord. The block is supported so that the cord is not stretched. The unstretched length of the cord is 0.500 m, and its mass is 5.00 g. The "spring constant" for the cord is 100 N/m. The block is released and stops at the lowest point. (a) Determine the tension in the cord when the block is at this lowest point. (b) What is the length of the cord in this "stretched" position? (c) Find the speed of a transverse wave in the cord if the block is held in this lowest position.

56. **Review Problem.** A block of mass M hangs from a rubber cord. The block is supported so that the cord is not stretched. The unstretched length of the cord is L_0, and its mass is m, much less than M. The "spring constant" for the cord is k. The block is released and stops at the lowest point. (a) Determine the tension in the cord when the block is at this lowest point. (b) What is the length of the cord in this "stretched" position? (c) Find the speed of a transverse wave in the cord if the block is held in this lowest position.

57. A sinusoidal wave in a rope is described by the wave function

$$y = (0.20 \text{ m}) \sin(0.75\pi x + 18\pi t)$$

where x and y are in meters and t is in seconds. The rope has a linear mass density of 0.250 kg/m. If the tension in the rope is provided by an arrangement like the one illustrated in Figure 16.12, what is the value of the suspended mass?

58. A wire of density ρ is tapered so that its cross-sectional area varies with x, according to the equation

$$A = (1.0 \times 10^{-3}x + 0.010) \text{ cm}^2$$

(a) If the wire is subject to a tension T, derive a relationship for the speed of a wave as a function of position. (b) If the wire is aluminum and is subject to a tension of 24.0 N, determine the speed at the origin and at $x = 10.0$ m.

59. A rope of total mass m and length L is suspended vertically. Show that a transverse wave pulse travels the length of the rope in a time $t = 2\sqrt{L/g}$. (*Hint:* First find an expression for the wave speed at any point a distance x from the lower end by considering the tension in the rope as resulting from the weight of the segment below that point.)

60. If mass M is suspended from the bottom of the rope in Problem 59, (a) show that the time for a transverse wave to travel the length of the rope is

$$t = 2\sqrt{\frac{L}{mg}}\left[\sqrt{(M + m)} - \sqrt{M}\right]$$

(b) Show that this reduces to the result of Problem 59 when $M = 0$. (c) Show that for $m \ll M$, the expression in part (a) reduces to

$$t = \sqrt{\frac{mL}{Mg}}$$

61. It is stated in Problem 59 that a wave pulse travels from the bottom to the top of a rope of length L in a time $t = 2\sqrt{L/g}$. Use this result to answer the following questions. (It is *not* necessary to set up any new integrations.) (a) How long does it take for a wave pulse to travel halfway up the rope? (Give your answer as a fraction of the quantity $2\sqrt{L/g}$.) (b) A pulse starts traveling up the rope. How far has it traveled after a time $\sqrt{L/g}$?

62. Determine the speed and direction of propagation of each of the following sinusoidal waves, assuming that x is measured in meters and t in seconds:
(a) $y = 0.60 \cos(3.0x - 15t + 2)$
(b) $y = 0.40 \cos(3.0x + 15t - 2)$
(c) $y = 1.2 \sin(15t + 2.0x)$
(d) $y = 0.20 \sin(12t - x/2 + \pi)$

63. **Review Problem.** An aluminum wire under zero tension at room temperature is clamped at each end. The tension in the wire is increased by reducing the temperature, which results in a decrease in the wire's equilibrium length. What strain $(\Delta L/L)$ results in a transverse wave speed of 100 m/s? Take the cross-sectional area of the wire to be 5.00×10^{-6} m^2, the density of the material to be 2.70×10^3 kg/m^3, and Young's modulus to be 7.00×10^{10} N/m^2.

64. (a) Show that the speed of longitudinal waves along a spring of force constant k is $v = \sqrt{kL/\mu}$, where L is the unstretched length of the spring and μ is the mass per unit length. (b) A spring with a mass of 0.400 kg has an unstretched length of 2.00 m and a force constant of 100 N/m. Using the result you obtained in (a), determine the speed of longitudinal waves along this spring.

65. A string of length L consists of two sections: The left half has mass per unit length $\mu = \mu_0/2$, whereas the right half has a mass per unit length $\mu' = 3\mu = 3\mu_0/2$. Tension in the string is T_0. Notice from the data given that this string has the same total mass as a uniform string of length L and of mass per unit length μ_0.
(a) Find the speeds v and v' at which transverse wave pulses travel in the two sections. Express the speeds in terms of T_0 and μ_0, and also as multiples of the speed $v_0 = (T_0/\mu_0)^{1/2}$. (b) Find the time required for a pulse to travel from one end of the string to the other. Give your result as a multiple of $t_0 = L/v_0$.

66. A wave pulse traveling along a string of linear mass density μ is described by the relationship

$$y = [A_0 e^{-bx}] \sin(kx - \omega t)$$

where the factor in brackets before the sine function is said to be the amplitude. (a) What is the power $\mathcal{P}(x)$ carried by this wave at a point x? (b) What is the power carried by this wave at the origin? (c) Compute the ratio $\mathcal{P}(x)/\mathcal{P}(0)$.

67. An earthquake on the ocean floor in the Gulf of Alaska produces a *tsunami* (sometimes called a "tidal wave") that reaches Hilo, Hawaii, 4 450 km away, in a time of 9 h 30 min. Tsunamis have enormous wavelengths (100–200 km), and the propagation speed of these waves is $v \approx \sqrt{gd}$, where d is the average depth of the water. From the information given, find the average wave speed and the average ocean depth between Alaska and Hawaii. (This method was used in 1856 to estimate the average depth of the Pacific Ocean long before soundings were made to obtain direct measurements.)

ANSWERS TO QUICK QUIZZES

16.1 (a) It is longitudinal because the disturbance (the shift of position) is parallel to the direction in which the wave travels. (b) It is transverse because the people stand up and sit down (vertical motion), whereas the wave moves either to the left or to the right (motion perpendicular to the disturbance).

16.2

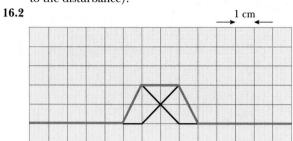

16.3 Only answers (f) and (h) are correct. (a) and (b) affect the transverse speed of a particle of the string, but not the wave speed along the string. (c) and (d) change the amplitude. (e) and (g) increase the time by decreasing the wave speed.

16.4 The transverse speed increases because $v_{y, \text{max}} = \omega A = 2\pi f A$. The wave speed does not change because it depends only on the tension and mass per length of the string, neither of which has been modified. The wavelength must decrease because the wave speed $v = \lambda f$ remains constant.

c h a p t e r

17

Sound Waves

Sound waves are the most important example of longitudinal waves. They can travel through any material medium with a speed that depends on the properties of the medium. As the waves travel, the particles in the medium vibrate to produce changes in density and pressure along the direction of motion of the wave. These changes result in a series of high-pressure and low-pressure regions. If the source of the sound waves vibrates sinusoidally, the pressure variations are also sinusoidal. We shall find that the mathematical description of sinusoidal sound waves is identical to that of sinusoidal string waves, which was discussed in the previous chapter.

Sound waves are divided into three categories that cover different frequency ranges. (1) *Audible waves* are waves that lie within the range of sensitivity of the human ear. They can be generated in a variety of ways, such as by musical instruments, human vocal cords, and loudspeakers. (2) *Infrasonic waves* are waves having frequencies below the audible range. Elephants can use infrasonic waves to communicate with each other, even when separated by many kilometers. (3) *Ultrasonic waves* are waves having frequencies above the audible range. You may have used a "silent" whistle to retrieve your dog. The ultrasonic sound it emits is easily heard by dogs, although humans cannot detect it at all. Ultrasonic waves are also used in medical imaging.

We begin this chapter by discussing the speed of sound waves and then wave intensity, which is a function of wave amplitude. We then provide an alternative description of the intensity of sound waves that compresses the wide range of intensities to which the ear is sensitive to a smaller range. Finally, we treat effects of the motion of sources and/or listeners.

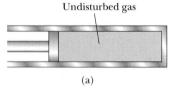

Undisturbed gas

(a)

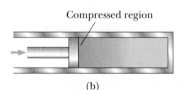

Compressed region

(b)

17.1 ▷ SPEED OF SOUND WAVES

Let us describe pictorially the motion of a one-dimensional longitudinal pulse moving through a long tube containing a compressible gas (Fig. 17.1). A piston at the left end can be moved to the right to compress the gas and create the pulse. Before the piston is moved, the gas is undisturbed and of uniform density, as represented by the uniformly shaded region in Figure 17.1a. When the piston is suddenly pushed to the right (Fig. 17.1b), the gas just in front of it is compressed (as represented by the more heavily shaded region); the pressure and density in this region are now higher than they were before the piston moved. When the piston comes to rest (Fig. 17.1c), the compressed region of the gas continues to move to the right, corresponding to a longitudinal pulse traveling through the tube with

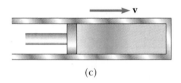

(c)

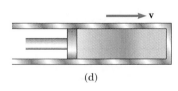

(d)

Figure 17.1   Motion of a longitudinal pulse through a compressible gas. The compression (darker region) is produced by the moving piston.

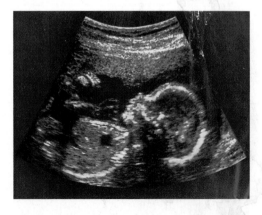

An ultrasound image of a human fetus in the womb after 20 weeks of development, showing the head, body, arms, and legs in profile. *(U.H.B. Trust/Tony Stone Images)*

speed v. Note that the piston speed does *not* equal v. Furthermore, the compressed region does not "stay with" the piston as the piston moves, because the speed of the wave may be greater than the speed of the piston.

The speed of sound waves depends on the compressibility and inertia of the medium. If the medium has a bulk modulus B (see Section 12.4) and density ρ, the speed of sound waves in that medium is

$$v = \sqrt{\frac{B}{\rho}}$$

(17.1)   Speed of sound

It is interesting to compare this expression with Equation 16.4 for the speed of transverse waves on a string, $v = \sqrt{T/\mu}$. In both cases, the wave speed depends on an elastic property of the medium—bulk modulus B or string tension T—and on an inertial property of the medium—ρ or μ. In fact, the speed of *all mechanical waves* follows an expression of the general form

$$v = \sqrt{\frac{\text{elastic property}}{\text{inertial property}}}$$

The speed of sound also depends on the temperature of the medium. For sound traveling through air, the relationship between wave speed and medium temperature is

$$v = (331 \text{ m/s}) \sqrt{1 + \frac{T_C}{273°C}}$$

where 331 m/s is the speed of sound in air at 0°C, and T_C is the temperature in degrees Celsius. Using this equation, one finds that at 20°C the speed of sound in air is approximately 343 m/s.

This information provides a convenient way to estimate the distance to a thunderstorm, as demonstrated in the QuickLab. During a lightning flash, the temperature of a long channel of air rises rapidly as the bolt passes through it. This temperature increase causes the air in the channel to expand rapidly, and this expansion creates a sound wave. The channel produces sound throughout its entire length at essentially the same instant. If the orientation of the channel is such that all of its parts are approximately the same distance from you, sounds from the different parts reach you at the same time, and you hear a short, intense thunderclap. However, if the distances between your ear and different portions of the channel vary, sounds from different portions arrive at your ears at different times. If the channel were a straight line, the resulting sound would be a steady roar, but the zigzag shape of the path produces variations in loudness.

QuickLab

The next time a thunderstorm approaches, count the seconds between a flash of lightning (which reaches you almost instantaneously) and the following thunderclap. Divide this time by 3 to determine the approximate number of kilometers (or by 5 to estimate the miles) to the storm.

To learn more about lightning, read E. Williams, "The Electrification of Thunderstorms" *Sci. Am.* 259(5):88–89, 1988.

Quick Quiz 17.1

The speed of sound in air is a function of (a) wavelength, (b) frequency, (c) temperature, (d) amplitude.

Quick Quiz 17.2

As a result of a distant explosion, an observer first senses a ground tremor and then hears the explosion later. Explain.

EXAMPLE 17.1   **Speed of Sound in a Solid**

If a solid bar is struck at one end with a hammer, a longitudinal pulse propagates down the bar with a speed $v = \sqrt{Y/\rho}$, where Y is the Young's modulus for the material (see Section 12.4). Find the speed of sound in an aluminum bar.

Solution   From Table 12.1 we obtain $Y = 7.0 \times 10^{10}$ N/m^2 for aluminum, and from Table 1.5 we obtain $\rho = 2.70 \times 10^3$ kg/m^3. Therefore,

$$v_{\text{Al}} = \sqrt{\frac{Y}{\rho}} = \sqrt{\frac{7.0 \times 10^{10} \text{ N/m}^2}{2.70 \times 10^3 \text{ kg/m}^3}} \approx \boxed{5.1 \text{ km/s}}$$

This typical value for the speed of sound in solids is much greater than the speed of sound in gases, as Table 17.1 shows. This difference in speeds makes sense because the molecules of a solid are bound together into a much more rigid structure than those in a gas and hence respond more rapidly to a disturbance.

EXAMPLE 17.2   **Speed of Sound in a Liquid**

(a) Find the speed of sound in water, which has a bulk modulus of 2.1×10^9 N/m^2 and a density of 1.00×10^3 kg/m^3.

Solution   Using Equation 17.1, we find that

$$v_{\text{water}} = \sqrt{\frac{B}{\rho}} = \sqrt{\frac{2.1 \times 10^9 \text{ N/m}^2}{1.00 \times 10^3 \text{ kg/m}^3}} = \boxed{1.4 \text{ km/s}}$$

In general, sound waves travel more slowly in liquids than in solids because liquids are more compressible than solids.

(b) Dolphins use sound waves to locate food. Experiments have shown that a dolphin can detect a 7.5-cm target 110 m away, even in murky water. For a bit of "dinner" at that distance, how much time passes between the moment the dolphin emits a sound pulse and the moment the dolphin hears its reflection and thereby detects the distant target?

Solution   The total distance covered by the sound wave as it travels from dolphin to target and back is 2×110 m = 220 m. From Equation 2.2, we have

$$\Delta t = \frac{\Delta x}{v_x} = \frac{220 \text{ m}}{1\,400 \text{ m/s}} = \boxed{0.16 \text{ s}}$$

Bottle-nosed dolphin. *(Stuart Westmoreland/Tony Stone Images)*

17.2   PERIODIC SOUND WAVES

This section will help you better comprehend the nature of sound waves. You will learn that pressure variations control what we hear—an important fact for understanding how our ears work.

One can produce a one-dimensional periodic sound wave in a long, narrow tube containing a gas by means of an oscillating piston at one end, as shown in Figure 17.2. The darker parts of the colored areas in this figure represent regions where the gas is compressed and thus the density and pressure are above their equilibrium values. A compressed region is formed whenever the piston is pushed into the tube. This compressed region, called a **condensation,** moves through the tube as a pulse, continuously compressing the region just in front of itself. When the piston is pulled back, the gas in front of it expands, and the pressure and density in this region fall below their equilibrium values (represented by the lighter parts of the colored areas in Fig. 17.2). These low-pressure regions, called **rarefactions,** also propagate along the tube, following the condensations. Both regions move with a speed equal to the speed of sound in the medium.

As the piston oscillates sinusoidally, regions of condensation and rarefaction are continuously set up. The distance between two successive condensations (or two successive rarefactions) equals the wavelength λ. As these regions travel through the tube, any small volume of the medium moves with simple harmonic motion parallel to the direction of the wave. If $s(x, t)$ is the displacement of a small volume element from its equilibrium position, we can express this harmonic displacement function as

$$s(x, t) = s_{max} \cos(kx - \omega t) \qquad (17.2)$$

where s_{max} **is the maximum displacement of the medium from equilibrium** (in other words, the **displacement amplitude** of the wave), k is the angular wavenumber, and ω is the angular frequency of the piston. Note that the displacement of the medium is along x, in the direction of motion of the sound wave, which means we are describing a longitudinal wave.

As we shall demonstrate shortly, the variation in the gas pressure ΔP, measured from the equilibrium value, is also periodic and for the displacement function in Equation 17.2 is given by

$$\Delta P = \Delta P_{max} \sin(kx - \omega t) \qquad (17.3)$$

where **the pressure amplitude** ΔP_{max}—which is the **maximum change in pres-**

TABLE 17.1
Speeds of Sound in Various Media

Medium	v (m/s)
Gases	
Hydrogen (0°C)	1 286
Helium (0°C)	972
Air (20°C)	343
Air (0°C)	331
Oxygen (0°C)	317
Liquids at 25°C	
Glycerol	1 904
Sea water	1 533
Water	1 493
Mercury	1 450
Kerosene	1 324
Methyl alcohol	1 143
Carbon tetrachloride	926
Solids	
Diamond	12 000
Pyrex glass	5 640
Iron	5 130
Aluminum	5 100
Brass	4 700
Copper	3 560
Gold	3 240
Lucite	2 680
Lead	1 322
Rubber	1 600

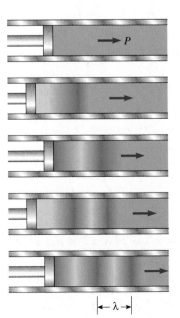

Figure 17.2 A sinusoidal longitudinal wave propagating through a gas-filled tube. The source of the wave is a sinusoidally oscillating piston at the left. The high-pressure and low-pressure regions are colored darkly and lightly, respectively.

sure from the equilibrium value—is given by

$$\Delta P_{max} = \rho v \omega s_{max} \qquad \textbf{(17.4)}$$

Thus, we see that a sound wave may be considered as either a displacement wave or a pressure wave. A comparison of Equations 17.2 and 17.3 shows that **the pressure wave is 90° out of phase with the displacement wave.** Graphs of these functions are shown in Figure 17.3. Note that the pressure variation is a maximum when the displacement is zero, and the displacement is a maximum when the pressure variation is zero.

Pressure amplitude

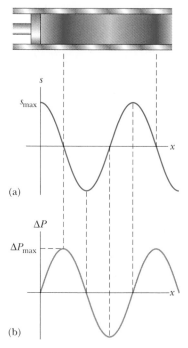

Figure 17.3 (a) Displacement amplitude versus position and (b) pressure amplitude versus position for a sinusoidal longitudinal wave. The displacement wave is 90° out of phase with the pressure wave.

> **Quick Quiz 17.3**

If you blow across the top of an empty soft-drink bottle, a pulse of air travels down the bottle. At the moment the pulse reaches the bottom of the bottle, compare the displacement of air molecules with the pressure variation.

Derivation of Equation 17.3

From the definition of bulk modulus (see Eq. 12.8), the pressure variation in the gas is

$$\Delta P = -B \frac{\Delta V}{V_i}$$

The volume of gas that has a thickness Δx in the horizontal direction and a cross-sectional area A is $V_i = A \, \Delta x$. The change in volume ΔV accompanying the pressure change is equal to $A \, \Delta s$, where Δs is the difference between the value of s at $x + \Delta x$ and the value of s at x. Hence, we can express ΔP as

$$\Delta P = -B \frac{\Delta V}{V_i} = -B \frac{A \, \Delta s}{A \, \Delta x} = -B \frac{\Delta s}{\Delta x}$$

As Δx approaches zero, the ratio $\Delta s / \Delta x$ becomes $\partial s / \partial x$. (The partial derivative indicates that we are interested in the variation of s with position at a *fixed* time.) Therefore,

$$\Delta P = -B \frac{\partial s}{\partial x}$$

If the displacement is the simple sinusoidal function given by Equation 17.2, we find that

$$\Delta P = -B \frac{\partial}{\partial x} \left[s_{max} \cos(kx - \omega t) \right] = Bks_{max} \sin(kx - \omega t)$$

Because the bulk modulus is given by $B = \rho v^2$ (see Eq. 17.1), the pressure variation reduces to

$$\Delta P = \rho v^2 k s_{max} \sin(kx - \omega t)$$

From Equation 16.13, we can write $k = \omega/v$; hence, ΔP can be expressed as

$$\Delta P = \rho v \omega s_{max} \sin(kx - \omega t)$$

Because the sine function has a maximum value of 1, we see that the maximum value of the pressure variation is $\Delta P_{max} = \rho v \omega s_{max}$ (see Eq. 17.4), and we arrive at Equation 17.3:

$$\Delta P = \Delta P_{max} \sin(kx - \omega t)$$

17.3 ▶ INTENSITY OF PERIODIC SOUND WAVES

In the previous chapter, we showed that a wave traveling on a taut string transports energy. The same concept applies to sound waves. Consider a volume of air of mass Δm and width Δx in front of a piston oscillating with a frequency ω, as shown in Figure 17.4. The piston transmits energy to this volume of air in the tube, and the energy is propagated away from the piston by the sound wave.[1] To evaluate the rate of energy transfer for the sound wave, we shall evaluate the kinetic energy of this volume of air, which is undergoing simple harmonic motion. We shall follow a procedure similar to that in Section 16.8, in which we evaluated the rate of energy transfer for a wave on a string.

As the sound wave propagates away from the piston, the displacement of any volume of air in front of the piston is given by Equation 17.2. To evaluate the kinetic energy of this volume of air, we need to know its speed. We find the speed by taking the time derivative of Equation 17.2:

$$v(x,\, t) = \frac{\partial}{\partial t}\, s(x,\, t) = \frac{\partial}{\partial t}\,[\,s_{max}\cos(kx - \omega t)\,] = \omega s_{max}\sin(kx - \omega t)$$

Imagine that we take a "snapshot" of the wave at $t = 0$. The kinetic energy of a given volume of air at this time is

$$\Delta K = \tfrac{1}{2}\,\Delta m v^2 = \tfrac{1}{2}\,\Delta m (\omega s_{max}\sin kx)^2 = \tfrac{1}{2}\rho A\,\Delta x(\omega s_{max}\sin kx)^2$$
$$= \tfrac{1}{2}\rho A\,\Delta x(\omega s_{max})^2\sin^2 kx$$

where A is the cross-sectional area of the moving air and $A\,\Delta x$ is its volume. Now, as in Section 16.8, we integrate this expression over a full wavelength to find the total kinetic energy in one wavelength. Letting the volume of air shrink to infinitesimal thickness, so that $\Delta x \rightarrow dx$, we have

$$K_\lambda = \int dK = \int_0^\lambda \tfrac{1}{2}\rho A(\omega s_{max})^2\sin^2 kx\, dx = \tfrac{1}{2}\rho A(\omega s_{max})^2 \int_0^\lambda \sin^2 kx\, dx$$
$$= \tfrac{1}{2}\rho A(\omega s_{max})^2 \left(\tfrac{1}{2}\lambda\right) = \tfrac{1}{4}\rho A(\omega s_{max})^2\lambda$$

As in the case of the string wave in Section 16.8, the total potential energy for one wavelength has the same value as the total kinetic energy; thus, the total mechani-

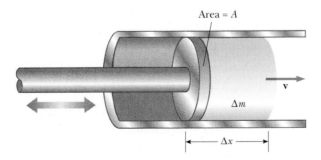

Figure 17.4   An oscillating piston transfers energy to the air in the tube, initially causing the volume of air of width Δx and mass Δm to oscillate with an amplitude s_{max}.

[1] Although it is not proved here, the work done by the piston equals the energy carried away by the wave. For a detailed mathematical treatment of this concept, see Chapter 4 in Frank S. Crawford, Jr., *Waves*, Berkeley Physics Course, vol. 3, New York, McGraw-Hill Book Company, 1968.

cal energy is

$$E_\lambda = K_\lambda + U_\lambda = \tfrac{1}{2}\rho A(\omega s_{max})^2 \lambda$$

As the sound wave moves through the air, this amount of energy passes by a given point during one period of oscillation. Hence, the rate of energy transfer is

$$\mathcal{P} = \frac{E_\lambda}{\Delta t} = \frac{\tfrac{1}{2}\rho A(\omega s_{max})^2 \lambda}{T} = \tfrac{1}{2}\rho A(\omega s_{max})^2 \left(\frac{\lambda}{T}\right) = \tfrac{1}{2}\rho A v(\omega s_{max})^2$$

where v is the speed of sound in air.

> We define the **intensity I** of a wave, or the power per unit area, to be the rate at which the energy being transported by the wave flows through a unit area A perpendicular to the direction of travel of the wave.

In the present case, therefore, the intensity is

Intensity of a sound wave

$$I = \frac{\mathcal{P}}{A} = \tfrac{1}{2}\rho v(\omega s_{max})^2 \tag{17.5}$$

Thus, we see that the intensity of a periodic sound wave is proportional to the square of the displacement amplitude and to the square of the angular frequency (as in the case of a periodic string wave). This can also be written in terms of the pressure amplitude ΔP_{max}; in this case, we use Equation 17.4 to obtain

$$I = \frac{\Delta P_{max}^2}{2\rho v} \tag{17.6}$$

EXAMPLE 17.3 ⟩ Hearing Limits

The faintest sounds the human ear can detect at a frequency of 1 000 Hz correspond to an intensity of about 1.00×10^{-12} W/m²—the so-called *threshold of hearing*. The loudest sounds the ear can tolerate at this frequency correspond to an intensity of about 1.00 W/m²—the *threshold of pain*. Determine the pressure amplitude and displacement amplitude associated with these two limits.

Solution First, consider the faintest sounds. Using Equation 17.6 and taking $v = 343$ m/s as the speed of sound waves in air and $\rho = 1.20$ kg/m³ as the density of air, we obtain

$$\Delta P_{max} = \sqrt{2\rho v I}$$
$$= \sqrt{2(1.20 \text{ kg/m}^3)(343 \text{ m/s})(1.00 \times 10^{-12} \text{ W/m}^2)}$$
$$= 2.87 \times 10^{-5} \text{ N/m}^2$$

Because atmospheric pressure is about 10^5 N/m², this result tells us that the ear can discern pressure fluctuations as small as 3 parts in 10^{10}!

We can calculate the corresponding displacement amplitude by using Equation 17.4, recalling that $\omega = 2\pi f$ (see Eqs. 16.10 and 16.12):

$$s_{max} = \frac{\Delta P_{max}}{\rho v \omega} = \frac{2.87 \times 10^{-5} \text{ N/m}^2}{(1.20 \text{ kg/m}^3)(343 \text{ m/s})(2\pi \times 1\,000 \text{ Hz})}$$
$$= 1.11 \times 10^{-11} \text{ m}$$

This is a remarkably small number! If we compare this result for s_{max} with the diameter of a molecule (about 10^{-10} m), we see that the ear is an extremely sensitive detector of sound waves.

In a similar manner, one finds that the loudest sounds the human ear can tolerate correspond to a pressure amplitude of 28.7 N/m² and a displacement amplitude equal to 1.11×10^{-5} m.

Sound Level in Decibels

The example we just worked illustrates the wide range of intensities the human ear can detect. Because this range is so wide, it is convenient to use a logarithmic scale, where the **sound level** β (Greek letter beta) is defined by the equation

$$\beta = 10 \log\left(\frac{I}{I_0}\right) \tag{17.7}$$

The constant I_0 is the *reference intensity,* taken to be at the threshold of hearing ($I_0 = 1.00 \times 10^{-12}$ W/m^2), and I is the intensity, in watts per square meter, at the sound level β, where β is measured in **decibels** (dB).[2] On this scale, the threshold of pain ($I = 1.00$ W/m^2) corresponds to a sound level of $\beta = 10 \log[(1 \text{ W/m}^2)/(10^{-12} \text{ W/m}^2)] = 10 \log(10^{12}) = 120$ dB, and the threshold of hearing corresponds to $\beta = 10 \log[(10^{-12} \text{ W/m}^2)/(10^{-12} \text{ W/m}^2)] = 0$ dB.

Prolonged exposure to high sound levels may seriously damage the ear. Ear plugs are recommended whenever sound levels exceed 90 dB. Recent evidence suggests that "noise pollution" may be a contributing factor to high blood pressure, anxiety, and nervousness. Table 17.2 gives some typical sound-level values.

TABLE 17.2
Sound Levels

Source of Sound	β (dB)
Nearby jet airplane	150
Jackhammer; machine gun	130
Siren; rock concert	120
Subway; power mower	100
Busy traffic	80
Vacuum cleaner	70
Normal conversation	50
Mosquito buzzing	40
Whisper	30
Rustling leaves	10
Threshold of hearing	0

EXAMPLE 17.4   Sound Levels

Two identical machines are positioned the same distance from a worker. The intensity of sound delivered by each machine at the location of the worker is 2.0×10^{-7} W/m^2. Find the sound level heard by the worker (a) when one machine is operating and (b) when both machines are operating.

Solution   (a) The sound level at the location of the worker with one machine operating is calculated from Equation 17.7:

$$\beta_1 = 10 \log\left(\frac{2.0 \times 10^{-7} \text{ W/m}^2}{1.00 \times 10^{-12} \text{ W/m}^2}\right) = 10 \log(2.0 \times 10^5)$$

$$= \boxed{53 \text{ dB}}$$

(b) When both machines are operating, the intensity is doubled to 4.0×10^{-7} W/m^2; therefore, the sound level now is

$$\beta_2 = 10 \log\left(\frac{4.0 \times 10^{-7} \text{ W/m}^2}{1.00 \times 10^{-12} \text{ W/m}^2}\right) = 10 \log(4.0 \times 10^5)$$

$$= \boxed{56 \text{ dB}}$$

From these results, we see that when the intensity is doubled, the sound level increases by only 3 dB.

Quick Quiz 17.4

A violin plays a melody line and is then joined by nine other violins, all playing at the same intensity as the first violin, in a repeat of the same melody. (a) When all of the violins are playing together, by how many decibels does the sound level increase? (b) If ten more violins join in, how much has the sound level increased over that for the single violin?

---

[2] The unit *bel* is named after the inventor of the telephone, Alexander Graham Bell (1847–1922). The prefix *deci-* is the SI prefix that stands for 10^{-1}.

17.4 ▶ SPHERICAL AND PLANE WAVES

If a spherical body oscillates so that its radius varies sinusoidally with time, a spherical sound wave is produced (Fig. 17.5). The wave moves outward from the source at a constant speed if the medium is uniform.

Because of this uniformity, we conclude that the energy in a spherical wave propagates equally in all directions. That is, no one direction is preferred over any other. If $\mathcal{P}_{av}$ is the average power emitted by the source, then this power at any distance r from the source must be distributed over a spherical surface of area $4\pi r^2$. Hence, the wave intensity at a distance r from the source is

$$I = \frac{\mathcal{P}_{av}}{A} = \frac{\mathcal{P}_{av}}{4\pi r^2} \tag{17.8}$$

Because $\mathcal{P}_{av}$ is the same for any spherical surface centered at the source, we see that the intensities at distances r_1 and r_2 are

$$I_1 = \frac{\mathcal{P}_{av}}{4\pi r_1{}^2} \quad \text{and} \quad I_2 = \frac{\mathcal{P}_{av}}{4\pi r_2{}^2}$$

Therefore, the ratio of intensities on these two spherical surfaces is

$$\frac{I_1}{I_2} = \frac{r_2{}^2}{r_1{}^2}$$

This inverse-square law states that the intensity decreases in proportion to the square of the distance from the source. Equation 17.5 tells us that the intensity is proportional to s_{max}^2. Setting the right side of Equation 17.5 equal to the right side

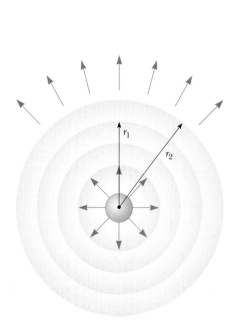

Figure 17.5   A spherical sound wave propagating radially outward from an oscillating spherical body. The intensity of the spherical wave varies as $1/r^2$.

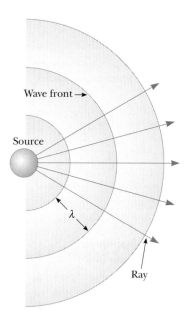

Figure 17.6   Spherical waves emitted by a point source. The circular arcs represent the spherical wave fronts that are concentric with the source. The rays are radial lines pointing outward from the source, perpendicular to the wave fronts.

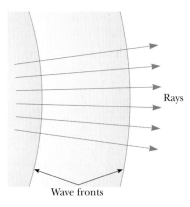

Figure 17.7 Far away from a point source, the wave fronts are nearly parallel planes, and the rays are nearly parallel lines perpendicular to the planes. Hence, a small segment of a spherical wave front is approximately a plane wave.

of Equation 17.8, we conclude that the displacement amplitude s_{max} of a spherical wave must vary as $1/r$. Therefore, we can write the wave function ψ (Greek letter psi) for an outgoing spherical wave in the form

$$\psi(r, t) = \frac{s_0}{r} \sin(kr - \omega t) \qquad \textbf{(17.9)}$$

where s_0, the displacement amplitude at unit distance from the source, is a constant parameter characterizing the whole wave.

It is useful to represent spherical waves with a series of circular arcs concentric with the source, as shown in Figure 17.6. Each arc represents a surface over which the phase of the wave is constant. We call such a surface of constant phase a **wave front.** The distance between adjacent wave fronts equals the wavelength λ. The radial lines pointing outward from the source are called **rays.**

Now consider a small portion of a wave front far from the source, as shown in Figure 17.7. In this case, the rays passing through the wave front are nearly parallel to one another, and the wave front is very close to being planar. Therefore, at distances from the source that are great compared with the wavelength, we can approximate a wave front with a plane. Any small portion of a spherical wave far from its source can be considered a **plane wave.**

Figure 17.8 illustrates a plane wave propagating along the x axis, which means that the wave fronts are parallel to the yz plane. In this case, the wave function depends only on x and t and has the form

$$\psi(x, t) = A \sin(kx - \omega t) \qquad \textbf{(17.10)}$$

That is, the wave function for a plane wave is identical in form to that for a one-dimensional traveling wave.

The intensity is the same at all points on a given wave front of a plane wave.

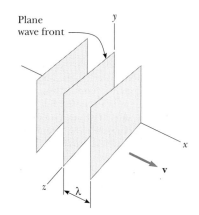

Figure 17.8 A representation of a plane wave moving in the positive x direction with a speed v. The wave fronts are planes parallel to the yz plane.

Representation of a plane wave

EXAMPLE 17.5 ▶ Intensity Variations of a Point Source

A point source emits sound waves with an average power output of 80.0 W. (a) Find the intensity 3.00 m from the source.

Solution A point source emits energy in the form of spherical waves (see Fig. 17.5). At a distance r from the source, the power is distributed over the surface area of a sphere, $4\pi r^2$. Therefore, the intensity at the distance r is given by Equation 17.8:

$$I = \frac{\mathcal{P}_{av}}{4\pi r^2} = \frac{80.0 \text{ W}}{4\pi(3.00 \text{ m})^2} = \boxed{0.707 \text{ W/m}^2}$$

an intensity that is close to the threshold of pain.

(b) Find the distance at which the sound level is 40 dB.

Solution We can find the intensity at the 40-dB sound level by using Equation 17.7 with $I_0 = 1.00 \times 10^{-12}$ W/m²:

$$10 \log\left(\frac{I}{I_0}\right) = 40 \text{ dB}$$

$$\log I - \log I_0 = \frac{40}{10} = 4$$

$$\log I = 4 + \log 10^{-12}$$

$$\log I = -8$$

$$I = 1.00 \times 10^{-8} \text{ W/m}^2$$

Using this value for I in Equation 17.8 and solving for r, we obtain

$$r = \sqrt{\frac{\mathcal{P}_{av}}{4\pi I}} = \sqrt{\frac{80.0 \text{ W}}{4\pi \times 1.00 \times 10^{-8} \text{ W/m}^2}}$$

$$= \boxed{2.52 \times 10^4 \text{ m}}$$

which equals about 16 miles!

17.5   THE DOPPLER EFFECT

QuickLab

(Before attempting to do this Quick-Lab, you should check to see whether it is legal to sound a horn in your area.) Sound your car horn while driving toward and away from a friend in a campus parking lot or on a country road. Try this at different speeds while driving toward and past the friend (not *at* the friend). Do the frequencies of the sounds your friend hears agree with what is described in the text?

Perhaps you have noticed how the sound of a vehicle's horn changes as the vehicle moves past you. The frequency of the sound you hear as the vehicle approaches you is higher than the frequency you hear as it moves away from you (see Quick-Lab). This is one example of the **Doppler effect.**[3]

To see what causes this apparent frequency change, imagine you are in a boat that is lying at anchor on a gentle sea where the waves have a period of $T = 3.0$ s. This means that every 3.0 s a crest hits your boat. Figure 17.9a shows this situation, with the water waves moving toward the left. If you set your watch to $t = 0$ just as one crest hits, the watch reads 3.0 s when the next crest hits, 6.0 s when the third crest hits, and so on. From these observations you conclude that the wave frequency is $f = 1/T = (1/3.0)$ Hz. Now suppose you start your motor and head directly into the oncoming waves, as shown in Figure 17.9b. Again you set your watch to $t = 0$ as a crest hits the front of your boat. Now, however, because you are moving toward the next wave crest as it moves toward you, it hits you less than 3.0 s after the first hit. In other words, the period you observe is shorter than the 3.0-s period you observed when you were stationary. Because $f = 1/T$, you observe a higher wave frequency than when you were at rest.

If you turn around and move in the same direction as the waves (see Fig. 17.9c), you observe the opposite effect. You set your watch to $t = 0$ as a crest hits the back of the boat. Because you are now moving away from the next crest, more than 3.0 s has elapsed on your watch by the time that crest catches you. Thus, you observe a lower frequency than when you were at rest.

These effects occur because the relative speed between your boat and the waves depends on the direction of travel and on the speed of your boat. When you are moving toward the right in Figure 17.9b, this relative speed is higher than that of the wave speed, which leads to the observation of an increased frequency. When you turn around and move to the left, the relative speed is lower, as is the observed frequency of the water waves.

Let us now examine an analogous situation with sound waves, in which the water waves become sound waves, the water becomes the air, and the person on the boat becomes an observer listening to the sound. In this case, an observer O is moving and a sound source S is stationary. For simplicity, we assume that the air is also stationary and that the observer moves directly toward the source. The observer moves with a speed v_O toward a stationary point source ($v_S = 0$) (Fig. 17.10). In general, *at rest* means at rest with respect to the medium, air.

---

[3] Named after the Austrian physicist Christian Johann Doppler (1803–1853), who discovered the effect for light waves.

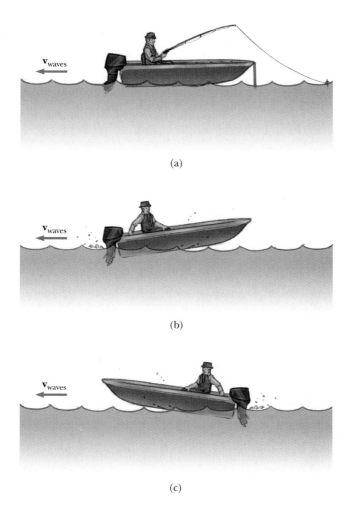

(a)

(b)

(c)

Figure 17.9 (a) Waves moving toward a stationary boat. The waves travel to the left, and their source is far to the right of the boat, out of the frame of the drawing. (b) The boat moving toward the wave source. (c) The boat moving away from the wave source.

We take the frequency of the source to be f, the wavelength to be λ, and the speed of sound to be v. If the observer were also stationary, he or she would detect f wave fronts per second. (That is, when $v_O = 0$ and $v_S = 0$, the observed frequency equals the source frequency.) When the observer moves toward the source,

Figure 17.10 An observer O (the cyclist) moves with a speed v_O toward a stationary point source S, the horn of a parked car. The observer hears a frequency f' that is greater than the source frequency.

the speed of the waves relative to the observer is $v' = v + v_O$, as in the case of the boat, but the wavelength λ is unchanged. Hence, using Equation 16.14, $v = \lambda f$, we can say that the frequency heard by the observer is *increased* and is given by

$$f' = \frac{v'}{\lambda} = \frac{v + v_O}{\lambda}$$

Because $\lambda = v/f$, we can express f' as

$$f' = \left(1 + \frac{v_O}{v}\right)f \qquad \text{(observer moving toward source)} \qquad \textbf{(17.11)}$$

If the observer is moving away from the source, the speed of the wave relative to the observer is $v' = v - v_O$. The frequency heard by the observer in this case is *decreased* and is given by

$$f' = \left(1 - \frac{v_O}{v}\right)f \qquad \text{(observer moving away from source)} \qquad \textbf{(17.12)}$$

In general, whenever an observer moves with a speed v_O relative to a stationary source, the frequency heard by the observer is

Frequency heard with an observer in motion

$$f' = \left(1 \pm \frac{v_O}{v}\right)f \qquad \textbf{(17.13)}$$

where the positive sign is used when the observer moves toward the source and the negative sign is used when the observer moves away from the source.

Now consider the situation in which the source is in motion and the observer is at rest. If the source moves directly toward observer A in Figure 17.11a, the wave fronts heard by the observer are closer together than they would be if the source were not moving. As a result, the wavelength λ' measured by observer A is shorter than the wavelength λ of the source. During each vibration, which lasts for a time T (the period), the source moves a distance $v_S T = v_S/f$ and the wavelength is

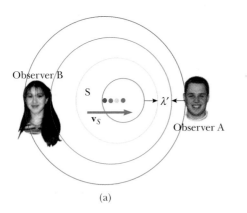

(a)

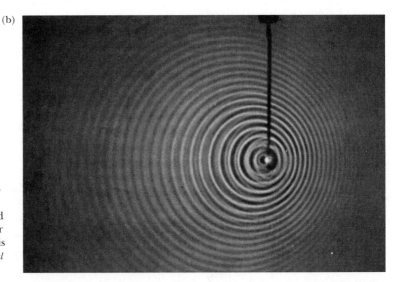

(b)

Figure 17.11   (a) A source S moving with a speed v_S toward a stationary observer A and away from a stationary observer B. Observer A hears an increased frequency, and observer B hears a decreased frequency. (b) The Doppler effect in water, observed in a ripple tank. A point source is moving to the right with speed v_S. *(Courtesy of the Educational Development Center, Newton, MA)*

shortened by this amount. Therefore, the observed wavelength λ' is

$$\lambda' = \lambda - \Delta\lambda = \lambda - \frac{v_S}{f}$$

Because $\lambda = v/f$, the frequency heard by observer A is

$$f' = \frac{v}{\lambda'} = \frac{v}{\lambda - \dfrac{v_S}{f}} = \frac{v}{\dfrac{v}{f} - \dfrac{v_S}{f}}$$

$$f' = \left(\frac{1}{1 - \dfrac{v_S}{v}}\right) f \qquad\qquad \textbf{(17.14)}$$

"I love hearing that lonesome wail of the train whistle as the magnitude of the frequency of the wave changes due to the Doppler effect." *(Sydney Harris)*

That is, the observed frequency is *increased* whenever the source is moving toward the observer.

When the source moves away from a stationary observer, as is the case for observer B in Figure 17.11a, the observer measures a wavelength λ' that is *greater* than λ and hears a *decreased* frequency:

$$f' = \left(\frac{1}{1 + \dfrac{v_S}{v}}\right) f \qquad\qquad \textbf{(17.15)}$$

Combining Equations 17.14 and 17.15, we can express the general relationship for the observed frequency when a source is moving and an observer is at rest as

$$f' = \left(\frac{1}{1 \mp \dfrac{v_S}{v}}\right) f \qquad\qquad \textbf{(17.16)}$$

Frequency heard with source in motion

Finally, if both source and observer are in motion, we find the following general relationship for the observed frequency:

$$f' = \left(\frac{v \pm v_O}{v \mp v_S}\right) f \qquad\qquad \textbf{(17.17)}$$

Frequency heard with observer and source in motion

In this expression, the upper signs ($+ v_O$ and $- v_S$) refer to motion of one toward the other, and the lower signs ($- v_O$ and $+ v_S$) refer to motion of one away from the other.

A convenient rule concerning signs for you to remember when working with all Doppler-effect problems is as follows:

The word *toward* is associated with an *increase* in observed frequency. The words *away from* are associated with a *decrease* in observed frequency.

Although the Doppler effect is most typically experienced with sound waves, it is a phenomenon that is common to all waves. For example, the relative motion of source and observer produces a frequency shift in light waves. The Doppler effect is used in police radar systems to measure the speeds of motor vehicles. Likewise, astronomers use the effect to determine the speeds of stars, galaxies, and other celestial objects relative to the Earth.

EXAMPLE 17.6   The Noisy Siren

As an ambulance travels east down a highway at a speed of 33.5 m/s (75 mi/h), its siren emits sound at a frequency of 400 Hz. What frequency is heard by a person in a car traveling west at 24.6 m/s (55 mi/h) (a) as the car approaches the ambulance and (b) as the car moves away from the ambulance?

Solution   (a) We can use Equation 17.17 in both cases, taking the speed of sound in air to be $v = 343$ m/s. As the ambulance and car approach each other, the person in the car hears the frequency

$$f' = \left(\frac{v + v_O}{v - v_S}\right)f = \left(\frac{343 \text{ m/s} + 24.6 \text{ m/s}}{343 \text{ m/s} - 33.5 \text{ m/s}}\right)(400 \text{ Hz})$$

$$= \boxed{475 \text{ Hz}}$$

(b) As the vehicles recede from each other, the person hears the frequency

$$f' = \left(\frac{v - v_O}{v + v_S}\right)f = \left(\frac{343 \text{ m/s} - 24.6 \text{ m/s}}{343 \text{ m/s} + 33.5 \text{ m/s}}\right)(400 \text{ Hz})$$

$$= \boxed{338 \text{ Hz}}$$

The *change* in frequency detected by the person in the car is $475 - 338 = 137$ Hz, which is more than 30% of the true frequency.

Exercise   Suppose the car is parked on the side of the highway as the ambulance speeds by. What frequency does the person in the car hear as the ambulance (a) approaches and (b) recedes?

Answer   (a) 443 Hz.   (b) 364 Hz.

Shock Waves

Now let us consider what happens when the speed v_S of a source *exceeds* the wave speed v. This situation is depicted graphically in Figure 17.12a. The circles represent spherical wave fronts emitted by the source at various times during its motion. At $t = 0$, the source is at S_0, and at a later time t, the source is at S_n. In the time t,

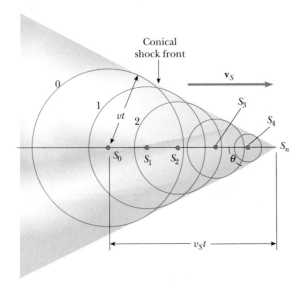

(a)

(b)

Figure 17.12   (a) A representation of a shock wave produced when a source moves from S_0 to S_n with a speed v_S, which is greater than the wave speed v in the medium. The envelope of the wave fronts forms a cone whose apex half-angle is given by $\sin \theta = v/v_S$. (b) A stroboscopic photograph of a bullet moving at supersonic speed through the hot air above a candle. Note the shock wave in the vicinity of the bullet. (*©1973 Kim Vandiver & Harold E. Edgerton/Courtesy of Palm Press, Inc.*)

Figure 17.13 The V-shaped bow wave of a boat is formed because the boat speed is greater than the speed of the water waves. A bow wave is analogous to a shock wave formed by an airplane traveling faster than sound. (©1994 Comstock)

the wave front centered at S_0 reaches a radius of vt. In this same amount of time, the source travels a distance $v_S t$ to S_n. At the instant the source is at S_n, waves are just beginning to be generated at this location, and hence the wave front has zero radius at this point. The tangent line drawn from S_n to the wave front centered on S_0 is tangent to all other wave fronts generated at intermediate times. Thus, we see that the envelope of these wave fronts is a cone whose apex half-angle θ is given by

$$\sin \theta = \frac{vt}{v_S t} = \frac{v}{v_S}$$

The ratio v_S/v is referred to as the *Mach number,* and the conical wave front produced when $v_S > v$ (supersonic speeds) is known as a *shock wave.* An interesting analogy to shock waves is the V-shaped wave fronts produced by a boat (the *bow wave*) when the boat's speed exceeds the speed of the surface-water waves (Fig. 17.13).

Jet airplanes traveling at supersonic speeds produce shock waves, which are responsible for the loud "sonic boom" one hears. The shock wave carries a great deal of energy concentrated on the surface of the cone, with correspondingly great pressure variations. Such shock waves are unpleasant to hear and can cause damage to buildings when aircraft fly supersonically at low altitudes. In fact, an airplane flying at supersonic speeds produces a double boom because two shock fronts are formed, one from the nose of the plane and one from the tail (Fig. 17.14). People near the path of the space shuttle as it glides toward its landing point often report hearing what sounds like two very closely spaced cracks of thunder.

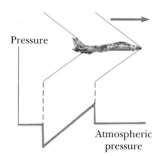

Pressure

Atmospheric pressure

Figure 17.14 The two shock waves produced by the nose and tail of a jet airplane traveling at supersonic speeds.

Quick Quiz 17.5

An airplane flying with a constant velocity moves from a cold air mass into a warm air mass. Does the Mach number increase, decrease, or stay the same?

Quick Quiz 17.6

Suppose that an observer and a source of sound are both at rest and that a strong wind blows from the source toward the observer. Describe the effect of the wind (if any) on

(a) the observed frequency of the sound waves, (b) the observed wave speed, and (c) the observed wavelength.

SUMMARY

Sound waves are longitudinal and travel through a compressible medium with a speed that depends on the compressibility and inertia of that medium. The speed of sound in a medium having a bulk modulus B and density ρ is

$$v = \sqrt{\frac{B}{\rho}} \tag{17.1}$$

With this formula you can determine the speed of a sound wave in many different materials.

For sinusoidal sound waves, the variation in the displacement is given by

$$s(x, t) = s_{max} \cos(kx - \omega t) \tag{17.2}$$

and the variation in pressure from the equilibrium value is

$$\Delta P = \Delta P_{max} \sin(kx - \omega t) \tag{17.3}$$

where ΔP_{max} is the **pressure amplitude.** The pressure wave is 90° out of phase with the displacement wave. The relationship between s_{max} and ΔP_{max} is given by

$$\Delta P_{max} = \rho v \omega s_{max} \tag{17.4}$$

The intensity of a periodic sound wave, which is the power per unit area, is

$$I = \tfrac{1}{2}\rho v (\omega s_{max})^2 = \frac{\Delta P_{max}^2}{2\rho v} \tag{17.5, 17.6}$$

The sound level of a sound wave, in decibels, is given by

$$\beta = 10 \log\left(\frac{I}{I_0}\right) \tag{17.7}$$

The constant I_0 is a reference intensity, usually taken to be at the threshold of hearing (1.00×10^{-12} W/m²), and I is the intensity of the sound wave in watts per square meter.

The intensity of a spherical wave produced by a point source is proportional to the average power emitted and inversely proportional to the square of the distance from the source:

$$I = \frac{\mathscr{P}_{av}}{4\pi r^2} \tag{17.8}$$

The change in frequency heard by an observer whenever there is relative motion between a source of sound waves and the observer is called the **Doppler effect.** The observed frequency is

$$f' = \left(\frac{v \pm v_O}{v \mp v_S}\right)f \tag{17.17}$$

The upper signs ($+ v_O$ and $- v_S$) are used with motion of one toward the other, and the lower signs ($- v_O$ and $+ v_S$) are used with motion of one away from the other. You can also use this formula when v_O or v_S is zero.

QUESTIONS

1. Why are sound waves characterized as longitudinal?
2. If an alarm clock is placed in a good vacuum and then activated, no sound is heard. Explain.
3. A sonic ranger is a device that determines the position of an object by sending out an ultrasonic sound pulse and measuring how long it takes for the sound wave to return after it reflects from the object. Typically, these devices cannot reliably detect an object that is less than half a meter from the sensor. Why is that?
4. In Example 17.5, we found that a point source with a power output of 80 W reduces to a sound level of 40 dB at a distance of about 16 miles. Why do you suppose you cannot normally hear a rock concert that is going on 16 miles away? (See Table 17.2.)
5. If the distance from a point source is tripled, by what factor does the intensity decrease?
6. Explain how the Doppler effect is used with microwaves to determine the speed of an automobile.
7. Explain what happens to the frequency of your echo as you move in a vehicle *toward* a canyon wall. What happens to the frequency as you move *away* from the wall?
8. Of the following sounds, which is most likely to have a sound level of 60 dB—a rock concert, the turning of a page in this text, normal conversation, or a cheering crowd at a football game?
9. Estimate the decibel level of each of the sounds in the previous question.
10. A binary star system consists of two stars revolving about their common center of mass. If we observe the light reaching us from one of these stars as it makes one complete revolution, what does the Doppler effect predict will happen to this light?
11. How can an object move with respect to an observer so that the sound from it is not shifted in frequency?
12. Why is it not possible to use sonar (sound waves) to determine the speed of an object traveling faster than the speed of sound in a given medium?
13. Why is it so quiet after a snowfall?
14. Why is the intensity of an echo less than that of the original sound?
15. If the wavelength of a sound source is reduced by a factor of 2, what happens to its frequency? Its speed?
16. In a recent discovery, a nearby star was found to have a large planet orbiting about it, although the planet could not be seen. In terms of the concept of a system rotating about its center of mass and the Doppler shift for light (which is in many ways similar to that for sound), explain how an astronomer could determine the presence of the invisible planet.
17. A friend sitting in her car far down the road waves to you and beeps her horn at the same time. How far away must her car be for you to measure the speed of sound to two significant figures by measuring the time it takes for the sound to reach you?

PROBLEMS

1, 2, 3 = straightforward, intermediate, challenging   ☐ = full solution available in the *Student Solutions Manual and Study Guide*
WEB = solution posted at **http://www.saunderscollege.com/physics/**   ▭ = Computer useful in solving problem   ▦ = Interactive Physics
☐ = paired numerical/symbolic problems

Section 17.1   Speed of Sound Waves

1. Suppose that you hear a clap of thunder 16.2 s after seeing the associated lightning stroke. The speed of sound waves in air is 343 m/s, and the speed of light in air is 3.00×10^8 m/s. How far are you from the lightning stroke?
2. Find the speed of sound in mercury, which has a bulk modulus of approximately 2.80×10^{10} N/m^2 and a density of 13 600 kg/m^3.
3. A flower pot is knocked off a balcony 20.0 m above the sidewalk and falls toward an unsuspecting 1.75-m-tall man who is standing below. How close to the sidewalk can the flower pot fall before it is too late for a shouted warning from the balcony to reach the man in time? Assume that the man below requires 0.300 s to respond to the warning.
4. You are watching a pier being constructed on the far shore of a saltwater inlet when some blasting occurs. You hear the sound in the water 4.50 s before it reaches you through the air. How wide is the inlet? (*Hint:* See Table 17.1. Assume that the air temperature is 20°C.)
5. Another approximation of the temperature dependence of the speed of sound in air (in meters per second) is given by the expression

$$v = 331.5 + 0.607T_{\mathrm{C}}$$

where T_{C} is the Celsius temperature. In dry air the temperature decreases about 1°C for every 150-m rise in altitude. (a) Assuming that this change is constant up to an altitude of 9 000 m, how long will it take the sound from an airplane flying at 9 000 m to reach the ground on a day when the ground temperature is 30°C? (b) Compare this to the time it would take if the air were at 30°C at all altitudes. Which interval is longer?
6. A bat can detect very small objects, such as an insect whose length is approximately equal to one wavelength

of the sound the bat makes. If bats emit a chirp at a frequency of 60.0 kHz, and if the speed of sound in air is 340 m/s, what is the smallest insect a bat can detect?

7. An airplane flies horizontally at a constant speed, piloted by rescuers who are searching for a disabled boat. When the plane is directly above the boat, the boat's crew blows a loud horn. By the time the plane's sound detector receives the horn's sound, the plane has traveled a distance equal to one-half its altitude above the ocean. If it takes the sound 2.00 s to reach the plane, determine (a) the speed of the plane and (b) its altitude. Take the speed of sound to be 343 m/s.

Section 17.2   Periodic Sound Waves

Note: In this section, use the following values as needed, unless otherwise specified. The equilibrium density of air is $\rho = 1.20$ kg/m^3; the speed of sound in air is $v = 343$ m/s. Pressure variations ΔP are measured relative to atmospheric pressure, 1.013×10^5 Pa.

8. A sound wave in air has a pressure amplitude equal to 4.00×10^{-3} Pa. Calculate the displacement amplitude of the wave at a frequency of 10.0 kHz.

9. A sinusoidal sound wave is described by the displacement

$$s(x, t) = (2.00 \ \mu\text{m}) \cos[(15.7 \ \text{m}^{-1})x - (858 \ \text{s}^{-1})t]$$

(a) Find the amplitude, wavelength, and speed of this wave. (b) Determine the instantaneous displacement of the molecules at the position $x = 0.050\ 0$ m at $t = 3.00$ ms. (c) Determine the maximum speed of a molecule's oscillatory motion.

10. As a sound wave travels through the air, it produces pressure variations (above and below atmospheric pressure) that are given by $\Delta P = 1.27 \sin(\pi x - 340\pi t)$ in SI units. Find (a) the amplitude of the pressure variations, (b) the frequency of the sound wave, (c) its wavelength in air, and (d) its speed.

11. Write an expression that describes the pressure variation as a function of position and time for a sinusoidal sound wave in air, if $\lambda = 0.100$ m and $\Delta P_{\text{max}} = 0.200$ Pa.

12. Write the function that describes the displacement wave corresponding to the pressure wave in Problem 11.

13. The tensile stress in a thick copper bar is 99.5% of its elastic breaking point of 13.0×10^{10} N/m^2. A 500-Hz sound wave is transmitted through the material. (a) What displacement amplitude will cause the bar to break? (b) What is the maximum speed of the particles at this moment?

14. Calculate the pressure amplitude of a 2.00-kHz sound wave in air if the displacement amplitude is equal to 2.00×10^{-8} m.

WEB 15. An experimenter wishes to generate in air a sound wave that has a displacement amplitude of 5.50×10^{-6} m. The pressure amplitude is to be limited to 8.40×10^{-1} Pa. What is the minimum wavelength the sound wave can have?

16. A sound wave in air has a pressure amplitude of 4.00 Pa and a frequency of 5.00 kHz. Take $\Delta P = 0$ at the point $x = 0$ when $t = 0$. (a) What is ΔP at $x = 0$ when $t = 2.00 \times 10^{-4}$ s? (b) What is ΔP at $x = 0.020\ 0$ m when $t = 0$?

Section 17.3   Intensity of Periodic Sound Waves

17. Calculate the sound level, in decibels, of a sound wave that has an intensity of 4.00 μW/m^2.

18. A vacuum cleaner has a measured sound level of 70.0 dB. (a) What is the intensity of this sound in watts per square meter? (b) What is the pressure amplitude of the sound?

19. The intensity of a sound wave at a fixed distance from a speaker vibrating at 1.00 kHz is 0.600 W/m^2. (a) Determine the intensity if the frequency is increased to 2.50 kHz while a constant displacement amplitude is maintained. (b) Calculate the intensity if the frequency is reduced to 0.500 kHz and the displacement amplitude is doubled.

20. The intensity of a sound wave at a fixed distance from a speaker vibrating at a frequency f is I. (a) Determine the intensity if the frequency is increased to f' while a constant displacement amplitude is maintained. (b) Calculate the intensity if the frequency is reduced to $f/2$ and the displacement amplitude is doubled.

WEB 21. A family ice show is held in an enclosed arena. The skaters perform to music with a sound level of 80.0 dB. This is too loud for your baby, who consequently yells at a level of 75.0 dB. (a) What total sound intensity engulfs you? (b) What is the combined sound level?

Section 17.4   Spherical and Plane Waves

22. For sound radiating from a point source, show that the difference in sound levels, β_1 and β_2, at two receivers is related to the ratio of the distances r_1 and r_2 from the source to the receivers by the expression

$$\beta_2 - \beta_1 = 20 \log\left(\frac{r_1}{r_2}\right)$$

23. A fireworks charge is detonated many meters above the ground. At a distance of 400 m from the explosion, the acoustic pressure reaches a maximum of 10.0 N/m^2. Assume that the speed of sound is constant at 343 m/s throughout the atmosphere over the region considered, that the ground absorbs all the sound falling on it, and that the air absorbs sound energy as described by the rate 7.00 dB/km. What is the sound level (in decibels) at 4.00 km from the explosion?

24. A loudspeaker is placed between two observers who are 110 m apart, along the line connecting them. If one observer records a sound level of 60.0 dB and the other records a sound level of 80.0 dB, how far is the speaker from each observer?

25. Two small speakers emit spherical sound waves of different frequencies. Speaker *A* has an output of 1.00 mW, and speaker *B* has an output of 1.50 mW. Determine the sound level (in decibels) at point *C* (Fig. P17.25) if (a) only speaker *A* emits sound, (b) only speaker *B* emits sound, (c) both speakers emit sound.

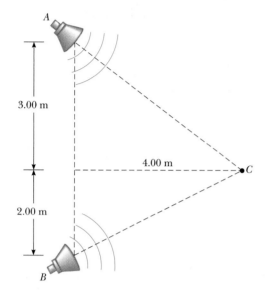

3.00 m

4.00 m

C

2.00 m

B

Figure P17.25

26. An experiment requires a sound intensity of 1.20 W/m² at a distance of 4.00 m from a speaker. What power output is required? Assume that the speaker radiates sound equally in all directions.

27. A source of sound (1 000 Hz) emits uniformly in all directions. An observer 3.00 m from the source measures a sound level of 40.0 dB. Calculate the average power output of the source.

28. A jackhammer, operated continuously at a construction site, behaves as a point source of spherical sound waves. A construction supervisor stands 50.0 m due north of this sound source and begins to walk due west. How far does she have to walk in order for the amplitude of the wave function to drop by a factor of 2.00?

29. The sound level at a distance of 3.00 m from a source is 120 dB. At what distances is the sound level (a) 100 dB and (b) 10.0 dB?

30. A fireworks rocket explodes 100 m above the ground. An observer directly under the explosion experiences an average sound intensity of 7.00×10^{-2} W/m² for 0.200 s. (a) What is the total sound energy of the explosion? (b) What sound level, in decibels, is heard by the observer?

31. As the people in a church sing on a summer morning, the sound level everywhere inside the church is 101 dB. The massive walls are opaque to sound, but all the windows and doors are open. Their total area is 22.0 m². (a) How much sound energy is radiated in 20.0 min? (b) Suppose the ground is a good reflector and sound

radiates uniformly in all horizontal and upward directions. Find the sound level 1.00 km away.

32. A spherical wave is radiating from a point source and is described by the wave function

$$\Delta P(r, t) = \left[\frac{25.0}{r} \right] \sin(1.25r - 1\ 870t)$$

where ΔP is in pascals, r in meters, and t in seconds. (a) What is the pressure amplitude 4.00 m from the source? (b) Determine the speed of the wave and hence the material the wave might be traveling through. (c) Find the sound level of the wave, in decibels, 4.00 m from the source. (d) Find the instantaneous pressure 5.00 m from the source at 0.080 0 s.

Section 17.5 The Doppler Effect

33. A commuter train passes a passenger platform at a constant speed of 40.0 m/s. The train horn is sounded at its characteristic frequency of 320 Hz. (a) What change in frequency is detected by a person on the platform as the train passes? (b) What wavelength is detected by a person on the platform as the train approaches?

34. A driver travels northbound on a highway at a speed of 25.0 m/s. A police car, traveling southbound at a speed of 40.0 m/s, approaches with its siren sounding at a frequency of 2 500 Hz. (a) What frequency does the driver observe as the police car approaches? (b) What frequency does the driver detect after the police car passes him? (c) Repeat parts (a) and (b) for the case in which the police car is northbound.

WEB 35. Standing at a crosswalk, you hear a frequency of 560 Hz from the siren of an approaching police car. After the police car passes, the observed frequency of the siren is 480 Hz. Determine the car's speed from these observations.

36. Expectant parents are thrilled to hear their unborn baby's heartbeat, revealed by an ultrasonic motion detector. Suppose the fetus's ventricular wall moves in simple harmonic motion with an amplitude of 1.80 mm and a frequency of 115 per minute. (a) Find the maximum linear speed of the heart wall. Suppose the motion detector in contact with the mother's abdomen produces sound at 2 000 000.0 Hz, which travels through tissue at 1.50 km/s. (b) Find the maximum frequency at which sound arrives at the wall of the baby's heart. (c) Find the maximum frequency at which reflected sound is received by the motion detector. (By electronically "listening" for echoes at a frequency different from the broadcast frequency, the motion detector can produce beeps of audible sound in synchronization with the fetal heartbeat.)

37. A tuning fork vibrating at 512 Hz falls from rest and accelerates at 9.80 m/s². How far below the point of release is the tuning fork when waves with a frequency of 485 Hz reach the release point? Take the speed of sound in air to be 340 m/s.

38. A block with a speaker bolted to it is connected to a spring having spring constant $k = 20.0$ N/m, as shown in Figure P17.38. The total mass of the block and speaker is 5.00 kg, and the amplitude of this unit's motion is 0.500 m. (a) If the speaker emits sound waves of frequency 440 Hz, determine the highest and lowest frequencies heard by the person to the right of the speaker. (b) If the maximum sound level heard by the person is 60.0 dB when he is closest to the speaker, 1.00 m away, what is the minimum sound level heard by the observer? Assume that the speed of sound is 343 m/s.

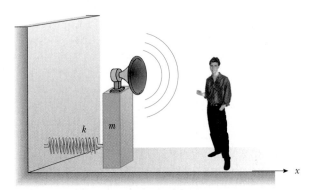

Figure P17.38

39. A train is moving parallel to a highway with a constant speed of 20.0 m/s. A car is traveling in the same direction as the train with a speed of 40.0 m/s. The car horn sounds at a frequency of 510 Hz, and the train whistle sounds at a frequency of 320 Hz. (a) When the car is behind the train, what frequency does an occupant of the car observe for the train whistle? (b) When the car is in front of the train, what frequency does a train passenger observe for the car horn just after the car passes?

40. At the Winter Olympics, an athlete rides her luge down the track while a bell just above the wall of the chute rings continuously. When her sled passes the bell, she hears the frequency of the bell fall by the musical interval called a minor third. That is, the frequency she hears drops to five sixths of its original value. (a) Find the speed of sound in air at the ambient temperature −10.0°C. (b) Find the speed of the athlete.

41. A jet fighter plane travels in horizontal flight at Mach 1.20 (that is, 1.20 times the speed of sound in air). At the instant an observer on the ground hears the shock wave, what is the angle her line of sight makes with the horizontal as she looks at the plane?

42. When high-energy charged particles move through a transparent medium with a speed greater than the speed of light in that medium, a shock wave, or bow wave, of light is produced. This phenomenon is called the *Cerenkov effect* and can be observed in the vicinity of the core of a swimming-pool nuclear reactor due to

high-speed electrons moving through the water. In a particular case, the Cerenkov radiation produces a wave front with an apex half-angle of 53.0°. Calculate the speed of the electrons in the water. (The speed of light in water is 2.25×10^8 m/s.)

WEB 43. A supersonic jet traveling at Mach 3.00 at an altitude of 20 000 m is directly over a person at time $t = 0$, as in Figure P17.43. (a) How long will it be before the person encounters the shock wave? (b) Where will the plane be when it is finally heard? (Assume that the speed of sound in air is 335 m/s.)

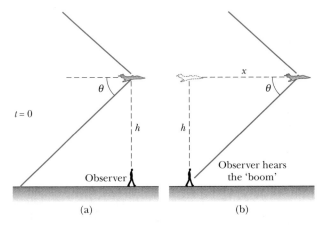

(a)   (b)

Figure P17.43

44. The tip of a circus ringmaster's whip travels at Mach 1.38 (that is, $v_S/v = 1.38$). What angle does the shock front make with the direction of the whip's motion?

ADDITIONAL PROBLEMS

45. A stone is dropped into a deep canyon and is heard to strike the bottom 10.2 s after release. The speed of sound waves in air is 343 m/s. How deep is the canyon? What would be the percentage error in the calculated depth if the time required for the sound to reach the canyon rim were ignored?

46. Unoccupied by spectators, a large set of football bleachers has solid seats and risers. You stand on the field in front of it and fire a starter's pistol or sharply clap two wooden boards together once. The sound pulse you produce has no frequency and no wavelength. You hear back from the bleachers a sound with definite pitch, which may remind you of a short toot on a trumpet, or of a buzzer or a kazoo. Account for this sound. Compute order-of-magnitude estimates for its frequency, wavelength, and duration on the basis of data that you specify.

47. Many artists sing very high notes in ornaments and cadenzas. The highest note written for a singer in a published score was F-sharp above high C, 1.480 kHz, sung

by Zerbinetta in the original version of Richard Strauss's opera *Ariadne auf Naxos*. (a) Find the wavelength of this sound in air. (b) Suppose that the people in the fourth row of seats hear this note with a level of 81.0 dB. Find the displacement amplitude of the sound. (c) In response to complaints, Strauss later transposed the note down to F above high C, 1.397 kHz. By what increment did the wavelength change?

48. A sound wave in a cylinder is described by Equations 17.2 through 17.4. Show that $\Delta P = \pm\, \rho v \omega \sqrt{s_{\max}^2 - s^2}$.

49. On a Saturday morning, pickup trucks carrying garbage to the town dump form a nearly steady procession on a country road, all traveling at 19.7 m/s. From this direction, two trucks arrive at the dump every three minutes. A bicyclist also is traveling toward the dump at 4.47 m/s. (a) With what frequency do the trucks pass him? (b) A hill does not slow the trucks but makes the out-of-shape cyclist's speed drop to 1.56 m/s. How often do the noisy trucks whiz past him now?

50. The ocean floor is underlain by a layer of basalt that constitutes the crust, or uppermost layer, of the Earth in that region. Below the crust is found denser peridotite rock, which forms the Earth's mantle. The boundary between these two layers is called the Mohorovicic discontinuity ("Moho" for short). If an explosive charge is set off at the surface of the basalt, it generates a seismic wave that is reflected back out at the Moho. If the speed of the wave in basalt is 6.50 km/s and the two-way travel time is 1.85 s, what is the thickness of this oceanic crust?

51. A worker strikes a steel pipeline with a hammer, generating both longitudinal and transverse waves. Reflected waves return 2.40 s apart. How far away is the reflection point? (For steel, $v_{\text{long}} = 6.20$ km/s and $v_{\text{trans}} = 3.20$ km/s.)

52. For a certain type of steel, stress is proportional to strain with Young's modulus as given in Table 12.1. The steel has the density listed for iron in Table 15.1. It bends permanently if subjected to compressive stress greater than its elastic limit, $\sigma = 400$ MPa, also called its *yield strength*. A rod 80.0 cm long, made of this steel, is projected at 12.0 m/s straight at a hard wall. (a) Find the speed of compressional waves moving along the rod. (b) After the front end of the rod hits the wall and stops, the back end of the rod keeps moving, as described by Newton's first law, until it is stopped by the excess pressure in a sound wave moving back through the rod. How much time elapses before the back end of the rod gets the message? (c) How far has the back end of the rod moved in this time? (d) Find the strain in the rod and (e) the stress. (f) If it is not to fail, show that the maximum impact speed a rod can have is given by the expression $\sigma/\sqrt{\rho Y}$.

53. To determine her own speed, a sky diver carries a buzzer that emits a steady tone at 1 800 Hz. A friend at the landing site on the ground directly below the sky diver listens to the amplified sound he receives from the buzzer. Assume that the air is calm and that the speed

of sound is 343 m/s, independent of altitude. While the sky diver is falling at terminal speed, her friend on the ground receives waves with a frequency of 2 150 Hz. (a) What is the sky diver's speed of descent? (b) Suppose the sky diver is also carrying sound-receiving equipment that is sensitive enough to detect waves reflected from the ground. What frequency does she receive?

54. A train whistle ($f = 400$ Hz) sounds higher or lower in pitch depending on whether it is approaching or receding. (a) Prove that the difference in frequency between the approaching and receding train whistle is

$$\Delta f = \frac{2(u/v)}{1 - (u^2/v^2)}\, f$$

where u is the speed of the train and v is the speed of sound. (b) Calculate this difference for a train moving at a speed of 130 km/h. Take the speed of sound in air to be 340 m/s.

55. A bat, moving at 5.00 m/s, is chasing a flying insect. If the bat emits a 40.0-kHz chirp and receives back an echo at 40.4 kHz, at what relative speed is the bat moving toward or away from the insect? (Take the speed of sound in air to be $v = 340$ m/s.)

Figure P17.55 *(Joe McDonald/Visuals Unlimited)*

56. A supersonic aircraft is flying parallel to the ground. When the aircraft is directly overhead, an observer on the ground sees a rocket fired from the aircraft. Ten seconds later the observer hears the sonic boom, which is followed 2.80 s later by the sound of the rocket engine. What is the Mach number of the aircraft?

57. A police car is traveling east at 40.0 m/s along a straight road, overtaking a car that is moving east at 30.0 m/s. The police car has a malfunctioning siren that is stuck at 1 000 Hz. (a) Sketch the appearance of the wave fronts of the sound produced by the siren. Show the wave fronts both to the east and to the west of the police car. (b) What would be the wavelength in air of the siren sound if the police car were at rest? (c) What is the wavelength in front of the car? (d) What is the wavelength behind the police car? (e) What frequency is heard by the driver being chased?

58. A copper bar is given a sharp compressional blow at one end. The sound of the blow, traveling through air at 0°C, reaches the opposite end of the bar 6.40 ms later than the sound transmitted through the metal of the bar. What is the length of the bar? (Refer to Table 17.1.)

59. The power output of a certain public address speaker is 6.00 W. Suppose it broadcasts equally in all directions. (a) Within what distance from the speaker would the sound be painful to the ear? (b) At what distance from the speaker would the sound be barely audible?

60. A jet flies toward higher altitude at a constant speed of 1 963 m/s in a direction that makes an angle θ with the horizontal (Fig. P17.60). An observer on the ground hears the jet for the first time when it is directly overhead. Determine the value of θ if the speed of sound in air is 340 m/s.

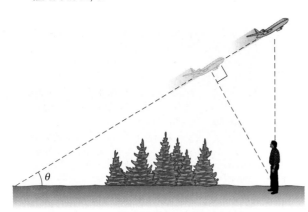

Figure P17.60

61. Two ships are moving along a line due east. The trailing vessel has a speed of 64.0 km/h relative to a land-based observation point, and the leading ship has a speed of 45.0 km/h relative to that point. The two ships are in a region of the ocean where the current is moving uniformly due west at 10.0 km/h. The trailing ship transmits a sonar signal at a frequency of 1 200.0 Hz. What frequency is monitored by the leading ship? (Use 1 520 m/s as the speed of sound in ocean water.)

62. A microwave oven generates a sound with intensity level 40.0 dB everywhere just outside it, when consuming 1.00 kW of power. Find the fraction of this power that is converted into the energy of sound waves. Assume the dimensions of the oven are 40.0 cm × 40.0 cm × 50.0 cm.

63. A meteoroid the size of a truck enters the Earth's atmosphere at a speed of 20.0 km/s and is not significantly slowed before entering the ocean. (a) What is the Mach angle of the shock wave from the meteoroid in the atmosphere? (Use 331 m/s as the sound speed.) (b) Assuming that the meteoroid survives the impact with the ocean surface, what is the (initial) Mach angle of the shock wave that the meteoroid produces in the water? (Use the wave speed for sea water given in Table 17.1.)

64. Consider a longitudinal (compressional) wave of wavelength λ traveling with speed v along the x direction through a medium of density ρ. The *displacement* of the molecules of the medium from their equilibrium position is

$$s = s_{max} \sin(kx - \omega t)$$

Show that the pressure variation in the medium is given by

$$\Delta P = -\left(\frac{2\pi\rho v^2}{\lambda} s_{max}\right) \cos(kx - \omega t)$$

WEB 65. By proper excitation, it is possible to produce both longitudinal and transverse waves in a long metal rod. A particular metal rod is 150 cm long and has a radius of 0.200 cm and a mass of 50.9 g. Young's modulus for the material is 6.80×10^{10} N/m². What must the tension in the rod be if the ratio of the speed of longitudinal waves to the speed of transverse waves is 8.00?

66. An interstate highway has been built through a neighborhood in a city. In the afternoon, the sound level in a rented room is 80.0 dB as 100 cars per minute pass outside the window. Late at night, the traffic flow on the freeway is only five cars per minute. What is the average late-night sound level in the room?

67. A siren creates a sound level of 60.0 dB at a location 500 m from the speaker. The siren is powered by a battery that delivers a total energy of 1.00 kJ. Assuming that the efficiency of the siren is 30.0% (that is, 30.0% of the supplied energy is transformed into sound energy), determine the total time the siren can sound.

68. A siren creates a sound level β at a distance d from the speaker. The siren is powered by a battery that delivers a total energy E. Assuming that the efficiency of the siren is e (that is, e is equal to the output sound energy divided by the supplied energy), determine the total time the siren can sound.

69. The Doppler equation presented in the text is valid when the motion between the observer and the source occurs on a straight line, so that the source and observer are moving either directly toward or directly away from each other. If this restriction is relaxed, one must use the more general Doppler equation

$$f' = \left(\frac{v + v_O \cos\theta_O}{v - v_S \cos\theta_S}\right) f$$

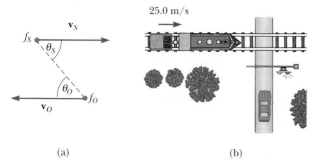

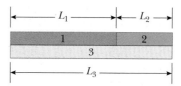

(a)   (b)

Figure P17.69

where θ_O and θ_S are defined in Figure P17.69a.
(a) Show that if the observer and source are moving away from each other, the preceding equation reduces to Equation 17.17 with lower signs. (b) Use the preceding equation to solve the following problem. A train moves at a constant speed of 25.0 m/s toward the intersection shown in Figure P17.69b. A car is stopped near the intersection, 30.0 m from the tracks. If the train's horn emits a frequency of 500 Hz, what frequency is heard by the passengers in the car when the train is 40.0 m from the intersection? Take the speed of sound to be 343 m/s.

70. Figure 17.5 illustrates that at distance r from a point source with power $\mathcal{P}_{av}$, the wave intensity is $I = \mathcal{P}_{av}/4\pi r^2$. Study Figure 17.11a and prove that at distance r straight in front of a point source with power $\mathcal{P}_{av}$, moving with constant speed v_S, the wave intensity is

$$I = \frac{\mathcal{P}_{av}}{4\pi r^2}\left(\frac{v - v_S}{v}\right)$$

71. Three metal rods are located relative to each other as shown in Figure P17.71, where $L_1 + L_2 = L_3$. The density values and Young's moduli for the three materials are $\rho_1 = 2.70 \times 10^3$ kg/m^3, $Y_1 = 7.00 \times 10^{10}$ N/m^2; $\rho_2 = 11.3 \times 10^3$ kg/m^3, $Y_2 = 1.60 \times 10^{10}$ N/m^2; $\rho_3 = 8.80 \times 10^3$ kg/m^3, $Y_3 = 11.0 \times 10^{10}$ N/m^2.
(a) If $L_3 = 1.50$ m, what must the ratio L_1/L_2 be if a sound wave is to travel the combined length of rods 1 and 2 in the same time it takes to travel the length of rod 3? (b) If the frequency of the source is 4.00 kHz, determine the phase difference between the wave traveling along rods 1 and 2 and the one traveling along rod 3.

Figure P17.71

72. The volume knob on a radio has what is known as a "logarithmic taper." The electrical device connected to the knob (called a potentiometer) has a resistance R whose logarithm is proportional to the angular position of the knob: that is, $\log R \propto \theta$. If the intensity of the sound I (in watts per square meter) produced by the speaker is proportional to the resistance R, show that the sound level β (in decibels) is a linear function of θ.

73. The smallest wavelength possible for a sound wave in air is on the order of the separation distance between air molecules. Find the order of magnitude of the highest-frequency sound wave possible in air, assuming a wave speed of 343 m/s, a density of 1.20 kg/m^3, and an average molecular mass of 4.82×10^{-26} kg.

ANSWERS TO QUICK QUIZZES

17.1 The only correct answer is (c). Although the speed of a wave is given by the product of its wavelength and frequency, it is not affected by changes in either one. For example, if the sound from a musical instrument increases in frequency, the wavelength decreases, and thus $v = \lambda f$ remains constant. The amplitude of a sound wave determines the size of the oscillations of air molecules but does not affect the speed of the wave through the air.

17.2 The ground tremor represents a sound wave moving through the Earth. Sound waves move faster through the Earth than through air because rock and other ground materials are much stiffer against compression. Therefore—the vibration through the ground and the sound in the air having started together—the vibration through the ground reaches the observer first.

17.3 Because the bottom of the bottle does not allow molecular motion, the displacement in this region is at its minimum value. Because the pressure variation is a maximum when the displacement is a minimum, the pressure variation at the bottom is a maximum.

17.4 (a) 10 dB. If we call the intensity of each violin I, the total intensity when all the violins are playing is $I + 9I = 10I$. Therefore, the addition of the nine violins increases the intensity of the sound over that of one violin by a factor of 10. From Equation 17.7 we see that an increase in intensity by a factor of 10 increases the sound level by 10 dB. (b) 13 dB. The intensity is now increased by a factor of 20 over that of a single violin.

17.5 The Mach number is the ratio of the plane's speed (which does not change) to the speed of sound, which is greater in the warm air than in the cold, as we learned in Section 17.1 (see Quick Quiz 17.1). The denominator of this fraction increases while the numerator stays constant. Therefore, the fraction as a whole—the Mach number—decreases.

17.6 (a) In the reference frame of the air, the observer is moving toward the source at the wind speed through stationary air, and the source is moving away from the observer with the same speed. In Equation 17.17, therefore, a plus sign is needed in both the numerator and the denominator:

$$f' = \left(\frac{v_{\text{sound}} + v_{\text{wind}}}{v_{\text{sound}} + v_{\text{wind}}} \right) f$$

meaning the observed frequency is the same as if no wind were blowing. (b) The observer "sees" the sound waves coming toward him at a higher speed $(v_{\text{sound}} + v_{\text{wind}})$. (c) At this higher speed, he attributes a greater wavelength $\lambda' = (v_{\text{sound}} + v_{\text{wind}})/f$ to the wave.

PUZZLER

A speaker for a stereo system operates even if the wires connecting it to the amplifier are reversed, that is, + for − and − for + (or red for black and black for red). Nonetheless, the owner's manual says that for best performance you should be careful to connect the two speakers properly, so that they are "in phase." Why is this such an important consideration for the quality of the sound you hear? *(George Semple)*

c h a p t e r

Superposition and Standing Waves

mportant in the study of waves is the combined effect of two or more waves traveling in the same medium. For instance, what happens to a string when a wave traveling along it hits a fixed end and is reflected back on itself? What is the air pressure variation at a particular seat in a theater when the instruments of an orchestra sound together?

When analyzing a linear medium—that is, one in which the restoring force acting on the particles of the medium is proportional to the displacement of the particles—we can apply the principle of superposition to determine the resultant disturbance. In Chapter 16 we discussed this principle as it applies to wave pulses. In this chapter we study the superposition principle as it applies to sinusoidal waves. If the sinusoidal waves that combine in a linear medium have the same frequency and wavelength, a stationary pattern—called a *standing wave*—can be produced at certain frequencies under certain circumstances. For example, a taut string fixed at both ends has a discrete set of oscillation patterns, called *modes of vibration,* that are related to the tension and linear mass density of the string. These modes of vibration are found in stringed musical instruments. Other musical instruments, such as the organ and the flute, make use of the natural frequencies of sound waves in hollow pipes. Such frequencies are related to the length and shape of the pipe and depend on whether the pipe is open at both ends or open at one end and closed at the other.

We also consider the superposition and interference of waves having different frequencies and wavelengths. When two sound waves having nearly the same frequency interfere, we hear variations in the loudness called *beats.* The beat frequency corresponds to the rate of alternation between constructive and destructive interference. Finally, we discuss how any non-sinusoidal periodic wave can be described as a sum of sine and cosine functions.

18.1 ▶ SUPERPOSITION AND INTERFERENCE OF SINUSOIDAL WAVES

Imagine that you are standing in a swimming pool and that a beach ball is floating a couple of meters away. You use your right hand to send a series of waves toward the beach ball, causing it to repeatedly move upward by 5 cm, return to its original position, and then move downward by 5 cm. After the water becomes still, you use your left hand to send an identical set of waves toward the beach ball and observe the same behavior. What happens if you use both hands at the same time to send two waves toward the beach ball? How the beach ball responds to the waves depends on whether the waves work together (that is, both waves make the beach ball go up at the same time and then down at the same time) or work against each other (that is, one wave tries to make the beach ball go up, while the other wave tries to make it go down). Because it is possible to have two or more waves in the same location at the same time, we have to consider how waves interact with each other and with their surroundings.

The superposition principle states that when two or more waves move in the same linear medium, the net displacement of the medium (that is, the resultant wave) at any point equals the algebraic sum of all the displacements caused by the individual waves. Let us apply this principle to two sinusoidal waves traveling in the same direction in a linear medium. If the two waves are traveling to the right and have the same frequency, wavelength, and amplitude but differ in phase, we can express their individual wave functions as

$$y_1 = A \sin(kx - \omega t) \qquad y_2 = A \sin(kx - \omega t + \phi)$$

where, as usual, $k = 2\pi/\lambda$, $\omega = 2\pi f$, and ϕ is the phase constant, which we introduced in the context of simple harmonic motion in Chapter 13. Hence, the resultant wave function y is

$$y = y_1 + y_2 = A[\sin(kx - \omega t) + \sin(kx - \omega t + \phi)]$$

To simplify this expression, we use the trigonometric identity

$$\sin a + \sin b = 2 \cos\left(\frac{a - b}{2}\right) \sin\left(\frac{a + b}{2}\right)$$

If we let $a = kx - \omega t$ and $b = kx - \omega t + \phi$, we find that the resultant wave function y reduces to

$$y = 2A \cos\left(\frac{\phi}{2}\right) \sin\left(kx - \omega t + \frac{\phi}{2}\right)$$

Resultant of two traveling sinusoidal waves

This result has several important features. The resultant wave function y also is sinusoidal and has the same frequency and wavelength as the individual waves, since the sine function incorporates the same values of k and ω that appear in the original wave functions. The amplitude of the resultant wave is $2A \cos(\phi/2)$, and its phase is $\phi/2$. If the phase constant ϕ equals 0, then $\cos(\phi/2) = \cos 0 = 1$, and the amplitude of the resultant wave is $2A$—twice the amplitude of either individual wave. In this case, in which $\phi = 0$, the waves are said to be everywhere *in phase* and thus **interfere constructively.** That is, the crests and troughs of the individual waves y_1 and y_2 occur at the same positions and combine to form the red curve y of amplitude $2A$ shown in Figure 18.1a. Because the individual waves are in phase, they are indistinguishable in Figure 18.1a, in which they appear as a single blue curve. In general, constructive interference occurs when $\cos(\phi/2) = \pm 1$. This is true, for example, when $\phi = 0, 2\pi, 4\pi, \ldots$ rad—that is, when ϕ is an *even* multiple of π.

Constructive interference

When ϕ is equal to π rad or to any *odd* multiple of π, then $\cos(\phi/2) = \cos(\pi/2) = 0$, and the crests of one wave occur at the same positions as the troughs of the second wave (Fig. 18.1b). Thus, the resultant wave has *zero* amplitude everywhere, as a consequence of **destructive interference.** Finally, when the phase constant has an arbitrary value other than 0 or other than an integer multiple of π rad (Fig. 18.1c), the resultant wave has an amplitude whose value is somewhere between 0 and $2A$.

Destructive interference

Interference of Sound Waves

One simple device for demonstrating interference of sound waves is illustrated in Figure 18.2. Sound from a loudspeaker S is sent into a tube at point P, where there is a T-shaped junction. Half of the sound power travels in one direction, and half travels in the opposite direction. Thus, the sound waves that reach the receiver R can travel along either of the two paths. The distance along any path from speaker to receiver is called the **path length** r. The lower path length r_1 is fixed, but the upper path length r_2 can be varied by sliding the U-shaped tube, which is similar to that on a slide trombone. When the difference in the path lengths $\Delta r = |r_2 - r_1|$ is either zero or some integer multiple of the wavelength λ (that is, $r = n\lambda$, where $n = 0, 1, 2, 3, \ldots$), the two waves reaching the receiver at any instant are in phase and interfere constructively, as shown in Figure 18.1a. For this case, a maximum in the sound intensity is detected at the receiver. If the path length r_2 is ad-

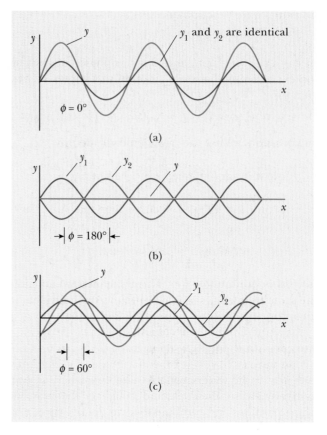

Figure 18.1   The superposition of two identical waves y_1 and y_2 (blue) to yield a resultant wave (red). (a) When y_1 and y_2 are in phase, the result is constructive interference. (b) When y_1 and y_2 are π rad out of phase, the result is destructive interference. (c) When the phase angle has a value other than 0 or π rad, the resultant wave y falls somewhere between the extremes shown in (a) and (b).

justed such that the path difference $\Delta r = \lambda/2, 3\lambda/2, \ldots, n\lambda/2$ (for n odd), the two waves are exactly π rad, or 180°, out of phase at the receiver and hence cancel each other. In this case of destructive interference, no sound is detected at the receiver. This simple experiment demonstrates that a phase difference may arise between two waves generated by the same source when they travel along paths of unequal lengths. This important phenomenon will be indispensable in our investigation of the interference of light waves in Chapter 37.

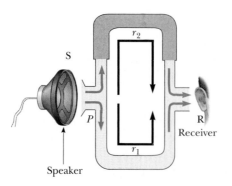

Figure 18.2   An acoustical system for demonstrating interference of sound waves. A sound wave from the speaker (S) propagates into the tube and splits into two parts at point P. The two waves, which superimpose at the opposite side, are detected at the receiver (R). The upper path length r_2 can be varied by sliding the upper section.

It is often useful to express the path difference in terms of the phase angle ϕ between the two waves. Because a path difference of one wavelength corresponds to a phase angle of 2π rad, we obtain the ratio $\phi/2\pi = \Delta r/\lambda$, or

$$\Delta r = \frac{\phi}{2\pi}\lambda \qquad (18.1)$$

Relationship between path difference and phase angle

Using the notion of path difference, we can express our conditions for constructive and destructive interference in a different way. If the path difference is any even multiple of $\lambda/2$, then the phase angle $\phi = 2n\pi$, where $n = 0, 1, 2, 3, \ldots$, and the interference is constructive. For path differences of odd multiples of $\lambda/2$, $\phi = (2n + 1)\pi$, where $n = 0, 1, 2, 3 \ldots$, and the interference is destructive. Thus, we have the conditions

$$\Delta r = (2n)\frac{\lambda}{2} \qquad \text{for constructive interference}$$

and $\qquad (18.2)$

$$\Delta r = (2n + 1)\frac{\lambda}{2} \qquad \text{for destructive interference}$$

EXAMPLE 18.1 ▸ Two Speakers Driven by the Same Source

A pair of speakers placed 3.00 m apart are driven by the same oscillator (Fig. 18.3). A listener is originally at point O, which is located 8.00 m from the center of the line connecting the two speakers. The listener then walks to point P, which is a perpendicular distance 0.350 m from O, before reaching the *first minimum* in sound intensity. What is the frequency of the oscillator?

Solution To find the frequency, we need to know the wavelength of the sound coming from the speakers. With this information, combined with our knowledge of the speed of sound, we can calculate the frequency. We can determine the wavelength from the interference information given. The first minimum occurs when the two waves reaching the listener at point P are 180° out of phase—in other words, when their path difference Δr equals $\lambda/2$. To calculate the path difference, we must first find the path lengths r_1 and r_2.

Figure 18.3 shows the physical arrangement of the speakers, along with two shaded right triangles that can be drawn on the basis of the lengths described in the problem. From

these triangles, we find that the path lengths are

$$r_1 = \sqrt{(8.00 \text{ m})^2 + (1.15 \text{ m})^2} = 8.08 \text{ m}$$

and

$$r_2 = \sqrt{(8.00 \text{ m})^2 + (1.85 \text{ m})^2} = 8.21 \text{ m}$$

Hence, the path difference is $r_2 - r_1 = 0.13$ m. Because we require that this path difference be equal to $\lambda/2$ for the first minimum, we find that $\lambda = 0.26$ m.

To obtain the oscillator frequency, we use Equation 16.14, $v = \lambda f$, where v is the speed of sound in air, 343 m/s:

$$f = \frac{v}{\lambda} = \frac{343 \text{ m/s}}{0.26 \text{ m}} = \boxed{1.3 \text{ kHz}}$$

Exercise If the oscillator frequency is adjusted such that the first location at which a listener hears no sound is at a distance of 0.75 m from O, what is the new frequency?

Answer 0.63 kHz.

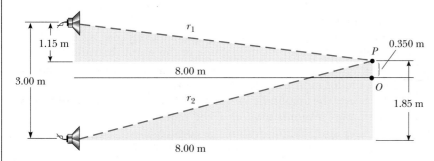

Figure 18.3

You can now understand why the speaker wires in a stereo system should be connected properly. When connected the wrong way—that is, when the positive (or red) wire is connected to the negative (or black) terminal—the speakers are said to be "out of phase" because the sound wave coming from one speaker destructively interferes with the wave coming from the other. In this situation, one speaker cone moves outward while the other moves inward. Along a line midway between the two, a rarefaction region from one speaker is superposed on a condensation region from the other speaker. Although the two sounds probably do not completely cancel each other (because the left and right stereo signals are usually not identical), a substantial loss of sound quality still occurs at points along this line.

18.2 STANDING WAVES

The sound waves from the speakers in Example 18.1 left the speakers in the forward direction, and we considered interference at a point in space in front of the speakers. Suppose that we turn the speakers so that they face each other and then have them emit sound of the same frequency and amplitude. We now have a situation in which two identical waves travel in opposite directions in the same medium. These waves combine in accordance with the superposition principle.

We can analyze such a situation by considering wave functions for two transverse sinusoidal waves having the same amplitude, frequency, and wavelength but traveling in opposite directions in the same medium:

$$y_1 = A \sin(kx - \omega t) \qquad y_2 = A \sin(kx + \omega t)$$

where y_1 represents a wave traveling to the right and y_2 represents one traveling to the left. Adding these two functions gives the resultant wave function y:

$$y = y_1 + y_2 = A \sin(kx - \omega t) + A \sin(kx + \omega t)$$

When we use the trigonometric identity $\sin(a \pm b) = \sin a \cos b \pm \cos a \sin b$, this expression reduces to

Wave function for a standing wave

$$y = (2A \sin kx) \cos \omega t \qquad \textbf{(18.3)}$$

which is the wave function of a standing wave. A **standing wave,** such as the one shown in Figure 18.4, is an oscillation pattern with a stationary outline that results from the superposition of two identical waves traveling in opposite directions.

Notice that Equation 18.3 does not contain a function of $kx \pm \omega t$. Thus, it is not an expression for a traveling wave. If we observe a standing wave, we have no sense of motion in the direction of propagation of either of the original waves. If we compare this equation with Equation 13.3, we see that Equation 18.3 describes a special kind of simple harmonic motion. Every particle of the medium oscillates in simple harmonic motion with the same frequency ω (according to the $\cos \omega t$ factor in the equation). However, the amplitude of the simple harmonic motion of a given particle (given by the factor $2A \sin kx$, the coefficient of the cosine function) depends on the location x of the particle in the medium. We need to distinguish carefully between the amplitude A of the individual waves and the amplitude $2A \sin kx$ of the simple harmonic motion of the particles of the medium. A given particle in a standing wave vibrates within the constraints of the *envelope* function $2A \sin kx$, where x is the particle's position in the medium. This is in contrast to the situation in a traveling sinusoidal wave, in which all particles oscillate with the

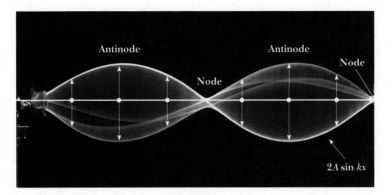

Figure 18.4   Multiflash photograph of a standing wave on a string. The time behavior of the vertical displacement from equilibrium of an individual particle of the string is given by $\cos \omega t$. That is, each particle vibrates at an angular frequency ω. The amplitude of the vertical oscillation of any particle on the string depends on the horizontal position of the particle. Each particle vibrates within the confines of the envelope function $2A \sin kx$. (©1991 Richard Megna/Fundamental Photographs)

same amplitude and the same frequency and in which the amplitude of the wave is the same as the amplitude of the simple harmonic motion of the particles.

The maximum displacement of a particle of the medium has a minimum value of zero when x satisfies the condition $\sin kx = 0$, that is, when

$$kx = \pi, 2\pi, 3\pi, \ldots$$

Because $k = 2\pi/\lambda$, these values for kx give

$$x = \frac{\lambda}{2}, \lambda, \frac{3\lambda}{2}, \ldots = \frac{n\lambda}{2} \qquad n = 0, 1, 2, 3, \ldots \qquad \textbf{(18.4)}$$

Position of nodes

These points of zero displacement are called **nodes.**

The particle with the greatest possible displacement from equilibrium has an amplitude of $2A$, and we define this as the amplitude of the standing wave. The positions in the medium at which this maximum displacement occurs are called **antinodes.** The antinodes are located at positions for which the coordinate x satisfies the condition $\sin kx = \pm 1$, that is, when

$$kx = \frac{\pi}{2}, \frac{3\pi}{2}, \frac{5\pi}{2}, \ldots$$

Thus, the positions of the antinodes are given by

$$x = \frac{\lambda}{4}, \frac{3\lambda}{4}, \frac{5\lambda}{4}, \ldots = \frac{n\lambda}{4} \qquad n = 1, 3, 5, \ldots \qquad \textbf{(18.5)}$$

Position of antinodes

In examining Equations 18.4 and 18.5, we note the following important features of the locations of nodes and antinodes:

The distance between adjacent antinodes is equal to $\lambda/2$.
The distance between adjacent nodes is equal to $\lambda/2$.
The distance between a node and an adjacent antinode is $\lambda/4$.

Displacement patterns of the particles of the medium produced at various times by two waves traveling in opposite directions are shown in Figure 18.5. The blue and green curves are the individual traveling waves, and the red curves are

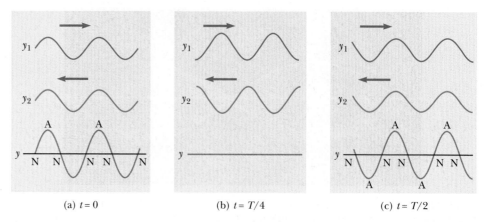

(a) $t = 0$ (b) $t = T/4$ (c) $t = T/2$

Figure 18.5 Standing-wave patterns produced at various times by two waves of equal amplitude traveling in opposite directions. For the resultant wave y, the nodes (N) are points of zero displacement, and the antinodes (A) are points of maximum displacement.

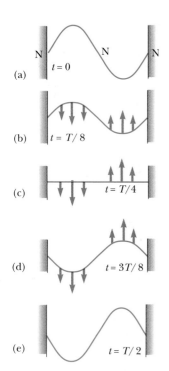

(a) $t = 0$
(b) $t = T/8$
(c) $t = T/4$
(d) $t = 3T/8$
(e) $t = T/2$

Figure 18.6 A standing-wave pattern in a taut string. The five "snapshots" were taken at half-cycle intervals. (a) At $t = 0$, the string is momentarily at rest; thus, $K = 0$, and all the energy is potential energy U associated with the vertical displacements of the string particles. (b) At $t = T/8$, the string is in motion, as indicated by the brown arrows, and the energy is half kinetic and half potential. (c) At $t = T/4$, the string is moving but horizontal (undeformed); thus, $U = 0$, and all the energy is kinetic. (d) The motion continues as indicated. (e) At $t = T/2$, the string is again momentarily at rest, but the crests and troughs of (a) are reversed. The cycle continues until ultimately, when a time interval equal to T has passed, the configuration shown in (a) is repeated.

the displacement patterns. At $t = 0$ (Fig. 18.5a), the two traveling waves are in phase, giving a displacement pattern in which each particle of the medium is experiencing its maximum displacement from equilibrium. One quarter of a period later, at $t = T/4$ (Fig. 18.5b), the traveling waves have moved one quarter of a wavelength (one to the right and the other to the left). At this time, the traveling waves are out of phase, and each particle of the medium is passing through the equilibrium position in its simple harmonic motion. The result is zero displacement for particles at all values of x—that is, the displacement pattern is a straight line. At $t = T/2$ (Fig. 18.5c), the traveling waves are again in phase, producing a displacement pattern that is inverted relative to the $t = 0$ pattern. In the standing wave, the particles of the medium alternate in time between the extremes shown in Figure 18.5a and c.

Energy in a Standing Wave

It is instructive to describe the energy associated with the particles of a medium in which a standing wave exists. Consider a standing wave formed on a taut string fixed at both ends, as shown in Figure 18.6. Except for the nodes, which are always stationary, all points on the string oscillate vertically with the same frequency but with different amplitudes of simple harmonic motion. Figure 18.6 represents snapshots of the standing wave at various times over one half of a period.

In a traveling wave, energy is transferred along with the wave, as we discussed in Chapter 16. We can imagine this transfer to be due to work done by one segment of the string on the next segment. As one segment moves upward, it exerts a force on the next segment, moving it through a displacement—that is, work is done. A particle of the string at a node, however, experiences no displacement. Thus, it cannot do work on the neighboring segment. As a result, no energy is transmitted along the string across a node, and energy does not propagate in a standing wave. For this reason, standing waves are often called **stationary waves.**

The energy of the oscillating string continuously alternates between elastic potential energy, when the string is momentarily stationary (see Fig. 18.6a), and kinetic energy, when the string is horizontal and the particles have their maximum speed (see Fig. 18.6c). At intermediate times (see Fig. 18.6b and d), the string particles have both potential energy and kinetic energy.

Quick Quiz 18.1

A standing wave described by Equation 18.3 is set up on a string. At what points on the string do the particles move the fastest?

EXAMPLE 18.2   Formation of a Standing Wave

Two waves traveling in opposite directions produce a standing wave. The individual wave functions $y = A \sin(kx - \omega t)$ are

$$y_1 = (4.0 \text{ cm}) \sin(3.0x - 2.0t)$$

and

$$y_2 = (4.0 \text{ cm}) \sin(3.0x + 2.0t)$$

where x and y are measured in centimeters. (a) Find the amplitude of the simple harmonic motion of the particle of the medium located at $x = 2.3$ cm.

Solution   The standing wave is described by Equation 18.3; in this problem, we have $A = 4.0$ cm, $k = 3.0$ rad/cm, and $\omega = 2.0$ rad/s. Thus,

$$y = (2A \sin kx) \cos \omega t = [(8.0 \text{ cm}) \sin 3.0x] \cos 2.0t$$

Thus, we obtain the amplitude of the simple harmonic motion of the particle at the position $x = 2.3$ cm by evaluating the coefficient of the cosine function at this position:

$$y_{max} = (8.0 \text{ cm}) \sin 3.0x \big|_{x=2.3}$$

$$= (8.0 \text{ cm}) \sin(6.9 \text{ rad}) = \boxed{4.6 \text{ cm}}$$

(b) Find the positions of the nodes and antinodes.

Solution   With $k = 2\pi/\lambda = 3.0$ rad/cm, we see that $\lambda = 2\pi/3$ cm. Therefore, from Equation 18.4 we find that the nodes are located at

$$x = n\frac{\lambda}{2} = \boxed{n\left(\frac{\pi}{3}\right) \text{cm}} \qquad n = 0, 1, 2, 3 \ldots$$

and from Equation 18.5 we find that the antinodes are located at

$$x = n\frac{\lambda}{4} = \boxed{n\left(\frac{\pi}{6}\right) \text{cm}} \qquad n = 1, 3, 5, \ldots$$

(c) What is the amplitude of the simple harmonic motion of a particle located at an antinode?

Solution   According to Equation 18.3, the maximum displacement of a particle at an antinode is the amplitude of the standing wave, which is twice the amplitude of the individual traveling waves:

$$y_{max} = 2A = 2(4.0 \text{ cm}) = \boxed{8.0 \text{ cm}}$$

Let us check this result by evaluating the coefficient of our standing-wave function at the positions we found for the antinodes:

$$y_{max} = (8.0 \text{ cm}) \sin 3.0x \big|_{x=n(\pi/6)}$$

$$= (8.0 \text{ cm}) \sin\left[3.0n\left(\frac{\pi}{6}\right) \text{rad}\right]$$

$$= (8.0 \text{ cm}) \sin\left[n\left(\frac{\pi}{2}\right) \text{rad}\right] = 8.0 \text{ cm}$$

In evaluating this expression, we have used the fact that n is an odd integer; thus, the sine function is equal to unity.

18.3 STANDING WAVES IN A STRING FIXED AT BOTH ENDS

Consider a string of length L fixed at both ends, as shown in Figure 18.7. Standing waves are set up in the string by a continuous superposition of waves incident on and reflected from the ends. Note that the ends of the string, because they are fixed and must necessarily have zero displacement, are nodes by definition. The string has a number of natural patterns of oscillation, called **normal modes,** each of which has a characteristic frequency that is easily calculated.

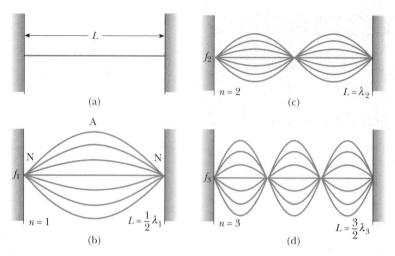

Figure 18.7   (a) A string of length L fixed at both ends. The normal modes of vibration form a harmonic series: (b) the fundamental, or first harmonic; (c) the second harmonic; (d) the third harmonic.

In general, the motion of an oscillating string fixed at both ends is described by the superposition of several normal modes. Exactly which normal modes are present depends on how the oscillation is started. For example, when a guitar string is plucked near its middle, the modes shown in Figure 18.7b and d, as well as other modes not shown, are excited.

In general, we can describe the normal modes of oscillation for the string by imposing the requirements that the ends be nodes and that the nodes and antinodes be separated by one fourth of a wavelength. The first normal mode, shown in Figure 18.7b, has nodes at its ends and one antinode in the middle. This is the longest-wavelength mode, and this is consistent with our requirements. This first normal mode occurs when the wavelength λ_1 is twice the length of the string, that is, $\lambda_1 = 2L$. The next normal mode, of wavelength λ_2 (see Fig. 18.7c), occurs when the wavelength equals the length of the string, that is, $\lambda_2 = L$. The third normal mode (see Fig. 18.7d) corresponds to the case in which $\lambda_3 = 2L/3$. In general, the wavelengths of the various normal modes for a string of length L fixed at both ends are

Wavelengths of normal modes

$$\lambda_n = \frac{2L}{n} \qquad n = 1, 2, 3, \ldots \qquad \textbf{(18.6)}$$

where the index n refers to the nth normal mode of oscillation. These are the *possible* modes of oscillation for the string. The *actual* modes that are excited by a given pluck of the string are discussed below.

The natural frequencies associated with these modes are obtained from the relationship $f = v/\lambda$, where the wave speed v is the same for all frequencies. Using Equation 18.6, we find that the natural frequencies f_n of the normal modes are

Frequencies of normal modes as functions of wave speed and length of string

$$f_n = \frac{v}{\lambda_n} = n\frac{v}{2L} \qquad n = 1, 2, 3, \ldots \qquad \textbf{(18.7)}$$

Because $v = \sqrt{T/\mu}$ (see Eq. 16.4), where T is the tension in the string and μ is its linear mass density, we can also express the natural frequencies of a taut string as

Frequencies of normal modes as functions of string tension and linear mass density

$$f_n = \frac{n}{2L}\sqrt{\frac{T}{\mu}} \qquad n = 1, 2, 3, \ldots \qquad \textbf{(18.8)}$$

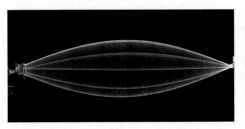 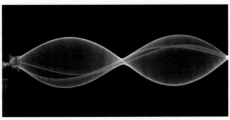 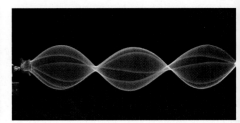

Multiflash photographs of standing-wave patterns in a cord driven by a vibrator at its left end. The single-loop pattern represents the first normal mode ($n = 1$). The double-loop pattern represents the second normal mode ($n = 2$), and the triple-loop pattern represents the third normal mode ($n = 3$). (*©1991 Richard Megna/Fundamental Photographs*)

The lowest frequency f_1, which corresponds to $n = 1$, is called either the **fundamental** or the **fundamental frequency** and is given by

$$f_1 = \frac{1}{2L}\sqrt{\frac{T}{\mu}}$$ **(18.9)**

> Fundamental frequency of a taut string

The frequencies of the remaining normal modes are integer multiples of the fundamental frequency. Frequencies of normal modes that exhibit an integer-multiple relationship such as this form a **harmonic series,** and the normal modes are called **harmonics.** The fundamental frequency f_1 is the frequency of the first harmonic; the frequency $f_2 = 2f_1$ is the frequency of the second harmonic; and the frequency $f_n = nf_1$ is the frequency of the nth harmonic. Other oscillating systems, such as a drumhead, exhibit normal modes, but the frequencies are not related as integer multiples of a fundamental. Thus, we do not use the term *harmonic* in association with these types of systems.

In obtaining Equation 18.6, we used a technique based on the separation distance between nodes and antinodes. We can obtain this equation in an alternative manner. Because we require that the string be fixed at $x = 0$ and $x = L$, the wave function $y(x, t)$ given by Equation 18.3 must be zero at these points for all times. That is, the *boundary conditions* require that $y(0, t) = 0$ and that $y(L, t) = 0$ for all values of t. Because the standing wave is described by $y = (2A \sin kx) \cos \omega t$, the first boundary condition, $y(0, t) = 0$, is automatically satisfied because $\sin kx = 0$ at $x = 0$. To meet the second boundary condition, $y(L, t) = 0$, we require that $\sin kL = 0$. This condition is satisfied when the angle kL equals an integer multiple of π rad. Therefore, the allowed values of k are given by[1]

$$k_n L = n\pi \qquad n = 1, 2, 3, \ldots$$ **(18.10)**

Because $k_n = 2\pi/\lambda_n$, we find that

$$\left(\frac{2\pi}{\lambda_n}\right)L = n\pi \qquad \text{or} \qquad \lambda_n = \frac{2L}{n}$$

which is identical to Equation 18.6.

Let us now examine how these various harmonics are created in a string. If we wish to excite just a single harmonic, we need to distort the string in such a way that its distorted shape corresponded to that of the desired harmonic. After being released, the string vibrates at the frequency of that harmonic. This maneuver is difficult to perform, however, and it is not how we excite a string of a musical in-

QuickLab ⟹

Compare the sounds of a guitar string plucked first near its center and then near one of its ends. More of the higher harmonics are present in the second situation. Can you hear the difference?

---

[1] We exclude $n = 0$ because this value corresponds to the trivial case in which no wave exists ($k = 0$).

strument. If the string is distorted such that its distorted shape is not that of just one harmonic, the resulting vibration includes various harmonics. Such a distortion occurs in musical instruments when the string is plucked (as in a guitar), bowed (as in a cello), or struck (as in a piano). When the string is distorted into a non-sinusoidal shape, only waves that satisfy the boundary conditions can persist on the string. These are the harmonics.

The frequency of a stringed instrument can be varied by changing either the tension or the string's length. For example, the tension in guitar and violin strings is varied by a screw adjustment mechanism or by tuning pegs located on the neck of the instrument. As the tension is increased, the frequency of the normal modes increases in accordance with Equation 18.8. Once the instrument is "tuned," players vary the frequency by moving their fingers along the neck, thereby changing the length of the oscillating portion of the string. As the length is shortened, the frequency increases because, as Equation 18.8 specifies, the normal-mode frequencies are inversely proportional to string length.

EXAMPLE 18.3   Give Me a C Note!

Middle C on a piano has a fundamental frequency of 262 Hz, and the first A above middle C has a fundamental frequency of 440 Hz. (a) Calculate the frequencies of the next two harmonics of the C string.

Solution   Knowing that the frequencies of higher harmonics are integer multiples of the fundamental frequency $f_1 = 262$ Hz, we find that

$$f_2 = 2f_1 = \boxed{524 \text{ Hz}}$$

$$f_3 = 3f_1 = \boxed{786 \text{ Hz}}$$

(b) If the A and C strings have the same linear mass density μ and length L, determine the ratio of tensions in the two strings.

Solution   Using Equation 18.8 for the two strings vibrating at their fundamental frequencies gives

$$f_{1A} = \frac{1}{2L}\sqrt{\frac{T_A}{\mu}} \quad \text{and} \quad f_{1C} = \frac{1}{2L}\sqrt{\frac{T_C}{\mu}}$$

Setting up the ratio of these frequencies, we find that

$$\frac{f_{1A}}{f_{1C}} = \sqrt{\frac{T_A}{T_C}}$$

$$\frac{T_A}{T_C} = \left(\frac{f_{1A}}{f_{1C}}\right)^2 = \left(\frac{440}{262}\right)^2 = \boxed{2.82}$$

(c) With respect to a real piano, the assumption we made in (b) is only partially true. The string densities are equal, but the length of the A string is only 64 percent of the length of the C string. What is the ratio of their tensions?

Solution   Using Equation 18.8 again, we set up the ratio of frequencies:

$$\frac{f_{1A}}{f_{1C}} = \frac{L_C}{L_A}\sqrt{\frac{T_A}{T_C}} = \left(\frac{100}{64}\right)\sqrt{\frac{T_A}{T_C}}$$

$$\frac{T_A}{T_C} = (0.64)^2\left(\frac{440}{262}\right)^2 = \boxed{1.16}$$

EXAMPLE 18.4   Guitar Basics

The high E string on a guitar measures 64.0 cm in length and has a fundamental frequency of 330 Hz. By pressing down on it at the first fret (Fig. 18.8), the string is shortened so that it plays an F note that has a frequency of 350 Hz. How far is the fret from the neck end of the string?

Solution   Equation 18.7 relates the string's length to the fundamental frequency. With $n = 1$, we can solve for the

speed of the wave on the string,

$$v = \frac{2L}{n}f_n = \frac{2(0.640 \text{ m})}{1}(330 \text{ Hz}) = 422 \text{ m/s}$$

Because we have not adjusted the tuning peg, the tension in the string, and hence the wave speed, remain constant. We can again use Equation 18.7, this time solving for L and sub-

Figure 18.8 Playing an F note on a guitar. *(Charles D. Winters)*

stituting the new frequency to find the shortened string length:

$$L = n\frac{v}{2f_n} = (1)\frac{422 \text{ m/s}}{2(350 \text{ Hz})} = 0.603 \text{ m}$$

The difference between this length and the measured length of 64.0 cm is the distance from the fret to the neck end of the string, or   3.70 cm.

18.4 RESONANCE

9.9 We have seen that a system such as a taut string is capable of oscillating in one or more normal modes of oscillation. **If a periodic force is applied to such a system, the amplitude of the resulting motion is greater than normal when the frequency of the applied force is equal to or nearly equal to one of the natural frequencies of the system.** We discussed this phenomenon, known as *resonance,* briefly in Section 13.7. Although a block–spring system or a simple pendulum has only one natural frequency, standing-wave systems can have a whole set of natural frequencies. Because an oscillating system exhibits a large amplitude when driven at any of its natural frequencies, these frequencies are often referred to as **resonance frequencies.**

Figure 18.9 shows the response of an oscillating system to various driving frequencies, where one of the resonance frequencies of the system is denoted by f_0. Note that the amplitude of oscillation of the system is greatest when the frequency of the driving force equals the resonance frequency. The maximum amplitude is limited by friction in the system. If a driving force begins to work on an oscillating system initially at rest, the input energy is used both to increase the amplitude of the oscillation and to overcome the frictional force. Once maximum amplitude is reached, the work done by the driving force is used only to overcome friction.

A system is said to be *weakly damped* when the amount of friction to be overcome is small. Such a system has a large amplitude of motion when driven at one of its resonance frequencies, and the oscillations persist for a long time after the driving force is removed. A system in which considerable friction must be overcome is said to be *strongly damped.* For a given driving force applied at a resonance frequency, the maximum amplitude attained by a strongly damped oscillator is smaller than that attained by a comparable weakly damped oscillator. Once the driving force in a strongly damped oscillator is removed, the amplitude decreases rapidly with time.

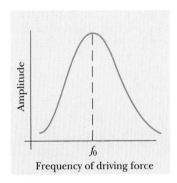

Figure 18.9 Graph of the amplitude (response) versus driving frequency for an oscillating system. The amplitude is a maximum at the resonance frequency f_0. Note that the curve is not symmetric.

Examples of Resonance

A playground swing is a pendulum having a natural frequency that depends on its length. Whenever we use a series of regular impulses to push a child in a swing, the swing goes higher if the frequency of the periodic force equals the natural fre-

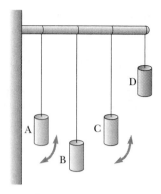

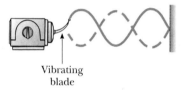

Figure 18.10 An example of resonance. If pendulum A is set into oscillation, only pendulum C, whose length matches that of A, eventually oscillates with large amplitude, or resonates. The arrows indicate motion perpendicular to the page.

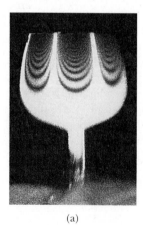

Vibrating blade

Figure 18.11 Standing waves are set up in a string when one end is connected to a vibrating blade. When the blade vibrates at one of the natural frequencies of the string, large-amplitude standing waves are created.

quency of the swing. We can demonstrate a similar effect by suspending pendulums of different lengths from a horizontal support, as shown in Figure 18.10. If pendulum A is set into oscillation, the other pendulums begin to oscillate as a result of the longitudinal waves transmitted along the beam. However, pendulum C, the length of which is close to the length of A, oscillates with a much greater amplitude than pendulums B and D, the lengths of which are much different from that of pendulum A. Pendulum C moves the way it does because its natural frequency is nearly the same as the driving frequency associated with pendulum A.

Next, consider a taut string fixed at one end and connected at the opposite end to an oscillating blade, as illustrated in Figure 18.11. The fixed end is a node, and the end connected to the blade is very nearly a node because the amplitude of the blade's motion is small compared with that of the string. As the blade oscillates, transverse waves sent down the string are reflected from the fixed end. As we learned in Section 18.3, the string has natural frequencies that are determined by its length, tension, and linear mass density (see Eq. 18.8). When the frequency of the blade equals one of the natural frequencies of the string, standing waves are produced and the string oscillates with a large amplitude. In this resonance case, the wave generated by the oscillating blade is in phase with the reflected wave, and the string absorbs energy from the blade. If the string is driven at a frequency that is not one of its natural frequencies, then the oscillations are of low amplitude and exhibit no stable pattern.

Once the amplitude of the standing-wave oscillations is a maximum, the mechanical energy delivered by the blade and absorbed by the system is lost because of the damping forces caused by friction in the system. If the applied frequency differs from one of the natural frequencies, energy is transferred to the string at first, but later the phase of the wave becomes such that it forces the blade to receive energy from the string, thereby reducing the energy in the string.

Quick Quiz 18.2

Some singers can shatter a wine glass by maintaining a certain frequency of their voice for several seconds. Figure 18.12a shows a side view of a wine glass vibrating because of a sound wave. Sketch the standing-wave pattern in the rim of the glass as seen from above. If an inte-

(a)   (b)

Figure 18.12 (a) Standing-wave pattern in a vibrating wine glass. The glass shatters if the amplitude of vibration becomes too great. *(Courtesy of Professor Thomas D. Rossing, Northern Illinois University)* (b) A wine glass shattered by the amplified sound of a human voice. *(©1992 Ben Rose/The Image Bank)*

gral number of waves "fit" around the circumference of the vibrating rim, how many wave-lengths fit around the rim in Figure 18.12a?

Quick Quiz 18.3

"Rumble strips" (Fig. 18.13) are sometimes placed across a road to warn drivers that they are approaching a stop sign, or laid along the sides of the road to alert drivers when they are drifting out of their lane. Why are these sets of small bumps so effective at getting a driver's attention?

Figure 18.13 Rumble strips along the side of a highway. *(Charles D. Winters)*

18.5 STANDING WAVES IN AIR COLUMNS

Standing waves can be set up in a tube of air, such as that in an organ pipe, as the result of interference between longitudinal sound waves traveling in opposite directions. The phase relationship between the incident wave and the wave reflected from one end of the pipe depends on whether that end is open or closed. This relationship is analogous to the phase relationships between incident and reflected transverse waves at the end of a string when the end is either fixed or free to move (see Figs. 16.13 and 16.14).

In a pipe closed at one end, **the closed end is a displacement node because the wall at this end does not allow longitudinal motion of the air molecules.** As a result, at a closed end of a pipe, the reflected sound wave is 180° out of phase with the incident wave. Furthermore, because the pressure wave is 90° out of phase with the displacement wave (see Section 17.2), **the closed end of an air column corresponds to a pressure antinode** (that is, a point of maximum pressure variation).

The open end of an air column is approximately a displacement antinode[2] **and a pressure node.** We can understand why no pressure variation occurs at an open end by noting that the end of the air column is open to the atmosphere; thus, the pressure at this end must remain constant at atmospheric pressure.

QuickLab

Snip off pieces at one end of a drinking straw so that the end tapers to a point. Chew on this end to flatten it, and you'll have created a double-reed instrument! Put your lips around the tapered end, press them tightly together, and blow through the straw. When you hear a steady tone, slowly snip off pieces of the straw from the other end. Be careful to maintain a constant pressure with your lips. How does the frequency change as the straw is shortened?

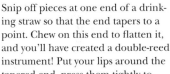

---

[2] Strictly speaking, the open end of an air column is not exactly a displacement antinode. A condensation reaching an open end does not reflect until it passes beyond the end. For a thin-walled tube of circular cross section, this end correction is approximately 0.6R, where R is the tube's radius. Hence, the effective length of the tube is longer than the true length L. We ignore this end correction in this discussion.

You may wonder how a sound wave can reflect from an open end, since there may not appear to be a change in the medium at this point. It is indeed true that the medium through which the sound wave moves is air both inside and outside the pipe. Remember that sound is a pressure wave, however, and a compression region of the sound wave is constrained by the sides of the pipe as long as the region is inside the pipe. As the compression region exits at the open end of the pipe, the constraint is removed and the compressed air is free to expand into the atmosphere. Thus, there is a change in the *character* of the medium between the inside of the pipe and the outside even though there is no change in the *material* of the medium. This change in character is sufficient to allow some reflection.

The first three normal modes of oscillation of a pipe open at both ends are shown in Figure 18.14a. When air is directed against an edge at the left, longitudinal standing waves are formed, and the pipe resonates at its natural frequencies. All normal modes are excited simultaneously (although not with the same amplitude). Note that both ends are displacement antinodes (approximately). In the first normal mode, the standing wave extends between two adjacent antinodes,

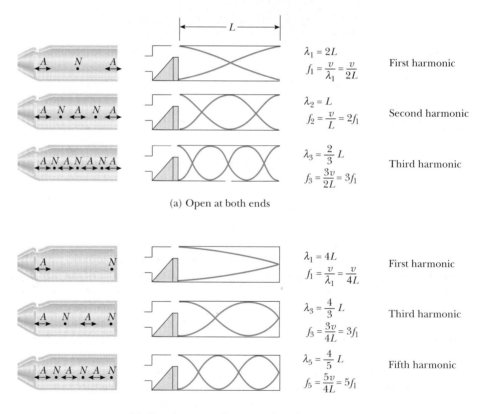

$\lambda_1 = 2L$
$f_1 = \dfrac{v}{\lambda_1} = \dfrac{v}{2L}$ First harmonic

$\lambda_2 = L$
$f_2 = \dfrac{v}{L} = 2f_1$ Second harmonic

$\lambda_3 = \dfrac{2}{3}L$
$f_3 = \dfrac{3v}{2L} = 3f_1$ Third harmonic

(a) Open at both ends

$\lambda_1 = 4L$
$f_1 = \dfrac{v}{\lambda_1} = \dfrac{v}{4L}$ First harmonic

$\lambda_3 = \dfrac{4}{3}L$
$f_3 = \dfrac{3v}{4L} = 3f_1$ Third harmonic

$\lambda_5 = \dfrac{4}{5}L$
$f_5 = \dfrac{5v}{4L} = 5f_1$ Fifth harmonic

(b) Closed at one end, open at the other

Figure 18.14   Motion of air molecules in standing longitudinal waves in a pipe, along with schematic representations of the waves. The graphs represent the displacement amplitudes, not the pressure amplitudes. (a) In a pipe open at both ends, the harmonic series created consists of all integer multiples of the fundamental frequency: $f_1, 2f_1, 3f_1, \ldots .$ (b) In a pipe closed at one end and open at the other, the harmonic series created consists of only odd-integer multiples of the fundamental frequency: $f_1, 3f_1, 5f_1, \ldots .$

which is a distance of half a wavelength. Thus, the wavelength is twice the length of the pipe, and the fundamental frequency is $f_1 = v/2L$. As Figure 18.14a shows, the frequencies of the higher harmonics are $2f_1, 3f_1, \ldots$. Thus, we can say that

> in a pipe open at both ends, the natural frequencies of oscillation form a harmonic series that includes all integral multiples of the fundamental frequency.

Because all harmonics are present, and because the fundamental frequency is given by the same expression as that for a string (see Eq. 18.7), we can express the natural frequencies of oscillation as

$$f_n = n\frac{v}{2L} \qquad n = 1, 2, 3 \ldots \qquad \text{(18.11)}$$

Natural frequencies of a pipe open at both ends

Despite the similarity between Equations 18.7 and 18.11, we must remember that v in Equation 18.7 is the speed of waves on the string, whereas v in Equation 18.11 is the speed of sound in air.

If a pipe is closed at one end and open at the other, the closed end is a displacement node (see Fig. 18.14b). In this case, the standing wave for the fundamental mode extends from an antinode to the adjacent node, which is one fourth of a wavelength. Hence, the wavelength for the first normal mode is $4L$, and the fundamental frequency is $f_1 = v/4L$. As Figure 18.14b shows, the higher-frequency waves that satisfy our conditions are those that have a node at the closed end and an antinode at the open end; this means that the higher harmonics have frequencies $3f_1, 5f_1, \ldots$:

QuickLab

Blow across the top of an empty soda-pop bottle. From a measurement of the height of the bottle, estimate the frequency of the sound you hear. Note that the cross-sectional area of the bottle is not constant; thus, this is not a perfect model of a cylindrical air column.

> In a pipe closed at one end and open at the other, the natural frequencies of oscillation form a harmonic series that includes only odd integer multiples of the fundamental frequency.

We express this result mathematically as

$$f_n = n\frac{v}{4L} \qquad n = 1, 3, 5, \ldots \qquad \text{(18.12)}$$

Natural frequencies of a pipe closed at one end and open at the other

It is interesting to investigate what happens to the frequencies of instruments based on air columns and strings during a concert as the temperature rises. The sound emitted by a flute, for example, becomes sharp (increases in frequency) as it warms up because the speed of sound increases in the increasingly warmer air inside the flute (consider Eq. 18.11). The sound produced by a violin becomes flat (decreases in frequency) as the strings expand thermally because the expansion causes their tension to decrease (see Eq. 18.8).

Quick Quiz 18.4

A pipe open at both ends resonates at a fundamental frequency f_{open}. When one end is covered and the pipe is again made to resonate, the fundamental frequency is f_{closed}. Which of the following expressions describes how these two resonant frequencies compare? (a) $f_{closed} = f_{open}$ (b) $f_{closed} = \frac{1}{2}f_{open}$ (c) $f_{closed} = 2f_{open}$ (d) $f_{closed} = \frac{3}{2}f_{open}$

EXAMPLE 18.5   Wind in a Culvert

A section of drainage culvert 1.23 m in length makes a howling noise when the wind blows. (a) Determine the frequencies of the first three harmonics of the culvert if it is open at both ends. Take $v = 343$ m/s as the speed of sound in air.

Solution   The frequency of the first harmonic of a pipe open at both ends is

$$f_1 = \frac{v}{2L} = \frac{343 \text{ m/s}}{2(1.23 \text{ m})} = \boxed{139 \text{ Hz}}$$

Because both ends are open, all harmonics are present; thus,

$$f_2 = 2f_1 = \boxed{278 \text{ Hz}} \quad \text{and} \quad f_3 = 3f_1 = \boxed{417 \text{ Hz}}.$$

  (b) What are the three lowest natural frequencies of the culvert if it is blocked at one end?

Solution   The fundamental frequency of a pipe closed at one end is

$$f_1 = \frac{v}{4L} = \frac{343 \text{ m/s}}{4(1.23 \text{ m})} = \boxed{69.7 \text{ Hz}}$$

In this case, only odd harmonics are present; hence, the next two harmonics have frequencies $f_3 = 3f_1 = \boxed{209 \text{ Hz}}$ and $f_5 = 5f_1 = \boxed{349 \text{ Hz}}.$

  (c) For the culvert open at both ends, how many of the harmonics present fall within the normal human hearing range (20 to 17 000 Hz)?

Solution   Because all harmonics are present, we can express the frequency of the highest harmonic heard as $f_n = nf_1$, where n is the number of harmonics that we can hear. For $f_n = 17\,000$ Hz, we find that the number of harmonics present in the audible range is

$$n = \frac{17\,000 \text{ Hz}}{139 \text{ Hz}} = \boxed{122}$$

Only the first few harmonics are of sufficient amplitude to be heard.

EXAMPLE 18.6   Measuring the Frequency of a Tuning Fork

A simple apparatus for demonstrating resonance in an air column is depicted in Figure 18.15. A vertical pipe open at both ends is partially submerged in water, and a tuning fork vibrating at an unknown frequency is placed near the top of the pipe. The length L of the air column can be adjusted by moving the pipe vertically. The sound waves generated by the fork are reinforced when L corresponds to one of the resonance frequencies of the pipe.

  For a certain tube, the smallest value of L for which a peak occurs in the sound intensity is 9.00 cm. What are (a) the frequency of the tuning fork and (b) the value of L for the next two resonance frequencies?

Solution   (a) Although the pipe is open at its lower end to allow the water to enter, the water's surface acts like a wall at one end. Therefore, this setup represents a pipe closed at one end, and so the fundamental frequency is $f_1 = v/4L$. Taking $v = 343$ m/s for the speed of sound in air and $L = 0.090\,0$ m, we obtain

$$f_1 = \frac{v}{4L} = \frac{343 \text{ m/s}}{4(0.090\,0 \text{ m})} = \boxed{953 \text{ Hz}}$$

Because the tuning fork causes the air column to resonate at this frequency, this must be the frequency of the tuning fork.

  (b) Because the pipe is closed at one end, we know from Figure 18.14b that the wavelength of the fundamental mode is $\lambda = 4L = 4(0.090\,0 \text{ m}) = 0.360$ m. Because the frequency of the tuning fork is constant, the next two normal modes (see Fig. 18.15b) correspond to lengths of $L = 3\lambda/4 = \boxed{0.270 \text{ m}}$ and $L = 5\lambda/4 = \boxed{0.450 \text{ m}}.$

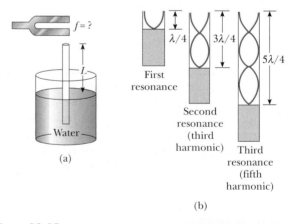

Figure 18.15   (a) Apparatus for demonstrating the resonance of sound waves in a tube closed at one end. The length L of the air column is varied by moving the tube vertically while it is partially submerged in water. (b) The first three normal modes of the system shown in part (a).

Optional Section

18.6 ▶ STANDING WAVES IN RODS AND PLATES

Standing waves can also be set up in rods and plates. A rod clamped in the middle and stroked at one end oscillates, as depicted in Figure 18.16a. The oscillations of the particles of the rod are longitudinal, and so the broken lines in Figure 18.16 represent *longitudinal* displacements of various parts of the rod. For clarity, we have drawn them in the transverse direction, just as we did for air columns. The midpoint is a displacement node because it is fixed by the clamp, whereas the ends are displacement antinodes because they are free to oscillate. The oscillations in this setup are analogous to those in a pipe open at both ends. The broken lines in Figure 18.16a represent the first normal mode, for which the wavelength is $2L$ and the frequency is $f = v/2L$, where v is the speed of longitudinal waves in the rod. Other normal modes may be excited by clamping the rod at different points. For example, the second normal mode (Fig. 18.16b) is excited by clamping the rod a distance $L/4$ away from one end.

Two-dimensional oscillations can be set up in a flexible membrane stretched over a circular hoop, such as that in a drumhead. As the membrane is struck at some point, wave pulses that arrive at the fixed boundary are reflected many times. The resulting sound is not harmonic because the oscillating drumhead and the drum's hollow interior together produce a set of standing waves having frequencies that are *not* related by integer multiples. Without this relationship, the sound may be more correctly described as *noise* than as music. This is in contrast to the situation in wind and stringed instruments, which produce sounds that we describe as musical.

Some possible normal modes of oscillation for a two-dimensional circular membrane are shown in Figure 18.17. The lowest normal mode, which has a frequency f_1, contains only one nodal curve; this curve runs around the outer edge of the membrane. The other possible normal modes show additional nodal curves that are circles and straight lines across the diameter of the membrane.

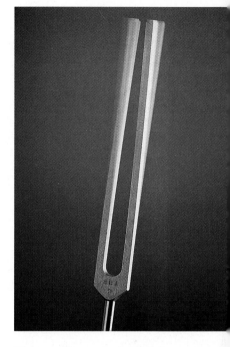

The sound from a tuning fork is produced by the vibrations of each of its prongs. *(Sam Dudgeon/Holt, Rinehart and Winston)*

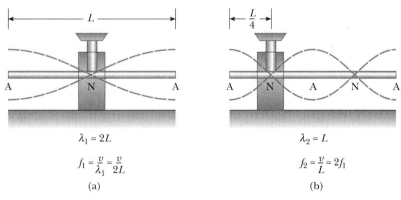

$$\lambda_1 = 2L$$

$$f_1 = \frac{v}{\lambda_1} = \frac{v}{2L}$$

(a)

$$\lambda_2 = L$$

$$f_2 = \frac{v}{L} = 2f_1$$

(b)

Figure 18.16 Normal-mode longitudinal vibrations of a rod of length L (a) clamped at the middle to produce the first normal mode and (b) clamped at a distance $L/4$ from one end to produce the second normal mode. Note that the dashed lines represent amplitudes parallel to the rod (longitudinal waves).

Wind chimes are usually designed so that the waves emanating from the vibrating rods blend into a harmonious sound. *(Joseph L. Fontenot/Visuals Unlimited)*

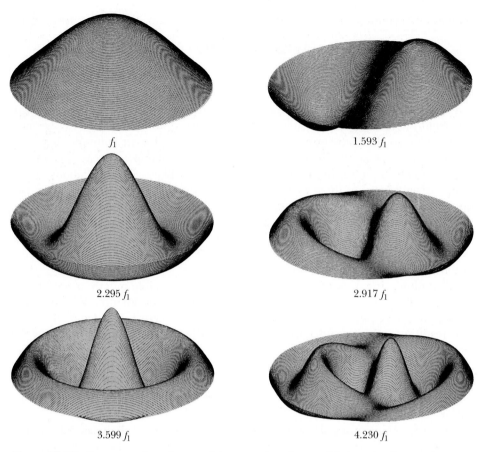

f_1

$1.593\,f_1$

$2.295\,f_1$

$2.917\,f_1$

$3.599\,f_1$

$4.230\,f_1$

Figure 18.17   Representation of some of the normal modes possible in a circular membrane fixed at its perimeter. The frequencies of oscillation do not form a harmonic series.   *(From M. L. Warren,* Introductory Physics, *New York, W. H. Freeman & Co., Publishers, 1979, with permission.)*

18.7 ▸ BEATS: INTERFERENCE IN TIME

The interference phenomena with which we have been dealing so far involve the superposition of two or more waves having the same frequency. Because the resultant wave depends on the coordinates of the disturbed medium, we refer to the phenomenon as *spatial interference.* Standing waves in strings and pipes are common examples of spatial interference.

We now consider another type of interference, one that results from the superposition of two waves having slightly *different* frequencies. In this case, when the two waves are observed at the point of superposition, they are periodically in and out of phase. That is, there is a *temporal* (time) alternation between constructive and destructive interference. Thus, we refer to this phenomenon as *interference in time* or *temporal interference.* For example, if two tuning forks of slightly different frequencies are struck, one hears a sound of periodically varying intensity. This phenomenon is called **beating**:

Definition of beating

> Beating is the periodic variation in intensity at a given point due to the superposition of two waves having slightly different frequencies.

The number of intensity maxima one hears per second, or the *beat frequency,* equals the difference in frequency between the two sources, as we shall show below. The maximum beat frequency that the human ear can detect is about 20 beats/s. When the beat frequency exceeds this value, the beats blend indistinguishably with the compound sounds producing them.

A piano tuner can use beats to tune a stringed instrument by "beating" a note against a reference tone of known frequency. The tuner can then adjust the string tension until the frequency of the sound it emits equals the frequency of the reference tone. The tuner does this by tightening or loosening the string until the beats produced by it and the reference source become too infrequent to notice.

Consider two sound waves of equal amplitude traveling through a medium with slightly different frequencies f_1 and f_2. We use equations similar to Equation 16.11 to represent the wave functions for these two waves at a point that we choose as $x = 0$:

$$y_1 = A \cos \omega_1 t = A \cos 2\pi f_1 t$$

$$y_2 = A \cos \omega_2 t = A \cos 2\pi f_2 t$$

Using the superposition principle, we find that the resultant wave function at this point is

$$y = y_1 + y_2 = A(\cos 2\pi f_1 t + \cos 2\pi f_2 t)$$

The trigonometric identity

$$\cos a + \cos b = 2 \cos\left(\frac{a-b}{2}\right) \cos\left(\frac{a+b}{2}\right)$$

allows us to write this expression in the form

$$y = \left[2 A \cos 2\pi\left(\frac{f_1 - f_2}{2}\right)t\right] \cos 2\pi\left(\frac{f_1 + f_2}{2}\right)t \qquad \textbf{(18.13)}$$

Resultant of two waves of different frequencies but equal amplitude

Graphs of the individual waves and the resultant wave are shown in Figure 18.18. From the factors in Equation 18.13, we see that the resultant sound for a listener standing at any given point has an effective frequency equal to the average frequency $(f_1 + f_2)/2$ and an amplitude given by the expression in the square

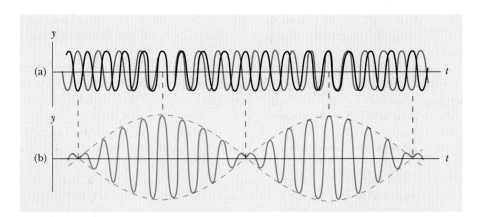

Figure 18.18 Beats are formed by the combination of two waves of slightly different frequencies. (a) The individual waves. (b) The combined wave has an amplitude (broken line) that oscillates in time.

brackets:

$$A_{\text{resultant}} = 2A \cos 2\pi\left(\frac{f_1 - f_2}{2}\right)t \tag{18.14}$$

That is, the **amplitude and therefore the intensity of the resultant sound vary in time.** The broken blue line in Figure 18.18b is a graphical representation of Equation 18.14 and is a sine wave varying with frequency $(f_1 - f_2)/2$.

Note that a maximum in the amplitude of the resultant sound wave is detected whenever

$$\cos 2\pi\left(\frac{f_1 - f_2}{2}\right)t = \pm 1$$

This means there are *two* maxima in each period of the resultant wave. Because the amplitude varies with frequency as $(f_1 - f_2)/2$, the number of beats per second, or the beat frequency f_b, is twice this value. That is,

$$f_b = |f_1 - f_2| \tag{18.15}$$

For instance, if one tuning fork vibrates at 438 Hz and a second one vibrates at 442 Hz, the resultant sound wave of the combination has a frequency of 440 Hz (the musical note A) and a beat frequency of 4 Hz. A listener would hear a 440-Hz sound wave go through an intensity maximum four times every second.

Optional Section

18.8 NON-SINUSOIDAL WAVE PATTERNS

The sound-wave patterns produced by the majority of musical instruments are non-sinusoidal. Characteristic patterns produced by a tuning fork, a flute, and a clarinet, each playing the same note, are shown in Figure 18.19. Each instrument has its own characteristic pattern. Note, however, that despite the differences in the patterns, each pattern is periodic. This point is important for our analysis of these waves, which we now discuss.

We can distinguish the sounds coming from a trumpet and a saxophone even when they are both playing the same note. On the other hand, we may have difficulty distinguishing a note played on a clarinet from the same note played on an oboe. We can use the pattern of the sound waves from various sources to explain these effects.

The wave patterns produced by a musical instrument are the result of the superposition of various harmonics. This superposition results in the corresponding richness of musical tones. The human perceptive response associated with various mixtures of harmonics is the *quality* or *timbre* of the sound. For instance, the sound of the trumpet is perceived to have a "brassy" quality (that is, we have learned to associate the adjective *brassy* with that sound); this quality enables us to distinguish the sound of the trumpet from that of the saxophone, whose quality is perceived as "reedy." The clarinet and oboe, however, are both straight air columns excited by reeds; because of this similarity, it is more difficult for the ear to distinguish them on the basis of their sound quality.

The problem of analyzing non-sinusoidal wave patterns appears at first sight to be a formidable task. However, if the wave pattern is periodic, it can be represented as closely as desired by the combination of a sufficiently large number of si-

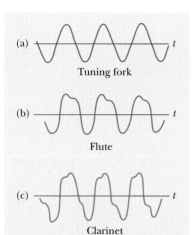

Figure 18.19 Sound wave patterns produced by (a) a tuning fork, (b) a flute, and (c) a clarinet, each at approximately the same frequency. *(Adapted from C. A. Culver, Musical Acoustics, 4th ed., New York, McGraw-Hill Book Company, 1956, p. 128.)*

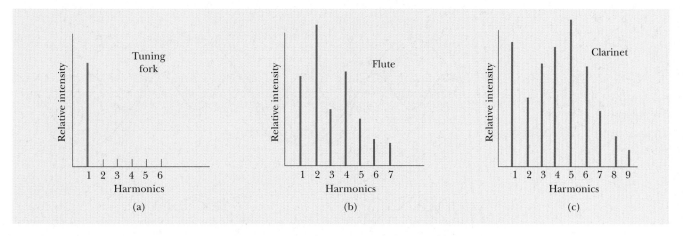

Figure 18.20   Harmonics of the wave patterns shown in Figure 18.19. Note the variations in intensity of the various harmonics. *(Adapted from C. A. Culver, Musical Acoustics, 4th ed., New York, McGraw-Hill Book Company, 1956.)*

nusoidal waves that form a harmonic series. In fact, we can represent any periodic function as a series of sine and cosine terms by using a mathematical technique based on **Fourier's theorem.**[3] The corresponding sum of terms that represents the periodic wave pattern is called a **Fourier series.**

Let $y(t)$ be any function that is periodic in time with period T, such that $y(t + T) = y(t)$. Fourier's theorem states that this function can be written as

$$y(t) = \sum_n (A_n \sin 2\pi f_n t + B_n \cos 2\pi f_n t) \qquad \textbf{(18.16)}$$

Fourier's theorem

where the lowest frequency is $f_1 = 1/T$. The higher frequencies are integer multiples of the fundamental, $f_n = nf_1$, and the coefficients A_n and B_n represent the amplitudes of the various waves. Figure 18.20 represents a harmonic analysis of the wave patterns shown in Figure 18.19. Note that a struck tuning fork produces only one harmonic (the first), whereas the flute and clarinet produce the first and many higher ones.

Note the variation in relative intensity of the various harmonics for the flute and the clarinet. In general, any musical sound consists of a fundamental frequency f plus other frequencies that are integer multiples of f, all having different intensities.

We have discussed the *analysis* of a wave pattern using Fourier's theorem. The analysis involves determining the coefficients of the harmonics in Equation 18.16 from a knowledge of the wave pattern. The reverse process, called *Fourier synthesis,* can also be performed. In this process, the various harmonics are added together to form a resultant wave pattern. As an example of Fourier synthesis, consider the building of a square wave, as shown in Figure 18.21. The symmetry of the square wave results in only odd multiples of the fundamental frequency combining in its synthesis. In Figure 18.21a, the orange curve shows the combination of f and $3f$. In Figure 18.21b, we have added $5f$ to the combination and obtained the green curve. Notice how the general shape of the square wave is approximated, even though the upper and lower portions are not flat as they should be.

---

[3] Developed by Jean Baptiste Joseph Fourier (1786–1830).

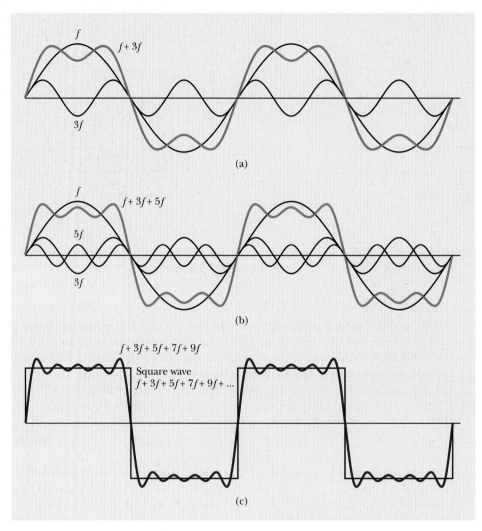

Figure 18.21 Fourier synthesis of a square wave, which is represented by the sum of odd multiples of the first harmonic, which has frequency f. (a) Waves of frequency f and $3f$ are added. (b) One more odd harmonic of frequency $5f$ is added. (c) The synthesis curve approaches the square wave when odd frequencies up to $9f$ are added.

This synthesizer can produce the characteristic sounds of different instruments by properly combining frequencies from electronic oscillators. *(Courtesy of Casio, Inc.)*

Figure 18.21c shows the result of adding odd frequencies up to $9f$. This approximation to the square wave (purple curve) is better than the approximations in parts a and b. To approximate the square wave as closely as possible, we would need to add all odd multiples of the fundamental frequency, up to infinite frequency.

Using modern technology, we can generate musical sounds electronically by mixing different amplitudes of any number of harmonics. These widely used electronic music synthesizers are capable of producing an infinite variety of musical tones.

SUMMARY

When two traveling waves having equal amplitudes and frequencies superimpose, the resultant wave has an amplitude that depends on the phase angle ϕ between the two waves. **Constructive interference** occurs when the two waves are in phase, corresponding to $\phi = 0, 2\pi, 4\pi, \ldots$ rad. **Destructive interference** occurs when the two waves are 180° out of phase, corresponding to $\phi = \pi, 3\pi, 5\pi, \ldots$ rad. Given two wave functions, you should be able to determine which if either of these two situations applies.

Standing waves are formed from the superposition of two sinusoidal waves having the same frequency, amplitude, and wavelength but traveling in opposite directions. The resultant standing wave is described by the wave function

$$y = (2A \sin kx) \cos \omega t \qquad \textbf{(18.3)}$$

Hence, the amplitude of the standing wave is $2A$, and the amplitude of the simple harmonic motion of any particle of the medium varies according to its position as $2A \sin kx$. The points of zero amplitude (called **nodes**) occur at $x = n\lambda/2$ ($n = 0$, 1, 2, 3, $\ldots$). The maximum amplitude points (called **antinodes**) occur at $x = n\lambda/4$ ($n = 1, 3, 5, \ldots$). Adjacent antinodes are separated by a distance $\lambda/2$. Adjacent nodes also are separated by a distance $\lambda/2$. You should be able to sketch the standing-wave pattern resulting from the superposition of two traveling waves.

The natural frequencies of vibration of a taut string of length L and fixed at both ends are

$$f_n = \frac{n}{2L}\sqrt{\frac{T}{\mu}} \qquad n = 1, 2, 3, \ldots \qquad \textbf{(18.8)}$$

where T is the tension in the string and μ is its linear mass density. The natural frequencies of vibration $f_1, 2f_1, 3f_1, \ldots$ form a **harmonic series.**

An oscillating system is in **resonance** with some driving force whenever the frequency of the driving force matches one of the natural frequencies of the system. When the system is resonating, it responds by oscillating with a relatively large amplitude.

Standing waves can be produced in a column of air inside a pipe. If the pipe is open at both ends, all harmonics are present and the natural frequencies of oscillation are

$$f_n = n\frac{v}{2L} \qquad n = 1, 2, 3, \ldots \qquad \textbf{(18.11)}$$

If the pipe is open at one end and closed at the other, only the odd harmonics are present, and the natural frequencies of oscillation are

$$f_n = n\frac{v}{4L} \qquad n = 1, 3, 5, \ldots \qquad \textbf{(18.12)}$$

The phenomenon of **beating** is the periodic variation in intensity at a given point due to the superposition of two waves having slightly different frequencies.

QUESTIONS

1. For certain positions of the movable section shown in Figure 18.2, no sound is detected at the receiver—a situation corresponding to destructive interference. This suggests that perhaps energy is somehow lost! What happens to the energy transmitted by the speaker?

2. Does the phenomenon of wave interference apply only to sinusoidal waves?

3. When two waves interfere constructively or destructively, is there any gain or loss in energy? Explain.

4. A standing wave is set up on a string, as shown in Figure 18.6. Explain why no energy is transmitted along the string.

5. What is common to *all* points (other than the nodes) on a string supporting a standing wave?

6. What limits the amplitude of motion of a real vibrating system that is driven at one of its resonant frequencies?

7. In Balboa Park in San Diego, CA, there is a huge outdoor organ. Does the fundamental frequency of a particular

pipe of this organ change on hot and cold days? How about on days with high and low atmospheric pressure?

8. Explain why your voice seems to sound better than usual when you sing in the shower.

9. What is the purpose of the slide on a trombone or of the valves on a trumpet?

10. Explain why all harmonics are present in an organ pipe open at both ends, but only the odd harmonics are present in a pipe closed at one end.

11. Explain how a musical instrument such as a piano may be tuned by using the phenomenon of beats.

12. An airplane mechanic notices that the sound from a twin-engine aircraft rapidly varies in loudness when both engines are running. What could be causing this variation from loudness to softness?

13. Why does a vibrating guitar string sound louder when placed on the instrument than it would if it were allowed to vibrate in the air while off the instrument?

14. When the base of a vibrating tuning fork is placed against a chalkboard, the sound that it emits becomes louder. This is due to the fact that the vibrations of the tuning fork are transmitted to the chalkboard. Because it has a larger area than that of the tuning fork, the vibrating

chalkboard sets a larger number of air molecules into vibration. Thus, the chalkboard is a better radiator of sound than the tuning fork. How does this affect the length of time during which the fork vibrates? Does this agree with the principle of conservation of energy?

15. To keep animals away from their cars, some people mount short thin pipes on the front bumpers. The pipes produce a high-frequency wail when the cars are moving. How do they create this sound?

16. Guitarists sometimes play a "harmonic" by lightly touching a string at the exact center and plucking the string. The result is a clear note one octave higher than the fundamental frequency of the string, even though the string is not pressed to the fingerboard. Why does this happen?

17. If you wet your fingers and lightly run them around the rim of a fine wine glass, a high-frequency sound is heard. Why? How could you produce various musical notes with a set of wine glasses, each of which contains a different amount of water?

18. Despite a reasonably steady hand, one often spills coffee when carrying a cup of it from one place to another. Discuss resonance as a possible cause of this difficulty, and devise a means for solving the problem.

PROBLEMS

1, 2, 3 = straightforward, intermediate, challenging   ☐ = full solution available in the *Student Solutions Manual and Study Guide*
WEB = solution posted at **http://www.saunderscollege.com/physics/**   🖥 = Computer useful in solving problem   📱 = Interactive Physics
☐ = paired numerical/symbolic problems

Section 18.1   Superposition and Interference of Sinusoidal Waves

WEB **1.** Two sinusoidal waves are described by the equations

$$y_1 = (5.00 \text{ m}) \sin[\pi(4.00x - 1\,200t)]$$

and

$$y_2 = (5.00 \text{ m}) \sin[\pi(4.00x - 1\,200t - 0.250)]$$

where x, y_1, and y_2 are in meters and t is in seconds. (a) What is the amplitude of the resultant wave? (b) What is the frequency of the resultant wave?

2. A sinusoidal wave is described by the equation

$$y_1 = (0.080\,0 \text{ m}) \sin[2\pi(0.100x - 80.0t)]$$

where y_1 and x are in meters and t is in seconds. Write an expression for a wave that has the same frequency, amplitude, and wavelength as y_1 but which, when added to y_1, gives a resultant with an amplitude of $8\sqrt{3}$ cm.

3. Two waves are traveling in the same direction along a stretched string. The waves are 90.0° out of phase. Each wave has an amplitude of 4.00 cm. Find the amplitude of the resultant wave.

4. Two identical sinusoidal waves with wavelengths of 3.00 m travel in the same direction at a speed of 2.00 m/s. The second wave originates from the same

point as the first, but at a later time. Determine the minimum possible time interval between the starting moments of the two waves if the amplitude of the resultant wave is the same as that of each of the two initial waves.

5. A tuning fork generates sound waves with a frequency of 246 Hz. The waves travel in opposite directions along a hallway, are reflected by walls, and return. The hallway is 47.0 m in length, and the tuning fork is located 14.0 m from one end. What is the phase difference between the reflected waves when they meet? The speed of sound in air is 343 m/s.

6. Two identical speakers 10.0 m apart are driven by the same oscillator with a frequency of $f = 21.5$ Hz (Fig. P18.6). (a) Explain why a receiver at point A records a minimum in sound intensity from the two speakers. (b) If the receiver is moved in the plane of the speakers, what path should it take so that the intensity remains at a minimum? That is, determine the relationship between x and y (the coordinates of the receiver) that causes the receiver to record a minimum in sound intensity. Take the speed of sound to be 343 m/s.

7. Two speakers are driven by the same oscillator with frequency of 200 Hz. They are located 4.00 m apart on a

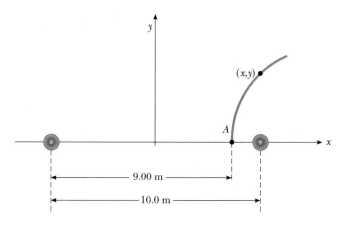

Figure P18.6

vertical pole. A man walks straight toward the lower speaker in a direction perpendicular to the pole, as shown in Figure P18.7. (a) How many times will he hear a minimum in sound intensity, and (b) how far is he from the pole at these moments? Take the speed of sound to be 330 m/s, and ignore any sound reflections coming off the ground.

8. Two speakers are driven by the same oscillator of frequency f. They are located a distance d from each other on a vertical pole. A man walks straight toward the lower speaker in a direction perpendicular to the pole, as shown in Figure P18.7. (a) How many times will he hear a minimum in sound intensity, and (b) how far is he from the pole at these moments? Take the speed of sound to be v, and ignore any sound reflections coming off the ground.

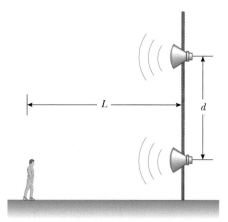

Figure P18.7   Problems 7 and 8.

Section 18.2   **Standing Waves**

9. Two sinusoidal waves traveling in opposite directions interfere to produce a standing wave described by the equation

$$y = (1.50 \text{ m}) \sin(0.400x) \cos(200t)$$

where x is in meters and t is in seconds. Determine the wavelength, frequency, and speed of the interfering waves.

10. Two waves in a long string are described by the equations

$$y_1 = (0.015\ 0 \text{ m}) \cos\left(\frac{x}{2} - 40t\right)$$

and

$$y_2 = (0.015\ 0 \text{ m}) \cos\left(\frac{x}{2} + 40t\right)$$

where y_1, y_2, and x are in meters and t is in seconds. (a) Determine the positions of the nodes of the resulting standing wave. (b) What is the maximum displacement at the position $x = 0.400$ m?

WEB 11. Two speakers are driven by a common oscillator at 800 Hz and face each other at a distance of 1.25 m. Locate the points along a line joining the two speakers where relative minima of sound pressure would be expected. (Use $v = 343$ m/s.)

12. Two waves that set up a standing wave in a long string are given by the expressions

$$y_1 = A \sin(kx - \omega t + \phi)$$

and

$$y_2 = A \sin(kx + \omega t)$$

Show (a) that the addition of the arbitrary phase angle changes only the position of the nodes, and (b) that the distance between the nodes remains constant in time.

13. Two sinusoidal waves combining in a medium are described by the equations

$$y_1 = (3.0 \text{ cm}) \sin \pi(x + 0.60t)$$

and

$$y_2 = (3.0 \text{ cm}) \sin \pi(x - 0.60t)$$

where x is in centimeters and t is in seconds. Determine the *maximum* displacement of the medium at (a) $x = 0.250$ cm, (b) $x = 0.500$ cm, and (c) $x = 1.50$ cm. (d) Find the three smallest values of x corresponding to antinodes.

14. A standing wave is formed by the interference of two traveling waves, each of which has an amplitude $A = \pi$ cm, angular wave number $k = (\pi/2)$ cm^{-1}, and angular frequency $\omega = 10\pi$ rad/s. (a) Calculate the distance between the first two antinodes. (b) What is the amplitude of the standing wave at $x = 0.250$ cm?

15. Verify by direct substitution that the wave function for a standing wave given in Equation 18.3, $y = 2A \sin kx \cos \omega t$, is a solution of the general linear

wave equation, Equation 16.26:

$$\frac{\partial^2 y}{\partial x^2} = \frac{1}{v^2}\frac{\partial^2 y}{\partial t^2}$$

Section 18.3 Standing Waves in a String Fixed at Both Ends

16. A 2.00-m-long wire having a mass of 0.100 kg is fixed at both ends. The tension in the wire is maintained at 20.0 N. What are the frequencies of the first three allowed modes of vibration? If a node is observed at a point 0.400 m from one end, in what mode and with what frequency is it vibrating?

17. Find the fundamental frequency and the next three frequencies that could cause a standing-wave pattern on a string that is 30.0 m long, has a mass per length of 9.00×10^{-3} kg/m, and is stretched to a tension of 20.0 N.

18. A standing wave is established in a 120-cm-long string fixed at both ends. The string vibrates in four segments when driven at 120 Hz. (a) Determine the wavelength. (b) What is the fundamental frequency of the string?

19. A cello A-string vibrates in its first normal mode with a frequency of 220 vibrations/s. The vibrating segment is 70.0 cm long and has a mass of 1.20 g. (a) Find the tension in the string. (b) Determine the frequency of vibration when the string vibrates in three segments.

20. A string of length L, mass per unit length μ, and tension T is vibrating at its fundamental frequency. Describe the effect that each of the following conditions has on the fundamental frequency: (a) The length of the string is doubled, but all other factors are held constant. (b) The mass per unit length is doubled, but all other factors are held constant. (c) The tension is doubled, but all other factors are held constant.

21. A 60.0-cm guitar string under a tension of 50.0 N has a mass per unit length of 0.100 g/cm. What is the highest resonance frequency of the string that can be heard by a person able to hear frequencies of up to 20 000 Hz?

22. A stretched wire vibrates in its first normal mode at a frequency of 400 Hz. What would be the fundamental frequency if the wire were half as long, its diameter were doubled, and its tension were increased four-fold?

23. A violin string has a length of 0.350 m and is tuned to concert G, with $f_G = 392$ Hz. Where must the violinist place her finger to play concert A, with $f_A = 440$ Hz? If this position is to remain correct to one-half the width of a finger (that is, to within 0.600 cm), what is the maximum allowable percentage change in the string's tension?

24. **Review Problem.** A sphere of mass M is supported by a string that passes over a light horizontal rod of length L (Fig. P18.24). Given that the angle is θ and that the fundamental frequency of standing waves in the section of the string above the horizontal rod is f, determine the mass of this section of the string.

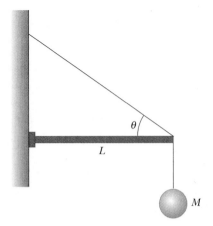

Figure P18.24

25. In the arrangement shown in Figure P18.25, a mass can be hung from a string (with a linear mass density of $\mu = 0.002\ 00$ kg/m) that passes over a light pulley. The string is connected to a vibrator (of constant frequency f), and the length of the string between point P and the pulley is $L = 2.00$ m. When the mass m is either 16.0 kg or 25.0 kg, standing waves are observed; however, no standing waves are observed with any mass between these values. (a) What is the frequency of the vibrator? (*Hint:* The greater the tension in the string, the smaller the number of nodes in the standing wave.) (b) What is the largest mass for which standing waves could be observed?

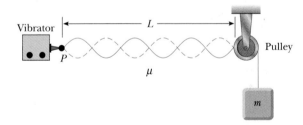

Figure P18.25

26. On a guitar, the fret closest to the bridge is a distance of 21.4 cm from it. The top string, pressed down at this last fret, produces the highest frequency that can be played on the guitar, 2 349 Hz. The next lower note has a frequency of 2 217 Hz. How far away from the last fret should the next fret be?

Section 18.4 Resonance

27. The chains suspending a child's swing are 2.00 m long. At what frequency should a big brother push to make the child swing with greatest amplitude?

28. Standing-wave vibrations are set up in a crystal goblet with four nodes and four antinodes equally spaced

around the 20.0-cm circumference of its rim. If transverse waves move around the glass at 900 m/s, an opera singer would have to produce a high harmonic with what frequency to shatter the glass with a resonant vibration?

29. An earthquake can produce a *seiche* (pronounced "saysh") in a lake, in which the water sloshes back and forth from end to end with a remarkably large amplitude and long period. Consider a seiche produced in a rectangular farm pond, as diagrammed in the cross-sectional view of Figure P18.29 (figure not drawn to scale). Suppose that the pond is 9.15 m long and of uniform depth. You measure that a wave pulse produced at one end reaches the other end in 2.50 s. (a) What is the wave speed? (b) To produce the seiche, you suggest that several people stand on the bank at one end and paddle together with snow shovels, moving them in simple harmonic motion. What must be the frequency of this motion?

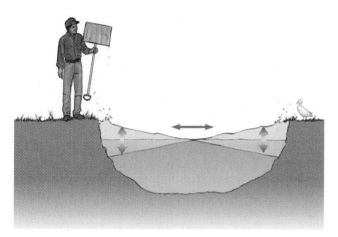

Figure P18.29

30. The Bay of Fundy, Nova Scotia, has the highest tides in the world. Assume that in mid-ocean and at the mouth of the bay, the Moon's gravity gradient and the Earth's rotation make the water surface oscillate with an amplitude of a few centimeters and a period of 12 h 24 min. At the head of the bay, the amplitude is several meters. Argue for or against the proposition that the tide is amplified by standing-wave resonance. Suppose that the bay has a length of 210 km and a depth everywhere of 36.1 m. The speed of long-wavelength water waves is given by $\sqrt{gd}$, where d is the water's depth.

Section 18.5 Standing Waves in Air Columns

Note: In this section, assume that the speed of sound in air is 343 m/s at 20°C and is described by the equation

$$v = (331 \text{ m/s})\sqrt{1 + \frac{T_C}{273°}}$$

at any Celsius temperature T_C.

31. Calculate the length of a pipe that has a fundamental frequency of 240 Hz if the pipe is (a) closed at one end and (b) open at both ends.

32. A glass tube (open at both ends) of length L is positioned near an audio speaker of frequency $f = 0.680$ kHz. For what values of L will the tube resonate with the speaker?

33. The overall length of a piccolo is 32.0 cm. The resonating air column vibrates as a pipe open at both ends. (a) Find the frequency of the lowest note that a piccolo can play, assuming that the speed of sound in air is 340 m/s. (b) Opening holes in the side effectively shortens the length of the resonant column. If the highest note that a piccolo can sound is 4 000 Hz, find the distance between adjacent antinodes for this mode of vibration.

34. The fundamental frequency of an open organ pipe corresponds to middle C (261.6 Hz on the chromatic musical scale). The third resonance of a closed organ pipe has the same frequency. What are the lengths of the two pipes?

35. Estimate the length of your ear canal, from its opening at the external ear to the eardrum. (Do not stick anything into your ear!) If you regard the canal as a tube that is open at one end and closed at the other, at approximately what fundamental frequency would you expect your hearing to be most sensitive? Explain why you can hear especially soft sounds just around this frequency.

36. An open pipe 0.400 m in length is placed vertically in a cylindrical bucket and nearly touches the bottom of the bucket, which has an area of 0.100 m². Water is slowly poured into the bucket until a sounding tuning fork of frequency 440 Hz, held over the pipe, produces resonance. Find the mass of water in the bucket at this moment.

WEB 37. A shower stall measures 86.0 cm × 86.0 cm × 210 cm. If you were singing in this shower, which frequencies would sound the richest (because of resonance)? Assume that the stall acts as a pipe closed at both ends, with nodes at opposite sides. Assume that the voices of various singers range from 130 Hz to 2 000 Hz. Let the speed of sound in the hot shower stall be 355 m/s.

38. When a metal pipe is cut into two pieces, the lowest resonance frequency in one piece is 256 Hz and that for the other is 440 Hz. (a) What resonant frequency would have been produced by the original length of pipe? (b) How long was the original pipe?

39. As shown in Figure P18.39, water is pumped into a long vertical cylinder at a rate of 18.0 cm³/s. The radius of the cylinder is 4.00 cm, and at the open top of the cylinder is a tuning fork vibrating with a frequency of 200 Hz. As the water rises, how much time elapses between successive resonances?

40. As shown in Figure P18.39, water is pumped into a long vertical cylinder at a volume flow rate R. The radius of

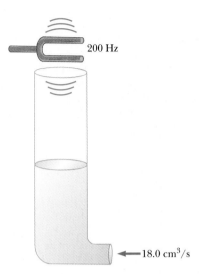

Figure P18.39   Problems 39 and 40.

the cylinder is r, and at the open top of the cylinder is a tuning fork vibrating with a frequency f. As the water rises, how much time elapses between successive resonances?

41. A tuning fork with a frequency of 512 Hz is placed near the top of the tube shown in Figure 18.15a. The water level is lowered so that the length L slowly increases from an initial value of 20.0 cm. Determine the next two values of L that correspond to resonant modes.

42. A student uses an audio oscillator of adjustable frequency to measure the depth of a water well. Two successive resonances are heard at 51.5 Hz and 60.0 Hz. How deep is the well?

43. A glass tube is open at one end and closed at the other by a movable piston. The tube is filled with air warmer than that at room temperature, and a 384-Hz tuning fork is held at the open end. Resonance is heard when the piston is 22.8 cm from the open end and again when it is 68.3 cm from the open end. (a) What speed of sound is implied by these data? (b) How far from the open end will the piston be when the next resonance is heard?

44. The longest pipe on an organ that has pedal stops is often 4.88 m. What is the fundamental frequency (at 0.00°C) if the nondriven end of the pipe is (a) closed and (b) open? (c) What are the frequencies at 20.0°C?

45. With a particular fingering, a flute sounds a note with a frequency of 880 Hz at 20.0°C. The flute is open at both ends. (a) Find the length of the air column. (b) Find the frequency it produces during the half-time performance at a late-season football game, when the ambient temperature is −5.00°C.

(Optional)
Section 18.6   Standing Waves in Rods and Plates

46. An aluminum rod is clamped one quarter of the way along its length and set into longitudinal vibration by a variable-frequency driving source. The lowest frequency that produces resonance is 4 400 Hz. The speed of sound in aluminum is 5 100 m/s. Determine the length of the rod.

47. An aluminum rod 1.60 m in length is held at its center. It is stroked with a rosin-coated cloth to set up a longitudinal vibration. (a) What is the fundamental frequency of the waves established in the rod? (b) What harmonics are set up in the rod held in this manner? (c) What would be the fundamental frequency if the rod were made of copper?

48. A 60.0-cm metal bar that is clamped at one end is struck with a hammer. If the speed of longitudinal (compressional) waves in the bar is 4 500 m/s, what is the lowest frequency with which the struck bar resonates?

Section 18.7   Beats: Interference in Time

WEB 49. In certain ranges of a piano keyboard, more than one string is tuned to the same note to provide extra loudness. For example, the note at 110 Hz has two strings that vibrate at this frequency. If one string slips from its normal tension of 600 N to 540 N, what beat frequency is heard when the hammer strikes the two strings simultaneously?

50. While attempting to tune the note C at 523 Hz, a piano tuner hears 2 beats/s between a reference oscillator and the string. (a) What are the possible frequencies of the string? (b) When she tightens the string slightly, she hears 3 beats/s. What is the frequency of the string now? (c) By what percentage should the piano tuner now change the tension in the string to bring it into tune?

51. A student holds a tuning fork oscillating at 256 Hz. He walks toward a wall at a constant speed of 1.33 m/s. (a) What beat frequency does he observe between the tuning fork and its echo? (b) How fast must he walk away from the wall to observe a beat frequency of 5.00 Hz?

(Optional)
Section 18.8   Non-Sinusoidal Wave Patterns

52. Suppose that a flutist plays a 523-Hz C note with first harmonic displacement amplitude $A_1 = 100$ nm. From Figure 18.20b, read, by proportion, the displacement amplitudes of harmonics 2 through 7. Take these as the values A_2 through A_7 in the Fourier analysis of the sound, and assume that $B_1 = B_2 = \ldots = B_7 = 0$. Construct a graph of the waveform of the sound. Your waveform will not look exactly like the flute waveform in Figure 18.19b because you simplify by ignoring cosine terms; nevertheless, it produces the same sensation to human hearing.

53. An A-major chord consists of the notes called A, C#, and E. It can be played on a piano by simultaneously striking strings that have fundamental frequencies of 440.00 Hz, 554.37 Hz, and 659.26 Hz. The rich consonance of the chord is associated with the near equality of the frequencies of some of the higher harmonics of the three tones. Consider the first five harmonics of each string and determine which harmonics show near equality.

ADDITIONAL PROBLEMS

54. **Review Problem.** For the arrangement shown in Figure P18.54, $\theta = 30.0°$, the inclined plane and the small pulley are frictionless, the string supports the mass M at the bottom of the plane, and the string has a mass m that is small compared with M. The system is in equilibrium, and the vertical part of the string has a length h. Standing waves are set up in the vertical section of the string. Find (a) the tension in the string, (b) the whole length of the string (ignoring the radius of curvature of the pulley), (c) the mass per unit length of the string, (d) the speed of waves on the string, (e) the lowest-frequency standing wave, (f) the period of the standing wave having three nodes, (g) the wavelength of the standing wave having three nodes, and (h) the frequency of the beats resulting from the interference of the sound wave of lowest frequency generated by the string with another sound wave having a frequency that is 2.00% greater.

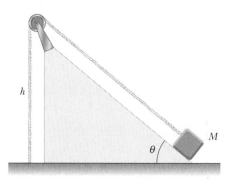

Figure P18.54

55. Two loudspeakers are placed on a wall 2.00 m apart. A listener stands 3.00 m from the wall directly in front of one of the speakers. The speakers are being driven by a single oscillator at a frequency of 300 Hz. (a) What is the phase difference between the two waves when they reach the observer? (b) What is the frequency closest to 300 Hz to which the oscillator may be adjusted such that the observer hears minimal sound?

56. On a marimba (Fig. P18.56), the wooden bar that sounds a tone when it is struck vibrates in a transverse standing wave having three antinodes and two nodes. The lowest-frequency note is 87.0 Hz; this note is produced by a bar 40.0 cm long. (a) Find the speed of transverse waves on the bar. (b) The loudness of the emitted sound is enhanced by a resonant pipe suspended vertically below the center of the bar. If the pipe is open at the top end only and the speed of sound in air is 340 m/s, what is the length of the pipe required to resonate with the bar in part (a)?

Figure P18.56 Marimba players in Mexico City. *(Murray Greenberg)*

57. Two train whistles have identical frequencies of 180 Hz. When one train is at rest in the station and is sounding its whistle, a beat frequency of 2.00 Hz is heard from a train moving nearby. What are the two possible speeds and directions that the moving train can have?

58. A speaker at the front of a room and an identical speaker at the rear of the room are being driven by the same oscillator at 456 Hz. A student walks at a uniform rate of 1.50 m/s along the length of the room. How many beats does the student hear per second?

59. While Jane waits on a railroad platform, she observes two trains approaching from the same direction at equal speeds of 8.00 m/s. Both trains are blowing their whistles (which have the same frequency), and one train is some distance behind the other. After the first train passes Jane, but before the second train passes her, she hears beats having a frequency of 4.00 Hz. What is the frequency of the trains' whistles?

60. A string fixed at both ends and having a mass of 4.80 g, a length of 2.00 m, and a tension of 48.0 N vibrates in its second ($n = 2$) natural mode. What is the wavelength in air of the sound emitted by this vibrating string?

61. A string 0.400 m in length has a mass per unit length of 9.00×10^{-3} kg/m. What must be the tension in the string if its second harmonic is to have the same frequency as the second resonance mode of a 1.75-m-long pipe open at one end?

62. In a major chord on the physical pitch musical scale, the frequencies are in the ratios $4:5:6:8$. A set of pipes, closed at one end, must be cut so that, when they are sounded in their first normal mode, they produce a major chord. (a) What is the ratio of the lengths of the pipes? (b) What are the lengths of the pipes needed if the lowest frequency of the chord is 256 Hz? (c) What are the frequencies of this chord?

63. Two wires are welded together. The wires are made of the same material, but the diameter of one wire is twice that of the other. They are subjected to a tension of 4.60 N. The thin wire has a length of 40.0 cm and a linear mass density of 2.00 g/m. The combination is fixed at both ends and vibrated in such a way that two antinodes are present, with the node between them being right at the weld. (a) What is the frequency of vibration? (b) How long is the thick wire?

64. Two identical strings, each fixed at both ends, are arranged near each other. If string A starts oscillating in its first normal mode, string B begins vibrating in its third ($n = 3$) natural mode. Determine the ratio of the tension of string B to the tension of string A.

65. A standing wave is set up in a string of variable length and tension by a vibrator of variable frequency. When the vibrator has a frequency f, in a string of length L and under a tension T, n antinodes are set up in the string. (a) If the length of the string is doubled, by what factor should the frequency be changed so that the same number of antinodes is produced? (b) If the frequency and length are held constant, what tension produces $n + 1$ antinodes? (c) If the frequency is tripled and the length of the string is halved, by what factor should the tension be changed so that twice as many antinodes are produced?

66. A 0.010 0-kg, 2.00-m-long wire is fixed at both ends and vibrates in its simplest mode under a tension of 200 N. When a tuning fork is placed near the wire, a beat frequency of 5.00 Hz is heard. (a) What could the frequency of the tuning fork be? (b) What should the tension in the wire be if the beats are to disappear?

WEB 67. If two adjacent natural frequencies of an organ pipe are determined to be 0.550 kHz and 0.650 kHz, calculate the fundamental frequency and length of the pipe. (Use $v = 340$ m/s.)

68. Two waves are described by the equations

$$y_1(x, t) = 5.0 \sin(2.0x - 10t)$$

and

$$y_2(x, t) = 10 \cos(2.0x - 10t)$$

where x is in meters and t is in seconds. Show that the resulting wave is sinusoidal, and determine the amplitude and phase of this sinusoidal wave.

69. The wave function for a standing wave is given in Equation 18.3 as $y = (2A \sin kx) \cos \omega t$. (a) Rewrite this wave function in terms of the wavelength λ and the wave speed v of the wave. (b) Write the wave function of the simplest standing-wave vibration of a stretched string of length L. (c) Write the wave function for the second harmonic. (d) Generalize these results, and write the wave function for the nth resonance vibration.

70. **Review Problem.**   A 12.0-kg mass hangs in equilibrium from a string with a total length of $L = 5.00$ m and a linear mass density of $\mu = 0.001\,00$ kg/m. The string is wrapped around two light, frictionless pulleys that are separated by a distance of $d = 2.00$ m (Fig. P18.70a). (a) Determine the tension in the string. (b) At what frequency must the string between the pulleys vibrate to form the standing-wave pattern shown in Figure P18.70b?

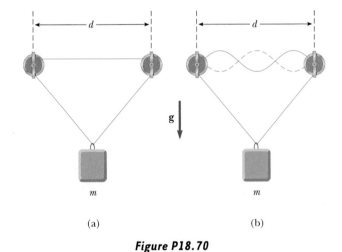

(a)   (b)

Figure P18.70

Answers to Quick Quizzes

18.1 At the antinodes. All particles have the same period $T = 2\pi/\omega$, but a particle at an antinode must travel through the greatest vertical distance in this amount of time and therefore must travel fastest.

18.2 For each natural frequency of the glass, the standing wave must "fit" exactly around the rim. In Figure 18.12a we see three antinodes on the near side of the glass, and thus there must be another three on the far side. This corresponds to three complete waves. In a top view, the wave pattern looks like this (although we have greatly exaggerated the amplitude):

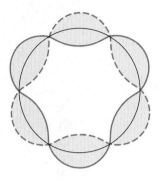

18.3 At highway speeds, a car crosses the ridges on the rumble strip at a rate that matches one of the car's natural frequencies of oscillation. This causes the car to oscillate substantially more than when it is traveling over the randomly spaced bumps of regular pavement. This sudden resonance oscillation alerts the driver that he or she must pay attention.

18.4 (b). With both ends open, the pipe has a fundamental frequency given by Equation 18.11: $f_{open} = v/2L$. With one end closed, the pipe has a fundamental frequency given by Equation 18.12:

$$f_{closed} = \frac{v}{4L} = \frac{1}{2} \frac{v}{2L} = \frac{1}{2} f_{open}$$

part 3

Thermodynamics

We now direct our attention to the study of thermodynamics, which deals with the concepts of heat and temperature. As we shall see, thermodynamics is very successful in explaining the bulk properties of matter and the correlation between these properties and the mechanics of both atoms and molecules.

Historically, the development of thermodynamics paralleled the development of the atomic theory of matter. By the 1820s, chemical experiments had provided solid evidence for the existence of atoms. At that time, scientists recognized that a connection between the theory of heat and temperature and the structure of matter must exist. In 1827, the botanist Robert Brown reported that grains of pollen suspended in a liquid move erratically from one place to another, as if under constant agitation. In 1905, Albert Einstein used kinetic theory to explain the cause of this erratic motion, which today is known as *Brownian motion*. Einstein explained this phenomenon by assuming that the grains are under constant bombardment by "invisible" molecules in the liquid, which themselves move erratically. This explanation gave scientists insight into molecular motion and gave credence to the idea that matter is made up of atoms. A connection was thus forged between the everyday world and the tiny, invisible building blocks that make up this world.

Thermodynamics also addresses more practical questions. Have you ever wondered how a refrigerator is able to cool its contents, what types of transformations occur in a power plant or in the engine of your automobile, or what happens to the kinetic energy of a moving object when the object comes to rest? The laws of thermodynamics and the concepts of heat and temperature provide explanations. In general, thermodynamics is concerned with transformations of matter in all of its states—solid, liquid, and gas.

◀ Richard A. Cooke III/Tony Stone Images.

⊞ P U Z Z L ∈ R

After this bottle of champagne was shaken, the cork was popped off and champagne spewed everywhere. Contrary to common belief, shaking a champagne bottle before opening it does not increase the pressure of the carbon dioxide (CO_2) inside. In fact, if you know the trick, you can open a thoroughly shaken bottle without spraying a drop. What's the secret? And why isn't the pressure inside the bottle greater after the bottle is shaken? *(Steve Niedorf/The Image Bank)*

c h a p t e r

19

Temperature

Chapter Outline

In our study of mechanics, we carefully defined such concepts as mass, force, and kinetic energy to facilitate our quantitative approach. Likewise, a quantitative description of thermal phenomena requires a careful definition of such important terms as *temperature, heat,* and *internal energy.* This chapter begins with a look at these three entities and with a description of one of the laws of thermodynamics (the poetically named "zeroth law"). We then discuss the three most common temperature scales—Celsius, Fahrenheit, and Kelvin.

Next, we consider why the composition of a body is an important factor when we are dealing with thermal phenomena. For example, gases expand appreciably when heated, whereas liquids and solids expand only slightly. If a gas is not free to expand as it is heated, its pressure increases. Certain substances may melt, boil, burn, or explode when they are heated, depending on their composition and structure.

This chapter concludes with a study of ideal gases on the macroscopic scale. Here, we are concerned with the relationships among such quantities as pressure, volume, and temperature. Later on, in Chapter 21, we shall examine gases on a microscopic scale, using a model that represents the components of a gas as small particles.

Molten lava flowing down a mountain in Kilauea, Hawaii. The temperature of the hot lava flowing from a central crater decreases until the lava is in thermal equilibrium with its surroundings. At that equilibrium temperature, the lava has solidified and formed the mountains. *(Ken Sakomoto/Black Star)*

19.1 TEMPERATURE AND THE ZEROTH LAW OF THERMODYNAMICS

We often associate the concept of temperature with how hot or cold an object feels when we touch it. Thus, our senses provide us with a qualitative indication of temperature. However, our senses are unreliable and often mislead us. For example, if we remove a metal ice tray and a cardboard box of frozen vegetables from the freezer, the ice tray feels colder than the box even though both are at the same temperature. The two objects feel different because metal is a better thermal conductor than cardboard is. What we need, therefore, is a reliable and reproducible method for establishing the relative hotness or coldness of bodies. Scientists have developed a variety of thermometers for making such quantitative measurements.

We are all familiar with the fact that two objects at different initial temperatures eventually reach some intermediate temperature when placed in contact with each other. For example, when a scoop of ice cream is placed in a room-temperature glass bowl, the ice cream melts and the temperature of the bowl decreases. Likewise, when an ice cube is dropped into a cup of hot coffee, it melts and the coffee's temperature decreases.

To understand the concept of temperature, it is useful to define two often-used phrases: *thermal contact* and *thermal equilibrium.* To grasp the meaning of thermal contact, let us imagine that two objects are placed in an insulated container such that they interact with each other but not with the rest of the world. If the objects are at different temperatures, energy is exchanged between them, even if they are initially not in physical contact with each other. **Heat is the transfer of energy from one object to another object as a result of a difference in temperature between the two.** We shall examine the concept of heat in greater detail in Chapter 20. For purposes of the current discussion, we assume that two objects are in **thermal contact** with each other if energy can be exchanged between them. **Thermal equilibrium** is a situation in which two objects in thermal contact with each other cease to exchange energy by the process of heat.

Let us consider two objects A and B, which are not in thermal contact, and a third object C, which is our thermometer. We wish to determine whether A and B

10.3 & 10.4

QuickLab

Fill three cups with tap water: one hot, one cold, and one lukewarm. Dip your left index finger into the hot water and your right index finger into the cold water. Slowly count to 20, then quickly dip both fingers into the lukewarm water. What do you feel?

are in thermal equilibrium with each other. The thermometer (object C) is first placed in thermal contact with object A until thermal equilibrium is reached. From that moment on, the thermometer's reading remains constant, and we record this reading. The thermometer is then removed from object A and placed in thermal contact with object B. The reading is again recorded after thermal equilibrium is reached. If the two readings are the same, then object A and object B are in thermal equilibrium with each other.

We can summarize these results in a statement known as the **zeroth law of thermodynamics** (the law of equilibrium):

Zeroth law of thermodynamics

> If objects A and B are separately in thermal equilibrium with a third object C, then objects A and B are in thermal equilibrium with each other.

This statement can easily be proved experimentally and is very important because it enables us to define temperature. We can think of **temperature** as the property that determines whether an object is in thermal equilibrium with other objects. **Two objects in thermal equilibrium with each other are at the same temperature.** Conversely, if two objects have different temperatures, then they are not in thermal equilibrium with each other.

19.2 THERMOMETERS AND THE CELSIUS TEMPERATURE SCALE

Thermometers are devices that are used to define and measure temperatures. All thermometers are based on the principle that some physical property of a system changes as the system's temperature changes. Some physical properties that change with temperature are (1) the volume of a liquid, (2) the length of a solid, (3) the pressure of a gas at constant volume, (4) the volume of a gas at constant pressure, (5) the electric resistance of a conductor, and (6) the color of an object. For a given substance and a given temperature range, a temperature scale can be established on the basis of any one of these physical properties.

A common thermometer in everyday use consists of a mass of liquid—usually mercury or alcohol—that expands into a glass capillary tube when heated (Fig. 19.1). In this case the physical property is the change in volume of a liquid. Any temperature change can be defined as being proportional to the change in length of the liquid column. The thermometer can be calibrated by placing it in thermal contact with some natural systems that remain at constant temperature. One such system is a mixture of water and ice in thermal equilibrium at atmospheric pressure. On the **Celsius temperature scale,** this mixture is defined to have a temperature of zero degrees Celsius, which is written as 0°C; this temperature is called the *ice point* of water. Another commonly used system is a mixture of water and steam in thermal equilibrium at atmospheric pressure; its temperature is 100°C, which is the *steam point* of water. Once the liquid levels in the thermometer have been established at these two points, the distance between the two points is divided into 100 equal segments to create the Celsius scale. Thus, each segment denotes a change in temperature of one Celsius degree. (This temperature scale used to be called the *centigrade scale* because there are 100 gradations between the ice and steam points of water.)

Thermometers calibrated in this way present problems when extremely accurate readings are needed. For instance, the readings given by an alcohol ther-

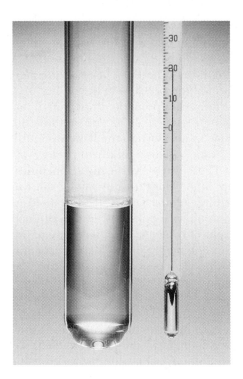

Figure 19.1   As a result of thermal expansion, the level of the mercury in the thermometer rises as the mercury is heated by water in the test tube.   *(Charles D. Winters)*

mometer calibrated at the ice and steam points of water might agree with those given by a mercury thermometer only at the calibration points. Because mercury and alcohol have different thermal expansion properties, when one thermometer reads a temperature of, for example, 50°C, the other may indicate a slightly different value. The discrepancies between thermometers are especially large when the temperatures to be measured are far from the calibration points.[1]

An additional practical problem of any thermometer is the limited range of temperatures over which it can be used. A mercury thermometer, for example, cannot be used below the freezing point of mercury, which is − 39°C, and an alcohol thermometer is not useful for measuring temperatures above 85°C, the boiling point of alcohol. To surmount this problem, we need a universal thermometer whose readings are independent of the substance used in it. The gas thermometer, discussed in the next section, approaches this requirement.

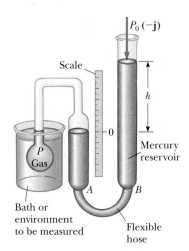

Figure 19.2   A constant-volume gas thermometer measures the pressure of the gas contained in the flask immersed in the bath. The volume of gas in the flask is kept constant by raising or lowering reservoir *B* to keep the mercury level in column *A* constant.

19.3 ▶ THE CONSTANT-VOLUME GAS THERMOMETER AND THE ABSOLUTE TEMPERATURE SCALE

The temperature readings given by a gas thermometer are nearly independent of the substance used in the thermometer. One version is the constant-volume gas thermometer shown in Figure 19.2. The physical change exploited in this device is the variation of pressure of a fixed volume of gas with temperature. When the constant-volume gas thermometer was developed, it was calibrated by using the ice

---

[1] Two thermometers that use the same liquid may also give different readings. This is due in part to difficulties in constructing uniform-bore glass capillary tubes.

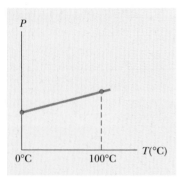

Figure 19.3 A typical graph of pressure versus temperature taken with a constant-volume gas thermometer. The two dots represent known reference temperatures (the ice and steam points of water).

10.3

and steam points of water, as follows (a different calibration procedure, which we shall discuss shortly, is now used): The flask was immersed in an ice bath, and mercury reservoir *B* was raised or lowered until the top of the mercury in column *A* was at the zero point on the scale. The height *h*, the difference between the mercury levels in reservoir *B* and column *A*, indicated the pressure in the flask at 0°C.

The flask was then immersed in water at the steam point, and reservoir *B* was readjusted until the top of the mercury in column *A* was again at zero on the scale; this ensured that the gas's volume was the same as it was when the flask was in the ice bath (hence, the designation "constant volume"). This adjustment of reservoir *B* gave a value for the gas pressure at 100°C. These two pressure and temperature values were then plotted, as shown in Figure 19.3. The line connecting the two points serves as a calibration curve for unknown temperatures. If we wanted to measure the temperature of a substance, we would place the gas flask in thermal contact with the substance and adjust the height of reservoir *B* until the top of the mercury column in *A* was at zero on the scale. The height of the mercury column would indicate the pressure of the gas; knowing the pressure, we could find the temperature of the substance using the graph in Figure 19.3.

Now let us suppose that temperatures are measured with gas thermometers containing different gases at different initial pressures. Experiments show that the thermometer readings are nearly independent of the type of gas used, as long as the gas pressure is low and the temperature is well above the point at which the gas liquefies (Fig. 19.4). The agreement among thermometers using various gases improves as the pressure is reduced.

If you extend the curves shown in Figure 19.4 toward negative temperatures, you find, in every case, that the pressure is zero when the temperature is − 273.15°C. This significant temperature is used as the basis for the **absolute temperature scale,** which sets − 273.15°C as its zero point. This temperature is often referred to as **absolute zero.** The size of a degree on the absolute temperature scale is identical to the size of a degree on the Celsius scale. Thus, the conversion between these temperatures is

$$T_C = T - 273.15 \qquad\qquad (19.1)$$

where T_C is the Celsius temperature and T is the absolute temperature.

Because the ice and steam points are experimentally difficult to duplicate, an absolute temperature scale based on a single fixed point was adopted in 1954 by the International Committee on Weights and Measures. From a list of fixed points associated with various substances (Table 19.1), the triple point of water was chosen as the reference temperature for this new scale. The **triple point of water** is the single combination of temperature and pressure at which liquid water, gaseous

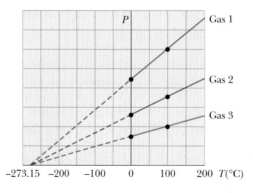

Figure 19.4 Pressure versus temperature for three dilute gases. Note that, for all gases, the pressure extrapolates to zero at the temperature − 273.15°C.

TABLE 19.1   Fixed-Point Temperatures[a]

Fixed Point	Temperature (°C)	Temperature (K)
Triple point of hydrogen	− 259.34	13.81
Boiling point of helium	− 268.93	4.215
Boiling point of hydrogen at 33.36 kPa pressure	− 256.108	17.042
Boiling point of hydrogen	− 252.87	20.28
Triple point of neon	− 246.048	27.102
Triple point of oxygen	− 218.789	54.361
Boiling point of oxygen	− 182.962	90.188
Triple point of water	0.01	273.16
Boiling point of water	100.00	373.15
Freezing point of tin	231.968 1	505.118 1
Freezing point of zinc	419.58	692.73
Freezing point of silver	961.93	1 235.08
Freezing point of gold	1 064.43	1 337.58

[a]All values are from National Bureau of Standards Special Publication 420; U. S. Department of Commerce, May 1975. All values are at standard atmospheric pressure except for triple points and as noted.

water, and ice (solid water) coexist in equilibrium. This triple point occurs at a temperature of approximately 0.01°C and a pressure of 4.58 mm of mercury. On the new scale, which uses the unit *kelvin,* the temperature of water at the triple point was set at 273.16 kelvin, abbreviated 273.16 K. (*Note:* no degree sign "°" is used with the unit kelvin.) This choice was made so that the old absolute temperature scale based on the ice and steam points would agree closely with the new scale based on the triple point. This new absolute temperature scale (also called the **Kelvin scale**) employs the SI unit of absolute temperature, the **kelvin,** which is defined to be **1/273.16 of the difference between absolute zero and the temperature of the triple point of water.**

Figure 19.5 shows the absolute temperature for various physical processes and structures. The temperature of absolute zero (0 K) cannot be achieved, although laboratory experiments incorporating the laser cooling of atoms have come very close.

What would happen to a gas if its temperature could reach 0 K? As Figure 19.4 indicates, the pressure it exerts on the walls of its container would be zero. In Section 19.5 we shall show that the pressure of a gas is proportional to the average kinetic energy of its molecules. Thus, according to classical physics, the kinetic energy of the gas molecules would become zero at absolute zero, and molecular motion would cease; hence, the molecules would settle out on the bottom of the container. Quantum theory modifies this model and shows that some residual energy, called the *zero-point energy,* would remain at this low temperature.

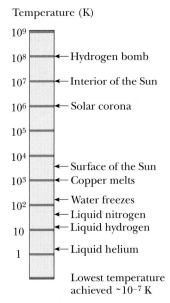

Figure 19.5 Absolute temperatures at which various physical processes occur. Note that the scale is logarithmic.

The Celsius, Fahrenheit, and Kelvin Temperature Scales[2]

Equation 19.1 shows that the Celsius temperature T_C is shifted from the absolute (Kelvin) temperature T by 273.15°. Because the size of a degree is the same on the

[2] Named after Anders Celsius (1701–1744), Gabriel Fahrenheit (1686–1736), and William Thomson, Lord Kelvin (1824–1907), respectively.

two scales, a temperature difference of 5°C is equal to a temperature difference of 5 K. The two scales differ only in the choice of the zero point. Thus, the ice-point temperature on the Kelvin scale, 273.15 K, corresponds to 0.00°C, and the Kelvin-scale steam point, 373.15 K, is equivalent to 100.00°C.

A common temperature scale in everyday use in the United States is the **Fahrenheit scale.** This scale sets the temperature of the ice point at 32°F and the temperature of the steam point at 212°F. The relationship between the Celsius and Fahrenheit temperature scales is

$$T_F = \tfrac{9}{5} T_C + 32°F \qquad \textbf{(19.2)}$$

Quick Quiz 19.1

What is the physical significance of the factor $\tfrac{9}{5}$ in Equation 19.2? Why is this factor missing in Equation 19.1?

Extending the ideas considered in Quick Quiz 19.1, we use Equation 19.2 to find a relationship between changes in temperature on the Celsius, Kelvin, and Fahrenheit scales:

$$\Delta T_C = \Delta T = \tfrac{5}{9} \Delta T_F \qquad \textbf{(19.3)}$$

EXAMPLE 19.1   Converting Temperatures

On a day when the temperature reaches 50°F, what is the temperature in degrees Celsius and in kelvins?

Solution   Substituting $T_F = 50°F$ into Equation 19.2, we obtain

$$T_C = \tfrac{5}{9}(T_F - 32) = \tfrac{5}{9}(50 - 32) = \boxed{10°C}$$

From Equation 19.1, we find that

$$T = T_C + 273.15 = 10°C + 273.15 = \boxed{283 \text{ K}}$$

A convenient set of weather-related temperature equivalents to keep in mind is that 0°C is (literally) freezing at 32°F, 10°C is cool at 50°F, 30°C is warm at 86°F, and 40°C is a hot day at 104°F.

EXAMPLE 19.2   Heating a Pan of Water

A pan of water is heated from 25°C to 80°C. What is the change in its temperature on the Kelvin scale and on the Fahrenheit scale?

Solution   From Equation 19.3, we see that the change in temperature on the Celsius scale equals the change on the Kelvin scale. Therefore,

$$\Delta T = \Delta T_C = 80°C - 25°C = 55°C = \boxed{55 \text{ K}}$$

From Equation 19.3, we also find that

$$\Delta T_F = \tfrac{9}{5} \Delta T_C = \tfrac{9}{5}(55°C) = \boxed{99°F}$$

19.4   THERMAL EXPANSION OF SOLIDS AND LIQUIDS

Our discussion of the liquid thermometer made use of one of the best-known changes in a substance: As its temperature increases, its volume almost always increases. (As we shall see shortly, in some substances the volume decreases when the temperature increases.) This phenomenon, known as **thermal expansion,** has

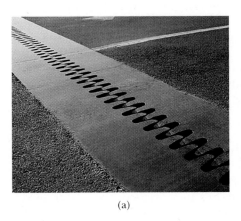
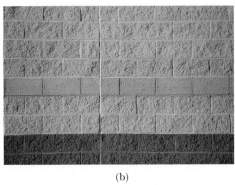

(a)   (b)

Figure 19.6   (a) Thermal-expansion joints are used to separate sections of roadways on bridges. Without these joints, the surfaces would buckle due to thermal expansion on very hot days or crack due to contraction on very cold days. (b) The long, vertical joint is filled with a soft material that allows the wall to expand and contract as the temperature of the bricks changes. *(a, Frank Siteman, Stock/Boston; b, George Semple)*

an important role in numerous engineering applications. For example, thermal-expansion joints, such as those shown in Figure 19.6, must be included in buildings, concrete highways, railroad tracks, brick walls, and bridges to compensate for dimensional changes that occur as the temperature changes.

Thermal expansion is a consequence of the change in the average separation between the constituent atoms in an object. To understand this, imagine that the atoms are connected by stiff springs, as shown in Figure 19.7. At ordinary temperatures, the atoms in a solid oscillate about their equilibrium positions with an amplitude of approximately 10^{-11} m and a frequency of approximately 10^{13} Hz. The average spacing between the atoms is about 10^{-10} m. As the temperature of the solid increases, the atoms oscillate with greater amplitudes; as a result, the average separation between them increases.[3] Consequently, the object expands.

If thermal expansion is sufficiently small relative to an object's initial dimensions, the change in any dimension is, to a good approximation, proportional to the first power of the temperature change. Suppose that an object has an initial length L_i along some direction at some temperature and that the length increases by an amount ΔL for a change in temperature ΔT. Because it is convenient to consider the fractional change in length per degree of temperature change, we define the **average coefficient of linear expansion** as

$$\alpha \equiv \frac{\Delta L / L_i}{\Delta T}$$

Experiments show that α is constant for small changes in temperature. For purposes of calculation, this equation is usually rewritten as

$$\Delta L = \alpha L_i \, \Delta T \qquad \textbf{(19.4)}$$

or as

$$L_f - L_i = \alpha L_i (T_f - T_i) \qquad \textbf{(19.5)}$$

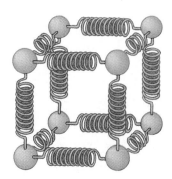

Figure 19.7   A mechanical model of the atomic configuration in a substance. The atoms (spheres) are imagined to be attached to each other by springs that reflect the elastic nature of the interatomic forces.

Average coefficient of linear expansion

The change in length of an object is proportional to the change in temperature

---

[3] More precisely, thermal expansion arises from the *asymmetrical* nature of the potential-energy curve for the atoms in a solid. If the oscillators were truly harmonic, the average atomic separations would not change regardless of the amplitude of vibration.

where L_f is the final length, T_i and T_f are the initial and final temperatures, and the proportionality constant α is the average coefficient of linear expansion for a given material and has units of $°C^{-1}$.

It may be helpful to think of thermal expansion as an effective magnification or as a photographic enlargement of an object. For example, as a metal washer is heated (Fig. 19.8), all dimensions, including the radius of the hole, increase according to Equation 19.4.

Table 19.2 lists the average coefficient of linear expansion for various materials. Note that for these materials α is positive, indicating an increase in length with increasing temperature. This is not always the case. Some substances—calcite ($CaCO_3$) is one example—expand along one dimension (positive α) and contract along another (negative α) as their temperatures are increased.

Because the linear dimensions of an object change with temperature, it follows that surface area and volume change as well. The change in volume at constant pressure is proportional to the initial volume V_i and to the change in temperature according to the relationship

> The change in volume of a solid at constant pressure is proportional to the change in temperature

$$\Delta V = \beta V_i \Delta T \tag{19.6}$$

where β is the **average coefficient of volume expansion.** For a solid, the average coefficient of volume expansion is approximately three times the average linear expansion coefficient: $\beta = 3\alpha$. (This assumes that the average coefficient of linear expansion of the solid is the same in all directions.)

To see that $\beta = 3\alpha$ for a solid, consider a box of dimensions ℓ, w, and h. Its volume at some temperature T_i is $V_i = \ell w h$. If the temperature changes to $T_i + \Delta T$, its volume changes to $V_i + \Delta V$, where each dimension changes according to Equation 19.4. Therefore,

$$
\begin{aligned}
V_i + \Delta V &= (\ell + \Delta \ell)(w + \Delta w)(h + \Delta h) \\
&= (\ell + \alpha \ell \,\Delta T)(w + \alpha w \,\Delta T)(h + \alpha h \,\Delta T) \\
&= \ell w h (1 + \alpha \,\Delta T)^3 \\
&= V_i [1 + 3\alpha \,\Delta T + 3(\alpha \,\Delta T)^2 + (\alpha \,\Delta T)^3]
\end{aligned}
$$

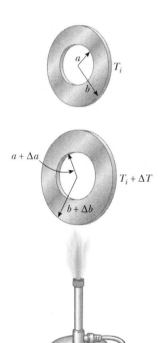

Figure 19.8 Thermal expansion of a homogeneous metal washer. As the washer is heated, all dimensions increase. (The expansion is exaggerated in this figure.)

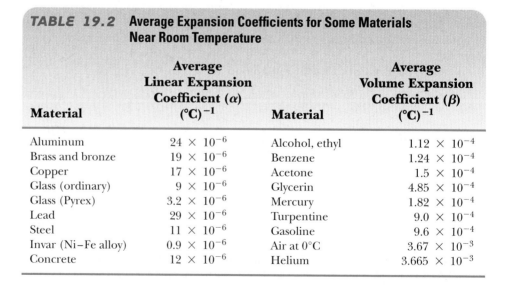

TABLE 19.2 Average Expansion Coefficients for Some Materials Near Room Temperature

Material	Average Linear Expansion Coefficient (α) $(°C)^{-1}$	Material	Average Volume Expansion Coefficient (β) $(°C)^{-1}$
Aluminum	24×10^{-6}	Alcohol, ethyl	1.12×10^{-4}
Brass and bronze	19×10^{-6}	Benzene	1.24×10^{-4}
Copper	17×10^{-6}	Acetone	1.5×10^{-4}
Glass (ordinary)	9×10^{-6}	Glycerin	4.85×10^{-4}
Glass (Pyrex)	3.2×10^{-6}	Mercury	1.82×10^{-4}
Lead	29×10^{-6}	Turpentine	9.0×10^{-4}
Steel	11×10^{-6}	Gasoline	9.6×10^{-4}
Invar (Ni–Fe alloy)	0.9×10^{-6}	Air at 0°C	3.67×10^{-3}
Concrete	12×10^{-6}	Helium	3.665×10^{-3}

If we now divide both sides by V_i and then isolate the term $\Delta V/V_i$, we obtain the fractional change in volume:

$$\frac{\Delta V}{V_i} = 3\alpha \, \Delta T + 3(\alpha \, \Delta T)^2 + (\alpha \, \Delta T)^3$$

Because $\alpha \, \Delta T \ll 1$ for typical values of ΔT $(< \sim 100°C)$, we can neglect the terms $3(\alpha \, \Delta T)^2$ and $(\alpha \, \Delta T)^3$. Upon making this approximation, we see that

$$\frac{\Delta V}{V_i} = 3\alpha \, \Delta T$$

$$3\alpha = \frac{1}{V_i} \frac{\Delta V}{\Delta T}$$

Equation 19.6 shows that the right side of this expression is equal to β, and so we have $3\alpha = \beta$, the relationship we set out to prove. In a similar way, you can show that the change in area of a rectangular plate is given by $\Delta A = 2\alpha A_i \, \Delta T$ (see Problem 53).

As Table 19.2 indicates, each substance has its own characteristic average coefficient of expansion. For example, when the temperatures of a brass rod and a steel rod of equal length are raised by the same amount from some common initial value, the brass rod expands more than the steel rod does because brass has a greater average coefficient of expansion than steel does. A simple mechanism called a *bimetallic strip* utilizes this principle and is found in practical devices such as thermostats. It consists of two thin strips of dissimilar metals bonded together. As the temperature of the strip increases, the two metals expand by different amounts and the strip bends, as shown in Figure 19.9.

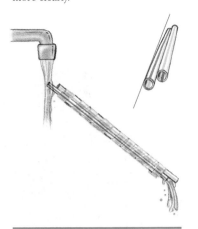

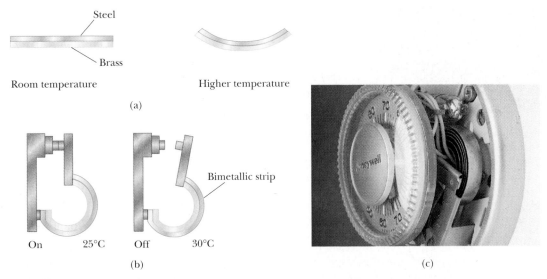

Figure 19.9 (a) A bimetallic strip bends as the temperature changes because the two metals have different expansion coefficients. (b) A bimetallic strip used in a thermostat to break or make electrical contact. (c) The interior of a thermostat, showing the coiled bimetallic strip. Why do you suppose the strip is coiled? *(c, Charles D. Winters)*

Quick Quiz 19.2

If you quickly plunge a room-temperature thermometer into very hot water, the mercury level will go *down* briefly before going up to a final reading. Why?

Quick Quiz 19.3

You are offered a prize for making the most sensitive glass thermometer using the materials in Table 19.2. Which glass and which working liquid would you choose?

EXAMPLE 19.3   **Expansion of a Railroad Track**

A steel railroad track has a length of 30.000 m when the temperature is 0.0°C. (a) What is its length when the temperature is 40.0°C?

Thermal expansion: The extreme temperature of a July day in Asbury Park, NJ, caused these railroad tracks to buckle and derail the train in the distance. *(AP/Wide World Photos)*

Solution   Making use of Table 19.2 and noting that the change in temperature is 40.0°C, we find that the increase in length is

$$\Delta L = \alpha L_i \Delta T = [11 \times 10^{-6} (°C)^{-1}](30.000 \text{ m})(40.0°C)$$
$$= 0.013 \text{ m}$$

If the track is 30.000 m long at 0.0°C, its length at 40.0°C is

30.013 m.

(b) Suppose that the ends of the rail are rigidly clamped at 0.0°C so that expansion is prevented. What is the thermal stress set up in the rail if its temperature is raised to 40.0°C?

Solution   From the definition of Young's modulus for a solid (see Eq. 12.6), we have

$$\text{Tensile stress} = \frac{F}{A} = Y \frac{\Delta L}{L_i}$$

Because Y for steel is $20 \times 10^{10} \text{ N/m}^2$ (see Table 12.1), we have

$$\frac{F}{A} = (20 \times 10^{10} \text{ N/m}^2)\left(\frac{0.013 \text{ m}}{30.000 \text{ m}}\right) = \boxed{8.7 \times 10^7 \text{ N/m}^2}$$

Exercise   If the rail has a cross-sectional area of 30.0 cm², what is the force of compression in the rail?

Answer $2.6 \times 10^5 \text{ N} = 58\,000 \text{ lb!}$

The Unusual Behavior of Water

Liquids generally increase in volume with increasing temperature and have average coefficients of volume expansion about ten times greater than those of solids. Water is an exception to this rule, as we can see from its density-versus-temperature curve shown in Figure 19.10. As the temperature increases from 0°C to 4°C, water contracts and thus its density increases. Above 4°C, water expands with increasing temperature, and so its density decreases. In other words, the density of water reaches a maximum value of 1 000 kg/m³ at 4°C.

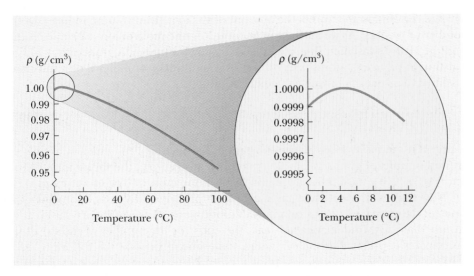

Figure 19.10 How the density of water at atmospheric pressure changes with temperature. The inset at the right shows that the maximum density of water occurs at 4°C.

We can use this unusual thermal-expansion behavior of water to explain why a pond begins freezing at the surface rather than at the bottom. When the atmospheric temperature drops from, for example, 7°C to 6°C, the surface water also cools and consequently decreases in volume. This means that the surface water is denser than the water below it, which has not cooled and decreased in volume. As a result, the surface water sinks, and warmer water from below is forced to the surface to be cooled. When the atmospheric temperature is between 4°C and 0°C, however, the surface water expands as it cools, becoming less dense than the water below it. The mixing process stops, and eventually the surface water freezes. As the water freezes, the ice remains on the surface because ice is less dense than water. The ice continues to build up at the surface, while water near the bottom remains at 4°C. If this were not the case, then fish and other forms of marine life would not survive.

19.5 MACROSCOPIC DESCRIPTION OF AN IDEAL GAS

In this section we examine the properties of a gas of mass m confined to a container of volume V at a pressure P and a temperature T. It is useful to know how these quantities are related. In general, the equation that interrelates these quantities, called the *equation of state,* is very complicated. However, if the gas is maintained at a very low pressure (or low density), the equation of state is quite simple and can be found experimentally. Such a low-density gas is commonly referred to as an *ideal gas.*[4]

---

[4] To be more specific, the assumption here is that the temperature of the gas must not be too low (the gas must not condense into a liquid) or too high, and that the pressure must be low. In reality, an ideal gas does not exist. However, the concept of an ideal gas is very useful in view of the fact that real gases at low pressures behave as ideal gases do. The concept of an ideal gas implies that the gas molecules do not interact except upon collision, and that the molecular volume is negligible compared with the volume of the container.

Figure 19.11   An ideal gas confined to a cylinder whose volume can be varied by means of a movable piston.

The universal gas constant

 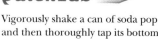
It is convenient to express the amount of gas in a given volume in terms of the number of moles n. As we learned in Section 1.3, one mole of any substance is that amount of the substance that contains Avogadro's number $N_A = 6.022 \times 10^{23}$ of constituent particles (atoms or molecules). The number of moles n of a substance is related to its mass m through the expression

$$n = \frac{m}{M} \tag{19.7}$$

where M is the molar mass of the substance (see Section 1.3), which is usually expressed in units of grams per mole (g/mol). For example, the molar mass of oxygen (O_2) is 32.0 g/mol. Therefore, the mass of one mole of oxygen is 32.0 g.

Now suppose that an ideal gas is confined to a cylindrical container whose volume can be varied by means of a movable piston, as shown in Figure 19.11. If we assume that the cylinder does not leak, the mass (or the number of moles) of the gas remains constant. For such a system, experiments provide the following information: First, when the gas is kept at a constant temperature, its pressure is inversely proportional to its volume (Boyle's law). Second, when the pressure of the gas is kept constant, its volume is directly proportional to its temperature (the law of Charles and Gay–Lussac). These observations are summarized by the **equation of state for an ideal gas:**

$$PV = nRT \tag{19.8}$$

In this expression, known as the **ideal gas law,** R is a universal constant that is the same for all gases and T is the absolute temperature in kelvins. Experiments on numerous gases show that as the pressure approaches zero, the quantity PV/nT approaches the same value R for all gases. For this reason, R is called the **universal gas constant.** In SI units, in which pressure is expressed in pascals (1 Pa = 1 N/m^2) and volume in cubic meters, the product PV has units of newton·meters, or joules, and R has the value

$$R = 8.315 \text{ J/mol·K} \tag{19.9}$$

If the pressure is expressed in atmospheres and the volume in liters (1 L = 10^3 cm^3 = 10^{-3} m^3), then R has the value

$$R = 0.082 \text{ 14 L·atm/mol·K}$$

Using this value of R and Equation 19.8, we find that the volume occupied by 1 mol of any gas at atmospheric pressure and at 0°C (273 K) is 22.4 L.

Now that we have presented the equation of state, we are ready for a formal definition of an ideal gas: **An ideal gas is one for which PV/nT is constant at all pressures.**

The ideal gas law states that if the volume and temperature of a fixed amount of gas do not change, then the pressure also remains constant. Consider the bottle of champagne shown at the beginning of this chapter. Because the temperature of the bottle and its contents remains constant, so does the pressure, as can be shown by replacing the cork with a pressure gauge. Shaking the bottle displaces some carbon dioxide gas from the "head space" to form bubbles within the liquid, and these bubbles become attached to the inside of the bottle. (No new gas is generated by shaking.) When the bottle is opened, the pressure is reduced; this causes the volume of the bubbles to increase suddenly. If the bubbles are attached to the bottle (beneath the liquid surface), their rapid expansion expels liquid from the bottle. If the sides and bottom of the bottle are first tapped until no bubbles remain beneath the surface, then when the champagne is opened, the drop in pressure will not force liquid from the bottle. Try the QuickLab, but practice before demonstrating to a friend!

The ideal gas law is often expressed in terms of the total number of molecules N. Because the total number of molecules equals the product of the number of moles n and Avogadro's number N_A, we can write Equation 19.8 as

$$PV = nRT = \frac{N}{N_A} RT$$

$$PV = Nk_B T \qquad\qquad \textbf{(19.10)}$$

where k_B is **Boltzmann's constant,** which has the value

$$k_B = \frac{R}{N_A} = 1.38 \times 10^{-23}\,\text{J/K} \qquad\qquad \textbf{(19.11)} \qquad \boxed{\text{Boltzmann's constant}}$$

It is common to call quantities such as P, V, and T the **thermodynamic variables** of an ideal gas. If the equation of state is known, then one of the variables can always be expressed as some function of the other two.

EXAMPLE 19.4 **How Many Gas Molecules in a Container?**

An ideal gas occupies a volume of 100 cm³ at 20°C and 100 Pa. Find the number of moles of gas in the container.

Solution The quantities given are volume, pressure, and temperature: $V = 100$ cm³ $= 1.00 \times 10^{-4}$ m³, $P = 100$ Pa, and $T = 20$°C $= 293$ K. Using Equation 19.8, we find that

$$n = \frac{PV}{RT} = \frac{(100\,\text{Pa})(10^{-4}\,\text{m}^3)}{(8.315\,\text{J/mol}\cdot\text{K})(293\,\text{K})} = \boxed{4.10 \times 10^{-6}\,\text{mol}}$$

Exercise How many molecules are in the container?

Answer 2.47×10^{18} molecules.

EXAMPLE 19.5 **Filling a Scuba Tank**

A certain scuba tank is designed to hold 66 ft³ of air when it is at atmospheric pressure at 22°C. When this volume of air is compressed to an absolute pressure of 3 000 lb/in.² and stored in a 10-L (0.35-ft³) tank, the air becomes so hot that the tank must be allowed to cool before it can be used. If the air does not cool, what is its temperature? (Assume that the air behaves like an ideal gas.)

Solution If no air escapes from the tank during filling, then the number of moles n remains constant; therefore, using $PV = nRT$, and with n and R being constant, we obtain for the initial and final values:

$$\frac{P_i V_i}{T_i} = \frac{P_f V_f}{T_f}$$

The initial pressure of the air is 14.7 lb/in.², its final pressure is 3 000 lb/in.², and the air is compressed from an initial volume of 66 ft³ to a final volume of 0.35 ft³. The initial temperature, converted to SI units, is 295 K. Solving for T_f, we obtain

$$T_f = \left(\frac{P_f V_f}{P_i V_i}\right) T_i = \frac{(3\,000\,\text{lb/in.}^2)(0.35\,\text{ft}^3)}{(14.7\,\text{lb/in.}^2)(66\,\text{ft}^3)}\,(295\,\text{K})$$

$$= \boxed{319\,\text{K}}$$

Exercise What is the air temperature in degrees Celsius and in degrees Fahrenheit?

Answer 45.9°C; 115°F.

Quick Quiz 19.4

In the previous example we used SI units for the temperature in our calculation step but not for the pressures or volumes. When working with the ideal gas law, how do you decide when it is necessary to use SI units and when it is not?

EXAMPLE 19.6   Heating a Spray Can

A spray can containing a propellant gas at twice atmospheric pressure (202 kPa) and having a volume of 125 cm^3 is at 22°C. It is then tossed into an open fire. When the temperature of the gas in the can reaches 195°C, what is the pressure inside the can? Assume any change in the volume of the can is negligible.

Solution   We employ the same approach we used in Example 19.5, starting with the expression

$$\frac{P_i V_i}{T_i} = \frac{P_f V_f}{T_f}$$

Because the initial and final volumes of the gas are assumed to be equal, this expression reduces to

$$\frac{P_i}{T_i} = \frac{P_f}{T_f}$$

Solving for P_f gives

$$P_f = \left(\frac{T_f}{T_i}\right)(P_i) = \left(\frac{468\ \text{K}}{295\ \text{K}}\right)(202\ \text{kPa}) = \boxed{320\ \text{kPa}}$$

Obviously, the higher the temperature, the higher the pressure exerted by the trapped gas. Of course, if the pressure increases high enough, the can will explode. Because of this possibility, you should never dispose of spray cans in a fire.

SUMMARY

Two bodies are in **thermal equilibrium** with each other if they have the same temperature.

The **zeroth law of thermodynamics** states that if objects A and B are separately in thermal equilibrium with a third object C, then objects A and B are in thermal equilibrium with each other.

The SI unit of absolute temperature is the **kelvin,** which is defined to be the fraction 1/273.16 of the temperature of the triple point of water.

When the temperature of an object is changed by an amount ΔT, its length changes by an amount ΔL that is proportional to ΔT and to its initial length L_i:

$$\Delta L = \alpha L_i \Delta T \tag{19.4}$$

where the constant α is the **average coefficient of linear expansion.** The **average volume expansion coefficient** β for a solid is approximately equal to 3α.

An **ideal gas** is one for which PV/nT is constant at all pressures. An ideal gas is described by the **equation of state,**

$$PV = nRT \tag{19.8}$$

where n equals the number of moles of the gas, V is its volume, R is the universal gas constant (8.315 J/mol·K), and T is the absolute temperature. A real gas behaves approximately as an ideal gas if it is far from liquefaction.

QUESTIONS

1. Is it possible for two objects to be in thermal equilibrium if they are not in contact with each other? Explain.

2. A piece of copper is dropped into a beaker of water. If the water's temperature increases, what happens to the temperature of the copper? Under what conditions are the water and copper in thermal equilibrium?

3. In principle, any gas can be used in a constant-volume gas thermometer. Why is it not possible to use oxygen for temperatures as low as 15 K? What gas would you use? (Refer to the data in Table 19.1.)

4. Rubber has a negative average coefficient of linear expansion. What happens to the size of a piece of rubber as it is warmed?

5. Why should the amalgam used in dental fillings have the same average coefficient of expansion as a tooth? What would occur if they were mismatched?

6. Explain why the thermal expansion of a spherical shell made of a homogeneous solid is equivalent to that of a solid sphere of the same material.

7. A steel ring bearing has an inside diameter that is 0.1 mm smaller than the diameter of an axle. How can it be made to fit onto the axle without removing any material?

8. Markings to indicate length are placed on a steel tape in a room that has a temperature of 22°C. Are measurements made with the tape on a day when the temperature is 27°C greater than, less than, or the same length as the object's length? Defend your answer.

9. Determine the number of grams in 1 mol of each of the following gases: (a) hydrogen, (b) helium, and (c) carbon monoxide.

10. An inflated rubber balloon filled with air is immersed in a flask of liquid nitrogen that is at 77 K. Describe what happens to the balloon, assuming that it remains flexible while being cooled.

11. Two identical cylinders at the same temperature each contain the same kind of gas and the same number of moles of gas. If the volume of cylinder A is three times greater than the volume of cylinder B, what can you say about the relative pressures in the cylinders?

12. The pendulum of a certain pendulum clock is made of brass. When the temperature increases, does the clock run too fast, run too slowly, or remain unchanged? Explain.

13. An automobile radiator is filled to the brim with water while the engine is cool. What happens to the water when the engine is running and the water is heated? What do modern automobiles have in their cooling systems to prevent the loss of coolants?

14. Metal lids on glass jars can often be loosened by running them under hot water. How is this possible?

15. When the metal ring and metal sphere shown in Figure Q19.15 are both at room temperature, the sphere can just be passed through the ring. After the sphere is heated, it cannot be passed through the ring. Explain.

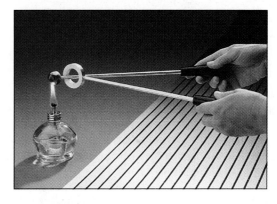

Figure Q19.15 *(Courtesy of Central Scientific Company)*

PROBLEMS

1, 2, 3 = straightforward, intermediate, challenging ☐ = full solution available in the *Student Solutions Manual and Study Guide*
WEB = solution posted at **http://www.saunderscollege.com/physics/** 🖥 = Computer useful in solving problem 🔧 = Interactive Physics
☐ = paired numerical/symbolic problems

Section 19.1 Temperature and the Zeroth Law of Thermodynamics
Section 19.2 Thermometers and the Celsius Temperature Scale
Section 19.3 The Constant-Volume Gas Thermometer and the Absolute Temperature Scale

Note: A pressure of 1 atm = 1.013×10^5 Pa = 101.3 kPa.

1. Convert the following to equivalent temperatures on the Celsius and Kelvin scales: (a) the normal human body temperature, 98.6°F; (b) the air temperature on a cold day, −5.00°F.

2. In a constant-volume gas thermometer, the pressure at 20.0°C is 0.980 atm. (a) What is the pressure at 45.0°C? (b) What is the temperature if the pressure is 0.500 atm?

WEB 3. A constant-volume gas thermometer is calibrated in dry ice (that is, carbon dioxide in the solid state, which has a temperature of − 80.0°C) and in boiling ethyl alcohol (78.0°C). The two pressures are 0.900 atm and

1.635 atm. (a) What Celsius value of absolute zero does the calibration yield? What is the pressure at (b) the freezing point of water and (c) the boiling point of water?

4. There is a temperature whose numerical value is the same on both the Celsius and Fahrenheit scales. What is this temperature?

5. Liquid nitrogen has a boiling point of − 195.81°C at atmospheric pressure. Express this temperature in (a) degrees Fahrenheit and (b) kelvins.

6. On a Strange temperature scale, the freezing point of water is − 15.0°S and the boiling point is + 60.0°S. Develop a *linear* conversion equation between this temperature scale and the Celsius scale.

7. The temperature difference between the inside and the outside of an automobile engine is 450°C. Express this temperature difference on the (a) Fahrenheit scale and (b) Kelvin scale.

8. The melting point of gold is 1 064°C , and the boiling point is 2 660°C. (a) Express these temperatures in kelvins. (b) Compute the difference between these temperatures in Celsius degrees and in kelvins.

Section 19.4 Thermal Expansion of Solids and Liquids

Note: When solving the problems in this section, use the data in Table 19.2.

9. A copper telephone wire has essentially no sag between poles 35.0 m apart on a winter day when the temperature is − 20.0°C. How much longer is the wire on a summer day when $T_C = 35.0$°C?

10. The concrete sections of a certain superhighway are designed to have a length of 25.0 m. The sections are poured and cured at 10.0°C. What minimum spacing should the engineer leave between the sections to eliminate buckling if the concrete is to reach a temperature of 50.0°C?

11. An aluminum tube is 3.000 0 m long at 20.0°C. What is its length at (a) 100.0°C and (b) 0.0°C?

12. A brass ring with a diameter of 10.00 cm at 20.0°C is heated and slipped over an aluminum rod with a diameter of 10.01 cm at 20.0°C. Assume that the average coefficients of linear expansion are constant. (a) To what temperature must this combination be cooled to separate them? Is this temperature attainable? (b) If the aluminum rod were 10.02 cm in diameter, what would be the required temperature?

13. A pair of eyeglass frames is made of epoxy plastic. At room temperature (20.0°C), the frames have circular lens holes 2.20 cm in radius. To what temperature must the frames be heated if lenses 2.21 cm in radius are to be inserted in them? The average coefficient of linear expansion for epoxy is 1.30×10^{-4} (°C)$^{-1}$.

14. The New River Gorge bridge in West Virginia is a steel arch bridge 518 m in length. How much does its length change between temperature extremes of − 20.0°C and 35.0°C?

15. A square hole measuring 8.00 cm along each side is cut in a sheet of copper. (a) Calculate the change in the area of this hole if the temperature of the sheet is increased by 50.0 K. (b) Does the result represent an increase or a decrease in the area of the hole?

16. The average coefficient of volume expansion for carbon tetrachloride is 5.81×10^{-4} (°C)$^{-1}$. If a 50.0-gal steel container is filled completely with carbon tetrachloride when the temperature is 10.0°C, how much will spill over when the temperature rises to 30.0°C?

WEB 17. The active element of a certain laser is a glass rod 30.0 cm long by 1.50 cm in diameter. If the temperature of the rod increases by 65.0°C, what is the increase in (a) its length, (b) its diameter, and (c) its volume? (Assume that $\alpha = 9.00 \times 10^{-6}$ (°C)$^{-1}$.)

18. A volumetric glass flask made of Pyrex is calibrated at 20.0°C. It is filled to the 100-mL mark with 35.0°C acetone with which it immediately comes to thermal equilibrium. (a) What is the volume of the acetone when it cools to 20.0°C? (b) How significant is the change in volume of the flask?

19. A concrete walk is poured on a day when the temperature is 20.0°C, in such a way that the ends are unable to move. (a) What is the stress in the cement on a hot day of 50.0°C? (b) Does the concrete fracture? Take Young's modulus for concrete to be 7.00×10^9 N/m^2 and the tensile strength to be 2.00×10^9 N/m^2.

20. Figure P19.20 shows a circular steel casting with a gap. If the casting is heated, (a) does the width of the gap increase or decrease? (b) The gap width is 1.600 cm when the temperature is 30.0°C. Determine the gap width when the temperature is 190°C.

Figure P19.20

21. A steel rod undergoes a stretching force of 500 N. Its cross-sectional area is 2.00 cm^2. Find the change in temperature that would elongate the rod by the same amount that the 500-N force does. (*Hint:* Refer to Tables 12.1 and 19.2.)

22. A steel rod 4.00 cm in diameter is heated so that its temperature increases by 70.0°C. It is then fastened between two rigid supports. The rod is allowed to cool to its original temperature. Assuming that Young's modulus for the steel is 20.6×10^{10} N/m^2 and that its average

coefficient of linear expansion is 11.0×10^{-6} (°C)$^{-1}$, calculate the tension in the rod.

23. A hollow aluminum cylinder 20.0 cm deep has an internal capacity of 2.000 L at 20.0°C. It is completely filled with turpentine and then warmed to 80.0°C. (a) How much turpentine overflows? (b) If the cylinder is then cooled back to 20.0°C, how far below the surface of the cylinder's rim does the turpentine's surface recede?

24. At 20.0°C, an aluminum ring has an inner diameter of 5.000 0 cm and a brass rod has a diameter of 5.050 0 cm. (a) To what temperature must the ring be heated so that it will just slip over the rod? (b) To what common temperature must the two be heated so that the ring just slips over the rod? Would this latter process work?

Section 19.5 Macroscopic Description of an Ideal Gas

25. Gas is contained in an 8.00-L vessel at a temperature of 20.0°C and a pressure of 9.00 atm. (a) Determine the number of moles of gas in the vessel. (b) How many molecules of gas are in the vessel?

26. A tank having a volume of 0.100 m^3 contains helium gas at 150 atm. How many balloons can the tank blow up if each filled balloon is a sphere 0.300 m in diameter at an absolute pressure of 1.20 atm?

27. An auditorium has dimensions 10.0 m × 20.0 m × 30.0 m. How many molecules of air fill the auditorium at 20.0°C and a pressure of 101 kPa?

28. Nine grams of water are placed in a 2.00-L pressure cooker and heated to 500°C. What is the pressure inside the container if no gas escapes?

WEB 29. The mass of a hot-air balloon and its cargo (not including the air inside) is 200 kg. The air outside is at 10.0°C and 101 kPa. The volume of the balloon is 400 m^3. To what temperature must the air in the balloon be heated before the balloon will lift off? (Air density at 10.0°C is 1.25 kg/m^3.)

30. One mole of oxygen gas is at a pressure of 6.00 atm and a temperature of 27.0°C. (a) If the gas is heated at constant volume until the pressure triples, what is the final temperature? (b) If the gas is heated until both the pressure and the volume are doubled, what is the final temperature?

31. (a) Find the number of moles in 1.00 m^3 of an ideal gas at 20.0°C and atmospheric pressure. (b) For air, Avogadro's number of molecules has a mass of 28.9 g. Calculate the mass of 1 m^3 of air. Compare the result with the tabulated density of air.

32. A cube 10.0 cm on each edge contains air (with equivalent molar mass 28.9 g/mol) at atmospheric pressure and temperature 300 K. Find (a) the mass of the gas, (b) its weight, and (c) the force it exerts on each face of the cube. (d) Comment on the underlying physical reason why such a small sample can exert such a great force.

33. An automobile tire is inflated with air originally at 10.0°C and normal atmospheric pressure. During the process, the air is compressed to 28.0% of its original volume and its temperature is increased to 40.0°C. (a) What is the tire pressure? (b) After the car is driven at high speed, the tire air temperature rises to 85.0°C and the interior volume of the tire increases by 2.00%. What is the new tire pressure (absolute) in pascals?

34. A spherical weather balloon is designed to expand to a maximum radius of 20.0 m when in flight at its working altitude, where the air pressure is 0.030 0 atm and the temperature is 200 K. If the balloon is filled at atmospheric pressure and 300 K, what is its radius at liftoff?

35. A room of volume 80.0 m^3 contains air having an equivalent molar mass of 28.9 g/mol. If the temperature of the room is raised from 18.0°C to 25.0°C, what mass of air (in kilograms) will leave the room? Assume that the air pressure in the room is maintained at 101 kPa.

36. A room of volume V contains air having equivalent molar mass M (in g/mol). If the temperature of the room is raised from T_1 to T_2, what mass of air will leave the room? Assume that the air pressure in the room is maintained at P_0.

37. At 25.0 m below the surface of the sea (density = 1 025 kg/m^3), where the temperature is 5.00°C, a diver exhales an air bubble having a volume of 1.00 cm^3. If the surface temperature of the sea is 20.0°C, what is the volume of the bubble right before it breaks the surface?

38. Estimate the mass of the air in your bedroom. State the quantities you take as data and the value you measure or estimate for each.

39. The pressure gauge on a tank registers the gauge pressure, which is the difference between the interior and exterior pressures. When the tank is full of oxygen (O_2), it contains 12.0 kg of the gas at a gauge pressure of 40.0 atm. Determine the mass of oxygen that has been withdrawn from the tank when the pressure reading is 25.0 atm. Assume that the temperature of the tank remains constant.

40. In state-of-the-art vacuum systems, pressures as low as 10^{-9} Pa are being attained. Calculate the number of molecules in a 1.00-m^3 vessel at this pressure if the temperature is 27°C.

41. Show that 1 mol of any gas (assumed to be ideal) at atmospheric pressure (101.3 kPa) and standard temperature (273 K) occupies a volume of 22.4 L.

42. A diving bell in the shape of a cylinder with a height of 2.50 m is closed at the upper end and open at the lower end. The bell is lowered from air into sea water (ρ = 1.025 g/cm^3). The air in the bell is initially at 20.0°C. The bell is lowered to a depth (measured to the bottom of the bell) of 45.0 fathoms, or 82.3 m. At this depth, the water temperature is 4.0°C, and the air in the bell is in thermal equilibrium with the water. (a) How high does sea water rise in the bell? (b) To what minimum pressure must the air in the bell be increased for the water that entered to be expelled?

ADDITIONAL PROBLEMS

43. A student measures the length of a brass rod with a steel tape at 20.0°C. The reading is 95.00 cm. What will the tape indicate for the length of the rod when the rod and the tape are at (a) $-15.0°C$ and (b) $55.0°C$?

44. The density of gasoline is 730 kg/m³ at 0°C. Its average coefficient of volume expansion is 9.60×10^{-4} (°C)$^{-1}$. If 1.00 gal of gasoline occupies 0.003 80 m³, how many extra kilograms of gasoline would you get if you bought 10.0 gal of gasoline at 0°C rather than at 20.0°C from a pump that is not temperature compensated?

45. A steel ball bearing is 4.000 cm in diameter at 20.0°C. A bronze plate has a hole in it that is 3.994 cm in diameter at 20.0°C. What common temperature must they have so that the ball just squeezes through the hole?

46. **Review Problem.** An aluminum pipe 0.655 m long at 20.0°C and open at both ends is used as a flute. The pipe is cooled to a low temperature but is then filled with air at 20.0°C as soon as it is played. By how much does its fundamental frequency change as the temperature of the metal increases from 5.00°C to 20.0°C?

47. A mercury thermometer is constructed as shown in Figure P19.47. The capillary tube has a diameter of 0.004 00 cm, and the bulb has a diameter of 0.250 cm. Neglecting the expansion of the glass, find the change in height of the mercury column that occurs with a temperature change of 30.0°C.

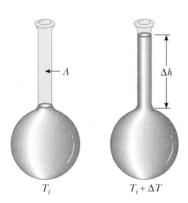

Figure P19.47   Problems 47 and 48.

48. A liquid with a coefficient of volume expansion β just fills a spherical shell of volume V_i at a temperature of T_i (see Fig. P19.47). The shell is made of a material that has an average coefficient of linear expansion α. The liquid is free to expand into an open capillary of area A projecting from the top of the sphere. (a) If the temperature increases by ΔT, show that the liquid rises in the capillary by the amount Δh given by the equation $\Delta h = (V_i/A)(\beta - 3\alpha)\,\Delta T$. (b) For a typical system, such as a mercury thermometer, why is it a good approximation to neglect the expansion of the shell?

WEB 49. A liquid has a density ρ. (a) Show that the fractional change in density for a change in temperature ΔT is $\Delta\rho/\rho = -\beta\,\Delta T$. What does the negative sign signify? (b) Fresh water has a maximum density of 1.000 0 g/cm³ at 4.0°C. At 10.0°C, its density is 0.999 7 g/cm³. What is β for water over this temperature interval?

50. A cylinder is closed by a piston connected to a spring of constant 2.00×10^3 N/m (Fig. P19.50). While the spring is relaxed, the cylinder is filled with 5.00 L of gas at a pressure of 1.00 atm and a temperature of 20.0°C. (a) If the piston has a cross-sectional area of 0.010 0 m² and a negligible mass, how high will it rise when the temperature is increased to 250°C? (b) What is the pressure of the gas at 250°C?

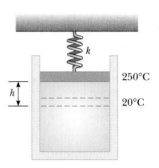

Figure P19.50

WEB 51. A vertical cylinder of cross-sectional area A is fitted with a tight-fitting, frictionless piston of mass m (Fig. P19.51). (a) If n moles of an ideal gas are in the cylinder at a temperature of T, what is the height h at which the piston is in equilibrium under its own weight? (b) What is the value for h if $n = 0.200$ mol, $T = 400$ K, $A = 0.008\ 00$ m², and $m = 20.0$ kg?

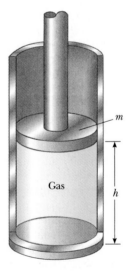

Figure P19.51

52. A bimetallic bar is made of two thin strips of dissimilar metals bonded together. As they are heated, the one with the greater average coefficient of expansion expands more than the other, forcing the bar into an arc, with the outer radius having a greater circumference (Fig. P19.52). (a) Derive an expression for the angle of bending θ as a function of the initial length of the strips, their average coefficients of linear expansion, the change in temperature, and the separation of the centers of the strips ($\Delta r = r_2 - r_1$). (b) Show that the angle of bending decreases to zero when ΔT decreases to zero or when the two average coefficients of expansion become equal. (c) What happens if the bar is cooled?

Figure P19.52

53. The rectangular plate shown in Figure P19.53 has an area A_i equal to ℓw. If the temperature increases by ΔT, show that the increase in area is $\Delta A = 2\alpha A_i \Delta T$, where α is the average coefficient of linear expansion. What approximation does this expression assume? (*Hint:* Note that each dimension increases according to the equation $\Delta L = \alpha L_i \Delta T$.)

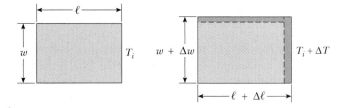

Figure P19.53

54. Precise temperature measurements are often made on the basis of the change in electrical resistance of a metal with temperature. The resistance varies approximately according to the expression $R = R_0(1 + A T_C)$, where R_0 and A are constants. A certain element has a resistance of 50.0 ohms (Ω) at 0°C and 71.5 Ω at the freezing point of tin (231.97°C). (a) Determine the constants A and R_0. (b) At what temperature is the resistance equal to 89.0 Ω?

55. **Review Problem.**   A clock with a brass pendulum has a period of 1.000 s at 20.0°C. If the temperature increases to 30.0°C, (a) by how much does the period change, and (b) how much time does the clock gain or lose in one week?

56. **Review Problem.**   Consider an object with any one of the shapes displayed in Table 10.2. What is the percentage increase in the moment of inertia of the object when it is heated from 0°C to 100°C if it is composed of (a) copper or (b) aluminum? (See Table 19.2. Assume that the average linear expansion coefficients do not vary between 0°C and 100°C.)

57. **Review Problem.**   (a) Derive an expression for the buoyant force on a spherical balloon that is submerged in water as a function of the depth below the surface, the volume (V_i) of the balloon at the surface, the pressure (P_0) at the surface, and the density of the water. (Assume that water temperature does not change with depth.) (b) Does the buoyant force increase or decrease as the balloon is submerged? (c) At what depth is the buoyant force one-half the surface value?

58. (a) Show that the density of an ideal gas occupying a volume V is given by $\rho = PM/RT$, where M is the molar mass. (b) Determine the density of oxygen gas at atmospheric pressure and 20.0°C.

59. Starting with Equation 19.10, show that the total pressure P in a container filled with a mixture of several ideal gases is $P = P_1 + P_2 + P_3 + \ldots$, where P_1, $P_2, \ldots$ are the pressures that each gas would exert if it alone filled the container. (These individual pressures are called the *partial pressures* of the respective gases.) This is known as *Dalton's law of partial pressures.*

60. A sample of dry air that has a mass of 100.00 g, collected at sea level, is analyzed and found to consist of the following gases:

$$\text{nitrogen (N}_2) = 75.52 \text{ g}$$
$$\text{oxygen (O}_2) = 23.15 \text{ g}$$
$$\text{argon (Ar)} = 1.28 \text{ g}$$
$$\text{carbon dioxide (CO}_2) = 0.05 \text{ g}$$

as well as trace amounts of neon, helium, methane, and other gases. (a) Calculate the partial pressure (see Problem 59) of each gas when the pressure is 101.3 kPa. (b) Determine the volume occupied by the 100-g sample at a temperature of 15.00°C and a pressure of 1.013×10^5 Pa. What is the density of the air for these conditions? (c) What is the effective molar mass of the air sample?

61. Steel rails for an interurban rapid transit system form a continuous track that is held rigidly in place in concrete. (a) If the track was laid when the temperature was 0°C, what is the stress in the rails on a warm day when the temperature is 25.0°C? (b) What fraction of the yield strength of 52.2×10^7 N/m² does this stress represent?

62. (a) Use the equation of state for an ideal gas and the definition of the average coefficient of volume expansion, in the form $\beta = (1/V)\,dV/dT$, to show that the average coefficient of volume expansion for an ideal gas at constant pressure is given by $\beta = 1/T$, where T is the absolute temperature. (b) What value does this expression predict for β at 0°C? Compare this with the experimental values for helium and air in Table 19.2.

63. Two concrete spans of a 250-m-long bridge are placed end to end so that no room is allowed for expansion (Fig. P19.63a). If a temperature increase of 20.0°C occurs, what is the height y to which the spans rise when they buckle (Fig. P19.63b)?

64. Two concrete spans of a bridge of length L are placed end to end so that no room is allowed for expansion (see Fig. P19.63a). If a temperature increase of ΔT occurs, what is the height y to which the spans rise when they buckle (see Fig. P19.63b)?

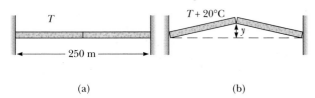

(a)   (b)

Figure P19.63   Problems 63 and 64.

65. A copper rod and a steel rod are heated. At 0°C the copper rod has length L_c, and the steel rod has length L_s. When the rods are being heated or cooled, the difference between their lengths stays constant at 5.00 cm. Determine the values of L_c and L_s.

66. A cylinder that has a 40.0-cm radius and is 50.0 cm deep is filled with air at 20.0°C and 1.00 atm (Fig. P19.66a). A 20.0-kg piston is now lowered into the cylinder, compressing the air trapped inside (Fig. P19.66b). Finally, a 75.0-kg man stands on the piston, further compressing the air, which remains at 20°C (Fig. P19.66c). (a) How far down (Δh) does the piston move when the man steps onto it? (b) To what temperature should the gas be heated to raise the piston and the man back to h_i?

67. The relationship $L_f = L_i(1 + \alpha\Delta T)$ is an approximation that works when the average coefficient of expansion is small. If α is large, one must integrate the relationship $dL/dT = \alpha L$ to determine the final length. (a) Assuming that the average coefficient of linear expansion is constant as L varies, determine a general expression for the final length. (b) Given a rod of length 1.00 m and a temperature change of 100.0 °C, determine the error caused by the approximation when $\alpha = 2.00 \times 10^{-5}\ (°C)^{-1}$ (a typical value for a metal) and

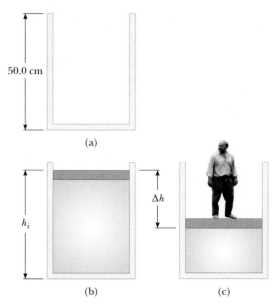

(a)

(b)   (c)

Figure P19.66

when $\alpha = 0.0200\ (°C)^{-1}$ (an unrealistically large value for comparison).

68. A steel wire and a copper wire, each of diameter 2.000 mm, are joined end to end. At 40.0°C, each has an unstretched length of 2.000 m; they are connected between two fixed supports 4.000 m apart on a tabletop, so that the steel wire extends from $x = -2.000$ m to $x = 0$, the copper wire extends from $x = 0$ to $x = 2.000$ m, and the tension is negligible. The temperature is then lowered to 20.0°C. At this lower temperature, what are the tension in the wire and the x coordinate of the junction between the wires? (Refer to Tables 12.1 and 19.2.)

69. **Review Problem.**   A steel guitar string with a diameter of 1.00 mm is stretched between supports 80.0 cm apart. The temperature is 0.0°C. (a) Find the mass per unit length of this string. (Use 7.86×10^3 kg/m³ as the mass density.) (b) The fundamental frequency of transverse oscillations of the string is 200 Hz. What is the tension in the string? (c) If the temperature is raised to 30.0°C, find the resulting values of the tension and the fundamental frequency. (Assume that both the Young's modulus [Table 12.1] and the average coefficient of linear expansion [Table 19.2] have constant values between 0.0°C and 30.0°C.)

70. A 1.00-km steel railroad rail is fastened securely at both ends when the temperature is 20.0°C. As the temperature increases, the rail begins to buckle. If its shape is an arc of a vertical circle, find the height h of the center of the buckle when the temperature is 25.0°C. (You will need to solve a transcendental equation.)

ANSWERS TO QUICK QUIZZES

19.1 The size of a degree on the Fahrenheit scale is $\frac{5}{9}$ the size of a degree on the Celsius scale. This is true because the Fahrenheit range of 32°F to 212°F is equivalent to the Celsius range of 0°C to 100°C. The factor $\frac{9}{5}$ in Equation 19.2 corrects for this difference. Equation 19.1 does not need this correction because the size of a Celsius degree is the same as the size of a kelvin.

19.2 The glass bulb containing most of the mercury warms up first because it is in direct thermal contact with the hot water. It expands slightly, and thus its volume increases. This causes the mercury level in the capillary tube to drop. As the mercury inside the bulb warms up, it also expands. Eventually, its increase in volume is sufficient to raise the mercury level in the capillary tube.

19.3 For the glass, choose Pyrex, which has a lower average coefficient of linear expansion than does ordinary glass. For the working liquid, choose gasoline, which has the largest average coefficient of volume expansion.

19.4 You do not have to convert the units for pressure and volume to SI units as long as the same units appear in the numerator and the denominator. This is not true for ratios of temperature units, as you can see by comparing the ratios 300 K/200 K and 26.85°C/(−73.15°C). You must always use absolute (kelvin) temperatures when applying the ideal gas law.

c h a p t e r

20

Heat and the First Law of Thermodynamics

U ntil about 1850, the fields of thermodynamics and mechanics were considered two distinct branches of science, and the law of conservation of energy seemed to describe only certain kinds of mechanical systems. However, mid–19th century experiments performed by the Englishman James Joule and others showed that energy may be added to (or removed from) a system either by heat or by doing work on the system (or having the system do work). Today we know that internal energy, which we formally define in this chapter, can be transformed to mechanical energy. Once the concept of energy was broadened to include internal energy, the law of conservation of energy emerged as a universal law of nature.

This chapter focuses on the concept of internal energy, the processes by which energy is transferred, the first law of thermodynamics, and some of the important applications of the first law. The first law of thermodynamics is the law of conservation of energy. It describes systems in which the only energy change is that of internal energy, which is due to transfers of energy by heat or work. Furthermore, the first law makes no distinction between the results of heat and the results of work. According to the first law, a system's internal energy can be changed either by an energy transfer by heat to or from the system or by work done on or by the system.

James Prescott Joule **British physicist (1818–1889)** Joule received some formal education in mathematics, philosophy, and chemistry but was in large part self-educated. His research led to the establishment of the principle of conservation of energy. His study of the quantitative relationship among electrical, mechanical, and chemical effects of heat culminated in his discovery in 1843 of the amount of work required to produce a unit of energy, called the mechanical equivalent of heat. *(By kind permission of the President and Council of the Royal Society)*

20.1 HEAT AND INTERNAL ENERGY

10.3 At the outset, it is important that we make a major distinction between internal energy and heat. **Internal energy is all the energy of a system that is associated with its microscopic components—atoms and molecules—when viewed from a reference frame at rest with respect to the object.** The last part of this sentence ensures that any bulk kinetic energy of the system due to its motion through space is not included in internal energy. Internal energy includes kinetic energy of translation, rotation, and vibration of molecules, potential energy within molecules, and potential energy between molecules. It is useful to relate internal energy to the temperature of an object, but this relationship is limited—we shall find in Section 20.3 that internal energy changes can also occur in the absence of temperature changes.

As we shall see in Chapter 21, the internal energy of a monatomic ideal gas is associated with the translational motion of its atoms. This is the only type of energy available for the microscopic components of this system. In this special case, the internal energy is simply the total kinetic energy of the atoms of the gas; the higher the temperature of the gas, the greater the average kinetic energy of the atoms and the greater the internal energy of the gas. More generally, in solids, liquids, and molecular gases, internal energy includes other forms of molecular energy. For example, a diatomic molecule can have rotational kinetic energy, as well as vibrational kinetic and potential energy.

Heat is defined as the transfer of energy across the boundary of a system due to a temperature difference between the system and its surroundings. When you *heat* a substance, you are transferring energy into it by placing it in contact with surroundings that have a higher temperature. This is the case, for example, when you place a pan of cold water on a stove burner—the burner is at a higher temperature than the water, and so the water gains energy. We shall also use the term *heat* to represent the amount of energy transferred by this method.

Scientists used to think of heat as a fluid called *caloric,* which they believed was transferred between objects; thus, they defined heat in terms of the temperature changes produced in an object during heating. Today we recognize the distinct difference between internal energy and heat. Nevertheless, we refer to quantities

Heat

using names that do not quite correctly define the quantities but which have become entrenched in physics tradition based on these early ideas. Examples of such quantities are *latent heat* and *heat capacity.*

As an analogy to the distinction between heat and internal energy, consider the distinction between work and mechanical energy discussed in Chapter 7. The work done on a system is a measure of the amount of energy transferred to the system from its surroundings, whereas the mechanical energy of the system (kinetic or potential, or both) is a consequence of the motion and relative positions of the members of the system. Thus, when a person does work on a system, energy is transferred from the person to the system. It makes no sense to talk about the work *of* a system—one can refer only to the work done *on* or *by* a system when some process has occurred in which energy has been transferred to or from the system. Likewise, it makes no sense to talk about the heat *of* a system—one can refer to *heat* only when energy has been transferred as a result of a temperature difference. Both heat and work are ways of changing the energy of a system.

It is also important to recognize that the internal energy of a system can be changed even when no energy is transferred by heat. For example, when a gas is compressed by a piston, the gas is warmed and its internal energy increases, but no transfer of energy by heat from the surroundings to the gas has occurred. If the gas then expands rapidly, it cools and its internal energy decreases, but no transfer of energy by heat from it to the surroundings has taken place. The temperature changes in the gas are due not to a difference in temperature between the gas and its surroundings but rather to the compression and the expansion. In each case, energy is transferred to or from the gas by *work,* and the energy change within the system is an increase or decrease of internal energy. The changes in internal energy in these examples are evidenced by corresponding changes in the temperature of the gas.

Units of Heat

The calorie

As we have mentioned, early studies of heat focused on the resultant increase in temperature of a substance, which was often water. The early notions of heat based on caloric suggested that the flow of this fluid from one body to another caused changes in temperature. From the name of this mythical fluid, we have an energy unit related to thermal processes, the **calorie (cal),** which is defined as **the amount of energy transfer necessary to raise the temperature of 1 g of water from 14.5°C to 15.5°C.**[1] (Note that the "Calorie," written with a capital "C" and used in describing the energy content of foods, is actually a kilocalorie.) The unit of energy in the British system is the **British thermal unit (Btu),** which is defined as **the amount of energy transfer required to raise the temperature of 1 lb of water from 63°F to 64°F.**

Scientists are increasingly using the SI unit of energy, the *joule,* when describing thermal processes. In this textbook, heat and internal energy are usually measured in joules. (Note that both heat and work are measured in energy units. Do not confuse these two means of energy *transfer* with energy itself, which is also measured in joules.)

---

[1] Originally, the calorie was defined as the "heat" necessary to raise the temperature of 1 g of water by 1°C. However, careful measurements showed that the amount of energy required to produce a 1°C change depends somewhat on the initial temperature; hence, a more precise definition evolved.

The Mechanical Equivalent of Heat

In Chapters 7 and 8, we found that whenever friction is present in a mechanical system, some mechanical energy is lost—in other words, mechanical energy is not conserved in the presence of nonconservative forces. Various experiments show that this lost mechanical energy does not simply disappear but is transformed into internal energy. We can perform such an experiment at home by simply hammering a nail into a scrap piece of wood. What happens to all the kinetic energy of the hammer once we have finished? Some of it is now in the nail as internal energy, as demonstrated by the fact that the nail is measurably warmer. Although this connection between mechanical and internal energy was first suggested by Benjamin Thompson, it was Joule who established the equivalence of these two forms of energy.

A schematic diagram of Joule's most famous experiment is shown in Figure 20.1. The system of interest is the water in a thermally insulated container. Work is done on the water by a rotating paddle wheel, which is driven by heavy blocks falling at a constant speed. The stirred water is warmed due to the friction between it and the paddles. If the energy lost in the bearings and through the walls is neglected, then the loss in potential energy associated with the blocks equals the work done by the paddle wheel on the water. If the two blocks fall through a distance h, the loss in potential energy is $2mgh$, where m is the mass of one block; it is this energy that causes the temperature of the water to increase. By varying the conditions of the experiment, Joule found that the loss in mechanical energy $2mgh$ is proportional to the increase in water temperature ΔT. The proportionality constant was found to be approximately 4.18 J/g·°C. Hence, 4.18 J of mechanical energy raises the temperature of 1 g of water by 1°C. More precise measurements taken later demonstrated the proportionality to be 4.186 J/g·°C when the temperature of the water was raised from 14.5°C to 15.5°C. We adopt this "15-degree calorie" value:

$$1 \text{ cal} \equiv 4.186 \text{ J} \qquad \textbf{(20.1)}$$

This equality is known, for purely historical reasons, as the **mechanical equivalent of heat.**

Benjamin Thompson (1753–1814). *(North Wind Picture Archives)*

Mechanical equivalent of heat

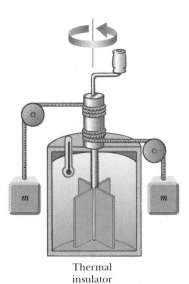

Thermal insulator

Figure 20.1 Joule's experiment for determining the mechanical equivalent of heat. The falling blocks rotate the paddles, causing the temperature of the water to increase.

EXAMPLE 20.1   **Losing Weight the Hard Way**

A student eats a dinner rated at 2 000 Calories. He wishes to do an equivalent amount of work in the gymnasium by lifting a 50.0-kg barbell. How many times must he raise the barbell to expend this much energy? Assume that he raises the barbell 2.00 m each time he lifts it and that he regains no energy when he drops the barbell to the floor.

Solution   Because 1 Calorie = 1.00×10^3 cal, the work required is 2.00×10^6 cal. Converting this value to joules, we have for the total work required:

$$W = (2.00 \times 10^6 \text{ cal})(4.186 \text{ J/cal}) = 8.37 \times 10^6 \text{ J}$$

The work done in lifting the barbell a distance h is equal to mgh, and the work done in lifting it n times is $nmgh$. We equate this to the total work required:

$$W = nmgh = 8.37 \times 10^6 \text{ J}$$

$$n = \frac{8.37 \times 10^6 \text{ J}}{(50.0 \text{ kg})(9.80 \text{ m/s}^2)(2.00 \text{ m})} = \boxed{8.54 \times 10^3 \text{ times}}$$

If the student is in good shape and lifts the barbell once every 5 s, it will take him about 12 h to perform this feat. Clearly, it is much easier for this student to lose weight by dieting.

20.2   HEAT CAPACITY AND SPECIFIC HEAT

When energy is added to a substance and no work is done, the temperature of the substance usually rises. (An exception to this statement is the case in which a substance undergoes a change of state—also called a *phase transition*—as discussed in the next section.) The quantity of energy required to raise the temperature of a given mass of a substance by some amount varies from one substance to another. For example, the quantity of energy required to raise the temperature of 1 kg of water by 1°C is 4 186 J, but the quantity of energy required to raise the temperature of 1 kg of copper by 1°C is only 387 J. In the discussion that follows, we shall use heat as our example of energy transfer, but we shall keep in mind that we could change the temperature of our system by doing work on it.

The **heat capacity** C of a particular sample of a substance is defined as the amount of energy needed to raise the temperature of that sample by 1°C. From this definition, we see that if heat Q produces a change ΔT in the temperature of a substance, then

Heat capacity

$$Q = C\Delta T \qquad \text{(20.2)}$$

The **specific heat** c of a substance is the heat capacity per unit mass. Thus, if energy Q transferred by heat to mass m of a substance changes the temperature of the sample by ΔT, then the specific heat of the substance is

Specific heat

$$c \equiv \frac{Q}{m\Delta T} \qquad \text{(20.3)}$$

Specific heat is essentially a measure of how thermally insensitive a substance is to the addition of energy. The greater a material's specific heat, the more energy must be added to a given mass of the material to cause a particular temperature change. Table 20.1 lists representative specific heats.

From this definition, we can express the energy Q transferred by heat between a sample of mass m of a material and its surroundings for a temperature change ΔT as

$$Q = mc\Delta T \qquad \text{(20.4)}$$

For example, the energy required to raise the temperature of 0.500 kg of water by 3.00°C is $(0.500 \text{ kg})(4\,186 \text{ J/kg} \cdot {}^\circ\text{C})(3.00^\circ\text{C}) = 6.28 \times 10^3$ J. Note that when the temperature increases, Q and ΔT are taken to be positive, and energy flows into

TABLE 20.1 Specific Heats of Some Substances at 25°C and Atmospheric Pressure

Substance	Specific Heat c	
	J/kg·°C	cal/g·°C
Elemental Solids		
Aluminum	900	0.215
Beryllium	1 830	0.436
Cadmium	230	0.055
Copper	387	0.092 4
Germanium	322	0.077
Gold	129	0.030 8
Iron	448	0.107
Lead	128	0.030 5
Silicon	703	0.168
Silver	234	0.056
Other Solids		
Brass	380	0.092
Glass	837	0.200
Ice (−5°C)	2 090	0.50
Marble	860	0.21
Wood	1 700	0.41
Liquids		
Alcohol (ethyl)	2 400	0.58
Mercury	140	0.033
Water (15°C)	4 186	1.00
Gas		
Steam (100°C)	2 010	0.48

the system. When the temperature decreases, Q and ΔT are negative, and energy flows out of the system.

Specific heat varies with temperature. However, if temperature intervals are not too great, the temperature variation can be ignored and c can be treated as a constant.[2] For example, the specific heat of water varies by only about 1% from 0°C to 100°C at atmospheric pressure. Unless stated otherwise, we shall neglect such variations.

Measured values of specific heats are found to depend on the conditions of the experiment. In general, measurements made at constant pressure are different from those made at constant volume. For solids and liquids, the difference between the two values is usually no greater than a few percent and is often neglected. Most of the values given in Table 20.1 were measured at atmospheric pressure and room temperature. As we shall see in Chapter 21, the specific heats for

---

[2] The definition given by Equation 20.3 assumes that the specific heat does not vary with temperature over the interval $\Delta T = T_f - T_i$. In general, if c varies with temperature over the interval, then the correct expression for Q is

$$Q = m \int_{T_i}^{T_f} c\, dT$$

gases measured at constant pressure are quite different from values measured at constant volume.

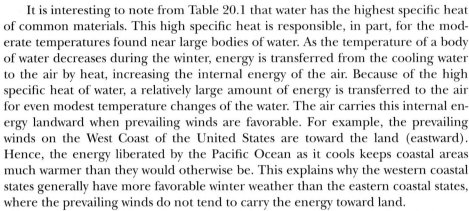

Quick Quiz 20.1

Imagine you have 1 kg each of iron, glass, and water, and that all three samples are at 10°C. (a) Rank the samples from lowest to highest temperature after 100 J of energy is added to each. (b) Rank them from least to greatest amount of energy transferred by heat if each increases in temperature by 20°C.

QuickLab

In an open area, such as a parking lot, use the flame from a match to pop an air-filled balloon. Now try the same thing with a water-filled balloon. Why doesn't the water-filled balloon pop?

It is interesting to note from Table 20.1 that water has the highest specific heat of common materials. This high specific heat is responsible, in part, for the moderate temperatures found near large bodies of water. As the temperature of a body of water decreases during the winter, energy is transferred from the cooling water to the air by heat, increasing the internal energy of the air. Because of the high specific heat of water, a relatively large amount of energy is transferred to the air for even modest temperature changes of the water. The air carries this internal energy landward when prevailing winds are favorable. For example, the prevailing winds on the West Coast of the United States are toward the land (eastward). Hence, the energy liberated by the Pacific Ocean as it cools keeps coastal areas much warmer than they would otherwise be. This explains why the western coastal states generally have more favorable winter weather than the eastern coastal states, where the prevailing winds do not tend to carry the energy toward land.

 A difference in specific heats causes the cheese topping on a slice of pizza to burn you more than a mouthful of crust at the same temperature. Both crust and cheese undergo the same change in temperature, starting at a high straight-from-the-oven value and ending at the temperature of the inside of your mouth, which is about 37°C. Because the cheese is much more likely to burn you, it must release much more energy as it cools than does the crust. If we assume roughly the same mass for both cheese and crust, then Equation 20.3 indicates that the specific heat of the cheese, which is mostly water, is greater than that of the crust, which is mostly air.

Conservation of Energy: Calorimetry

One technique for measuring specific heat involves heating a sample to some known temperature T_x, placing it in a vessel containing water of known mass and temperature $T_w < T_x$, and measuring the temperature of the water after equilibrium has been reached. Because a negligible amount of mechanical work is done in the process, the law of the conservation of energy requires that the amount of energy that leaves the sample (of unknown specific heat) equal the amount of energy that enters the water.[3] This technique is called **calorimetry,** and devices in which this energy transfer occurs are called **calorimeters.**

Conservation of energy allows us to write the equation

$$Q_{cold} = - Q_{hot} \qquad (20.5)$$

which simply states that the energy leaving the hot part of the system by heat is equal to that entering the cold part of the system. The negative sign in the equation is necessary to maintain consistency with our sign convention for heat. The

---

[3] For precise measurements, the water container should be included in our calculations because it also exchanges energy with the sample. However, doing so would require a knowledge of its mass and composition. If the mass of the water is much greater than that of the container, we can neglect the effects of the container.

heat Q_{hot} is negative because energy is leaving the hot sample. The negative sign in the equation ensures that the right-hand side is positive and thus consistent with the left-hand side, which is positive because energy is entering the cold water.

Suppose m_x is the mass of a sample of some substance whose specific heat we wish to determine. Let us call its specific heat c_x and its initial temperature T_x. Likewise, let m_w, c_w, and T_w represent corresponding values for the water. If T_f is the final equilibrium temperature after everything is mixed, then from Equation 20.4, we find that the energy transfer for the water is $m_w c_w(T_f - T_w)$, which is positive because $T_f > T_w$, and that the energy transfer for the sample of unknown specific heat is $m_x c_x(T_f - T_x)$, which is negative. Substituting these expressions into Equation 20.5 gives

$$m_w c_w(T_f - T_w) = -m_x c_x(T_f - T_x)$$

Solving for c_x gives

$$c_x = \frac{m_w c_w(T_f - T_w)}{m_x(T_x - T_f)}$$

EXAMPLE 20.2   Cooling a Hot Ingot

A 0.050 0-kg ingot of metal is heated to 200.0°C and then dropped into a beaker containing 0.400 kg of water initially at 20.0°C. If the final equilibrium temperature of the mixed system is 22.4°C, find the specific heat of the metal.

Solution   According to Equation 20.5, we can write

$$m_w c_w(T_f - T_w) = -m_x c_x(T_f - T_x)$$
$$(0.400 \text{ kg})(4\ 186 \text{ J/kg} \cdot °C)(22.4°C - 20.0°C) =$$
$$-(0.050\ 0 \text{ kg})(c_x)(22.4°C - 200.0°C)$$

From this we find that

$$c_x = \boxed{453 \text{ J/kg} \cdot °C}$$

The ingot is most likely iron, as we can see by comparing this result with the data given in Table 20.1. Note that the temperature of the ingot is initially above the steam point. Thus, some of the water may vaporize when we drop the ingot into the water. We assume that we have a sealed system and thus that this steam cannot escape. Because the final equilibrium temperature is lower than the steam point, any steam that does result recondenses back into water.

Exercise   What is the amount of energy transferred to the water as the ingot is cooled?

Answer   4 020 J.

EXAMPLE 20.3   Fun Time for a Cowboy

A cowboy fires a silver bullet with a mass of 2.00 g and with a muzzle speed of 200 m/s into the pine wall of a saloon. Assume that all the internal energy generated by the impact remains with the bullet. What is the temperature change of the bullet?

Solution   The kinetic energy of the bullet is

$$\tfrac{1}{2}mv^2 = \tfrac{1}{2}(2.00 \times 10^{-3} \text{ kg})(200 \text{ m/s})^2 = 40.0 \text{ J}$$

Because nothing in the environment is hotter than the bullet, the bullet gains no energy by heat. Its temperature increases because the 40.0 J of kinetic energy becomes 40.0 J of extra internal energy. The temperature change is the same as that which would take place if 40.0 J of energy were transferred by

heat from a stove to the bullet. If we imagine this latter process taking place, we can calculate ΔT from Equation 20.4. Using 234 J/kg · °C as the specific heat of silver (see Table 20.1), we obtain

$$\Delta T = \frac{Q}{mc} = \frac{40.0 \text{ J}}{(2.00 \times 10^{-3} \text{ kg})(234 \text{ J/kg} \cdot °C)} = \boxed{85.5°C}$$

Exercise   Suppose that the cowboy runs out of silver bullets and fires a lead bullet of the same mass and at the same speed into the wall. What is the temperature change of the bullet?

Answer   156°C.

20.3 ▸ LATENT HEAT

A substance often undergoes a change in temperature when energy is transferred between it and its surroundings. There are situations, however, in which the transfer of energy does not result in a change in temperature. This is the case whenever the physical characteristics of the substance change from one form to another; such a change is commonly referred to as a **phase change.** Two common phase changes are from solid to liquid (melting) and from liquid to gas (boiling); another is a change in the crystalline structure of a solid. All such phase changes involve a change in internal energy but no change in temperature. The increase in internal energy in boiling, for example, is represented by the breaking of bonds between molecules in the liquid state; this bond breaking allows the molecules to move farther apart in the gaseous state, with a corresponding increase in intermolecular potential energy.

As you might expect, different substances respond differently to the addition or removal of energy as they change phase because their internal molecular arrangements vary. Also, the amount of energy transferred during a phase change depends on the amount of substance involved. (It takes less energy to melt an ice cube than it does to thaw a frozen lake.) If a quantity Q of energy transfer is required to change the phase of a mass m of a substance, the ratio $L \equiv Q/m$ characterizes an important thermal property of that substance. Because this added or removed energy does not result in a temperature change, the quantity L is called the **latent heat** (literally, the "hidden" heat) of the substance. The value of L for a substance depends on the nature of the phase change, as well as on the properties of the substance.

From the definition of latent heat, and again choosing heat as our energy transfer mechanism, we find that the energy required to change the phase of a given mass m of a pure substance is

$$Q = mL \tag{20.6}$$

Latent heat of fusion L_f is the term used when the phase change is from solid to liquid (*to fuse* means "to combine by melting"), and **latent heat of vaporization**

TABLE 20.2 Latent Heats of Fusion and Vaporization

Substance	Melting Point (°C)	Latent Heat of Fusion (J/kg)	Boiling Point (°C)	Latent Heat of Vaporization (J/kg)
Helium	−269.65	5.23×10^3	−268.93	2.09×10^4
Nitrogen	−209.97	2.55×10^4	−195.81	2.01×10^5
Oxygen	−218.79	1.38×10^4	−182.97	2.13×10^5
Ethyl alcohol	−114	1.04×10^5	78	8.54×10^5
Water	0.00	3.33×10^5	100.00	2.26×10^6
Sulfur	119	3.81×10^4	444.60	3.26×10^5
Lead	327.3	2.45×10^4	1 750	8.70×10^5
Aluminum	660	3.97×10^5	2 450	1.14×10^7
Silver	960.80	8.82×10^4	2 193	2.33×10^6
Gold	1 063.00	6.44×10^4	2 660	1.58×10^6
Copper	1 083	1.34×10^5	1 187	5.06×10^6

L_v is the term used when the phase change is from liquid to gas (the liquid "vaporizes").[4] The latent heats of various substances vary considerably, as data in Table 20.2 show.

Which is more likely to cause a serious burn, 100°C liquid water or an equal mass of 100°C steam?

To understand the role of latent heat in phase changes, consider the energy required to convert a 1.00-g block of ice at −30.0°C to steam at 120.0°C. Figure 20.2 indicates the experimental results obtained when energy is gradually added to the ice. Let us examine each portion of the red curve.

Part A. On this portion of the curve, the temperature of the ice changes from −30.0°C to 0.0°C. Because the specific heat of ice is 2 090 J/kg·°C, we can calculate the amount of energy added by using Equation 20.4:

$$Q = m_i c_i \Delta T = (1.00 \times 10^{-3} \, \text{kg})(2\,090 \, \text{J/kg} \cdot °\text{C})(30.0°\text{C}) = 62.7 \, \text{J}$$

Part B. When the temperature of the ice reaches 0.0°C, the ice–water mixture remains at this temperature—even though energy is being added—until all the ice melts. The energy required to melt 1.00 g of ice at 0.0°C is, from Equation 20.6,

$$Q = mL_f = (1.00 \times 10^{-3} \, \text{kg})(3.33 \times 10^5 \, \text{J/kg}) = 333 \, \text{J}$$

Thus, we have moved to the 396 J (= 62.7 J + 333 J) mark on the energy axis.

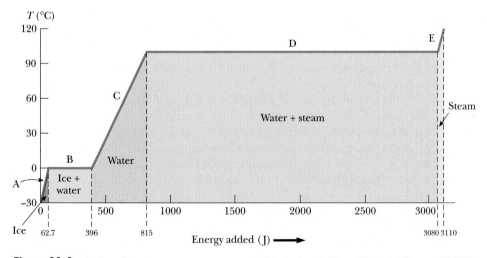

Figure 20.2 A plot of temperature versus energy added when 1.00 g of ice initially at −30.0°C is converted to steam at 120.0°C.

[4] When a gas cools, it eventually *condenses*—that is, it returns to the liquid phase. The energy given up per unit mass is called the *latent heat of condensation* and is numerically equal to the latent heat of vaporization. Likewise, when a liquid cools, it eventually solidifies, and the *latent heat of solidification* is numerically equal to the latent heat of fusion.

Part C. Between 0.0°C and 100.0°C, nothing surprising happens. No phase change occurs, and so all energy added to the water is used to increase its temperature. The amount of energy necessary to increase the temperature from 0.0°C to 100.0°C is

$$Q = m_w c_w \Delta T = (1.00 \times 10^{-3} \text{ kg})(4.19 \times 10^3 \text{ J/kg} \cdot °\text{C})(100.0°\text{C}) = 419 \text{ J}$$

Part D. At 100.0°C, another phase change occurs as the water changes from water at 100.0°C to steam at 100.0°C. Similar to the ice–water mixture in part B, the water–steam mixture remains at 100.0°C—even though energy is being added—until all of the liquid has been converted to steam. The energy required to convert 1.00 g of water to steam at 100.0°C is

$$Q = mL_v = (1.00 \times 10^{-3} \text{ kg})(2.26 \times 10^6 \text{ J/kg}) = 2.26 \times 10^3 \text{ J}$$

Part E. On this portion of the curve, as in parts A and C, no phase change occurs; thus, all energy added is used to increase the temperature of the steam. The energy that must be added to raise the temperature of the steam from 100.0°C to 120.0°C is

$$Q = m_s c_s \Delta T = (1.00 \times 10^{-3} \text{ kg})(2.01 \times 10^3 \text{ J/kg} \cdot °\text{C})(20.0°\text{C}) = 40.2 \text{ J}$$

The total amount of energy that must be added to change 1 g of ice at −30.0°C to steam at 120.0°C is the sum of the results from all five parts of the curve, which is 3.11×10^3 J. Conversely, to cool 1 g of steam at 120.0°C to ice at −30.0°C, we must remove 3.11×10^3 J of energy.

We can describe phase changes in terms of a rearrangement of molecules when energy is added to or removed from a substance. (For elemental substances in which the atoms do not combine to form molecules, the following discussion should be interpreted in terms of atoms. We use the general term *molecules* to refer to both molecular substances and elemental substances.) Consider first the liquid-to-gas phase change. The molecules in a liquid are close together, and the forces between them are stronger than those between the more widely separated molecules of a gas. Therefore, work must be done on the liquid against these attractive molecular forces if the molecules are to separate. The latent heat of vaporization is the amount of energy per unit mass that must be added to the liquid to accomplish this separation.

Similarly, for a solid, we imagine that the addition of energy causes the amplitude of vibration of the molecules about their equilibrium positions to become greater as the temperature increases. At the melting point of the solid, the amplitude is great enough to break the bonds between molecules and to allow molecules to move to new positions. The molecules in the liquid also are bound to each other, but less strongly than those in the solid phase. The latent heat of fusion is equal to the energy required per unit mass to transform the bonds among all molecules from the solid-type bond to the liquid-type bond.

As you can see from Table 20.2, the latent heat of vaporization for a given substance is usually somewhat higher than the latent heat of fusion. This is not surprising if we consider that the average distance between molecules in the gas phase is much greater than that in either the liquid or the solid phase. In the solid-to-liquid phase change, we transform solid-type bonds between molecules into liquid-type bonds between molecules, which are only slightly less strong. In the liquid-to-gas phase change, however, we break liquid-type bonds and create a situation in which the molecules of the gas essentially are not bonded to each other. Therefore, it is not surprising that more energy is required to vaporize a given mass of substance than is required to melt it.

Quick Quiz 20.3

Calculate the slopes for the A, C, and E portions of Figure 20.2. Rank the slopes from least to greatest and explain what this ordering means.

Problem-Solving Hints

Calorimetry Problems

If you are having difficulty in solving calorimetry problems, be sure to consider the following points:

- Units of measure must be consistent. For instance, if you are using specific heats measured in cal/g · °C, be sure that masses are in grams and temperatures are in Celsius degrees.
- Transfers of energy are given by the equation $Q = mc\Delta T$ only for those processes in which no phase changes occur. Use the equations $Q = mL_f$ and $Q = mL_v$ only when phase changes *are* taking place.
- Often, errors in sign are made when the equation $Q_{cold} = -Q_{hot}$ is used. Make sure that you use the negative sign in the equation, and remember that ΔT is always the final temperature minus the initial temperature.

EXAMPLE 20.4   Cooling the Steam

What mass of steam initially at 130°C is needed to warm 200 g of water in a 100-g glass container from 20.0°C to 50.0°C?

Solution The steam loses energy in three stages. In the first stage, the steam is cooled to 100°C. The energy transfer in the process is

$$Q_1 = m_s c_s \, \Delta T = m_s(2.01 \times 10^3 \, \text{J/kg} \cdot °\text{C})\,(-30.0°\text{C})$$
$$= -m_s(6.03 \times 10^4 \, \text{J/kg})$$

where m_s is the unknown mass of the steam.

In the second stage, the steam is converted to water. To find the energy transfer during this phase change, we use $Q = -mL_v$, where the negative sign indicates that energy is leaving the steam:

$$Q_2 = -m_s(2.26 \times 10^6 \, \text{J/kg})$$

In the third stage, the temperature of the water created from the steam is reduced to 50.0°C. This change requires an energy transfer of

$$Q_3 = m_s c_w \, \Delta T = m_s(4.19 \times 10^3 \, \text{J/kg} \cdot °\text{C})\,(-50.0°\text{C})$$
$$= -m_s(2.09 \times 10^5 \, \text{J/kg})$$

Adding the energy transfers in these three stages, we obtain

$$Q_{hot} = Q_1 + Q_2 + Q_3$$
$$= -m_s(6.03 \times 10^4 \, \text{J/kg} + 2.26 \times 10^6 \, \text{J/kg}$$
$$+ 2.09 \times 10^5 \, \text{J/kg})$$
$$= -m_s(2.53 \times 10^6 \, \text{J/kg})$$

Now, we turn our attention to the temperature increase of the water and the glass. Using Equation 20.4, we find that

$$Q_{cold} = (0.200 \, \text{kg})\,(4.19 \times 10^3 \, \text{J/kg} \cdot °\text{C})\,(30.0°\text{C})$$
$$+ (0.100 \, \text{kg})\,(837 \, \text{J/kg} \cdot °\text{C})\,(30.0°\text{C})$$
$$= 2.77 \times 10^4 \, \text{J}$$

Using Equation 20.5, we can solve for the unknown mass:

$$Q_{cold} = -Q_{hot}$$
$$2.77 \times 10^4 \, \text{J} = -[-m_s(2.53 \times 10^6 \, \text{J/kg})]$$
$$m_s = \boxed{1.09 \times 10^{-2} \, \text{kg} = 10.9 \, \text{g}}$$

EXAMPLE 20.5   **Boiling Liquid Helium**

Liquid helium has a very low boiling point, 4.2 K, and a very low latent heat of vaporization, 2.09×10^4 J/kg. If energy is transferred to a container of boiling liquid helium from an immersed electric heater at a rate of 10.0 W, how long does it take to boil away 1.00 kg of the liquid?

Solution   Because $L_v = 2.09 \times 10^4$ J/kg, we must supply 2.09×10^4 J of energy to boil away 1.00 kg. Because 10.0 W = 10.0 J/s, 10.0 J of energy is transferred to the helium each second. Therefore, the time it takes to transfer 2.09×10^4 J

of energy is

$$t = \frac{2.09 \times 10^4 \text{ J}}{10.0 \text{ J/s}} = 2.09 \times 10^3 \text{ s} \approx \boxed{35 \text{ min}}$$

Exercise   If 10.0 W of power is supplied to 1.00 kg of water at 100°C, how long does it take for the water to completely boil away?

Answer   62.8 h.

20.4   WORK AND HEAT IN THERMODYNAMIC PROCESSES

10.6   In the macroscopic approach to thermodynamics, we describe the *state* of a system using such variables as pressure, volume, temperature, and internal energy. The number of macroscopic variables needed to characterize a system depends on the nature of the system. For a homogeneous system, such as a gas containing only one type of molecule, usually only two variables are needed. However, it is important to note that a *macroscopic state* of an isolated system can be specified only if the system is in thermal equilibrium internally. In the case of a gas in a container, internal thermal equilibrium requires that every part of the gas be at the same pressure and temperature.

Consider a gas contained in a cylinder fitted with a movable piston (Fig. 20.3). At equilibrium, the gas occupies a volume V and exerts a uniform pressure P on the cylinder's walls and on the piston. If the piston has a cross-sectional area A, the

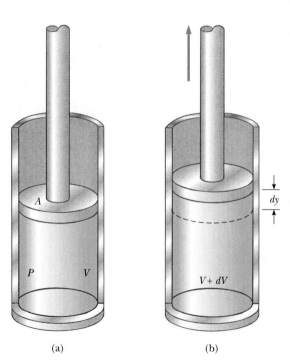

(a)   (b)

Figure 20.3   Gas contained in a cylinder at a pressure P does work on a moving piston as the system expands from a volume V to a volume $V + dV$.

force exerted by the gas on the piston is $F = PA$. Now let us assume that the gas expands **quasi-statically,** that is, slowly enough to allow the system to remain essentially in thermal equilibrium at all times. As the piston moves up a distance dy, the work done by the gas on the piston is

$$dW = F \, dy = PA \, dy$$

Because $A \, dy$ is the increase in volume of the gas dV, we can express the work done by the gas as

$$dW = P \, dV \qquad \qquad \textbf{(20.7)}$$

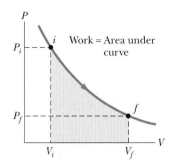

Figure 20.4 A gas expands quasi-statically (slowly) from state i to state f. The work done by the gas equals the area under the PV curve.

Because the gas expands, dV is positive, and so the work done by the gas is positive. If the gas were compressed, dV would be negative, indicating that the work done by the gas (which can be interpreted as work done *on* the gas) was negative.

In the thermodynamics problems that we shall solve, we shall identify the system of interest as a substance that is exchanging energy with the environment. In many problems, this will be a gas contained in a vessel; however, we will also consider problems involving liquids and solids. It is an unfortunate fact that, because of the separate historical development of thermodynamics and mechanics, positive work for a thermodynamic system is commonly defined as the work done *by* the system, rather than that done *on* the system. This is the reverse of the case for our study of work in mechanics. Thus, **in thermodynamics, positive work represents a transfer of energy out of the system.** We will use this convention to be consistent with common treatments of thermodynamics.

The total work done by the gas as its volume changes from V_i to V_f is given by the integral of Equation 20.7:

$$W = \int_{V_i}^{V_f} P \, dV \qquad \qquad \textbf{(20.8)}$$

To evaluate this integral, it is not enough that we know only the initial and final values of the pressure. We must also know the pressure at every instant during the expansion; we would know this if we had a functional dependence of P with respect to V. This important point is true for any process—the expansion we are discussing here, or any other. To fully specify a process, we must know the values of the thermodynamic variables at every state through which the system passes between the initial and final states. In the expansion we are considering here, we can plot the pressure and volume at each instant to create a PV diagram like the one shown in Figure 20.4. The value of the integral in Equation 20.8 is the area bounded by such a curve. Thus, we can say that

the work done by a gas in the expansion from an initial state to a final state is the area under the curve connecting the states in a PV diagram.

Work equals area under the curve in a PV diagram.

As Figure 20.4 shows, the work done in the expansion from the initial state i to the final state f depends on the path taken between these two states, where the *path* on a PV diagram is a description of the thermodynamic process through which the system is taken. To illustrate this important point, consider several paths connecting i and f (Fig. 20.5). In the process depicted in Figure 20.5a, the pressure of the gas is first reduced from P_i to P_f by cooling at constant volume V_i. The gas then expands from V_i to V_f at constant pressure P_f. The value of the work done along this path is equal to the area of the shaded rectangle, which is equal to

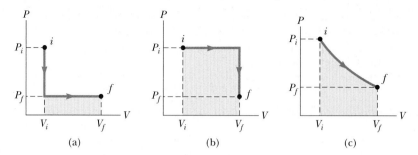

Figure 20.5   The work done by a gas as it is taken from an initial state to a final state depends on the path between these states.

$P_f(V_f - V_i)$. In Figure 20.5b, the gas first expands from V_i to V_f at constant pressure P_i. Then, its pressure is reduced to P_f at constant volume V_f. The value of the work done along this path is $P_i(V_f - V_i)$, which is greater than that for the process described in Figure 20.5a. Finally, for the process described in Figure 20.5c, where both P and V change continuously, the work done has some value intermediate between the values obtained in the first two processes. Therefore, we see that **the work done by a system depends on the initial and final states and on the path followed by the system between these states.**

The energy transfer by heat Q into or out of a system also depends on the process. Consider the situations depicted in Figure 20.6. In each case, the gas has the same initial volume, temperature, and pressure and is assumed to be ideal. In Figure 20.6a, the gas is thermally insulated from its surroundings except at the bottom of the gas-filled region, where it is in thermal contact with an energy reservoir. An *energy reservoir* is a source of energy that is considered to be so great that a finite transfer of energy from the reservoir does not change its temperature. The piston is held at its initial position by an external agent—a hand, for instance. When the force with which the piston is held is reduced slightly, the piston rises very slowly to its final position. Because the piston is moving upward, the gas is doing work on

> Work done depends on the path between the initial and final states.

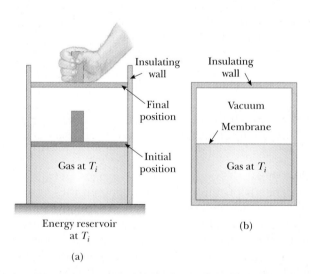

Figure 20.6   (a) A gas at temperature T_i expands slowly while absorbing energy from a reservoir in order to maintain a constant temperature. (b) A gas expands rapidly into an evacuated region after a membrane is broken.

the piston. During this expansion to the final volume V_f, just enough energy is transferred by heat from the reservoir to the gas to maintain a constant temperature T_i.

Now consider the completely thermally insulated system shown in Figure 20.6b. When the membrane is broken, the gas expands rapidly into the vacuum until it occupies a volume V_f and is at a pressure P_f. In this case, the gas does no work because there is no movable piston on which the gas applies a force. Furthermore, no energy is transferred by heat through the insulating wall.

The initial and final states of the ideal gas in Figure 20.6a are identical to the initial and final states in Figure 20.6b, but the paths are different. In the first case, the gas does work on the piston, and energy is transferred slowly to the gas. In the second case, no energy is transferred, and the value of the work done is zero. Therefore, we conclude that **energy transfer by heat, like work done, depends on the initial, final, and intermediate states of the system.** In other words, because heat and work depend on the path, neither quantity is determined solely by the end points of a thermodynamic process.

This device, called *Hero's engine*, was invented around 150 B.C. by Hero in Alexandria. When water is boiled in the flask, which is suspended by a cord, steam exits through two tubes at the sides (in opposite directions), creating a torque that rotates the flask. *(Courtesy of Central Scientific Company)*

20.5 THE FIRST LAW OF THERMODYNAMICS

When we introduced the law of conservation of mechanical energy in Chapter 8, we stated that the mechanical energy of a system is constant in the absence of nonconservative forces such as friction. That is, we did not include changes in the internal energy of the system in this mechanical model. The first law of thermodynamics is a generalization of the law of conservation of energy that encompasses changes in internal energy. It is a universally valid law that can be applied to many processes and provides a connection between the microscopic and macroscopic worlds.

We have discussed two ways in which energy can be transferred between a system and its surroundings. One is work done by the system, which requires that there be a macroscopic displacement of the point of application of a force (or pressure). The other is heat, which occurs through random collisions between the molecules of the system. Both mechanisms result in a change in the internal energy of the system and therefore usually result in measurable changes in the macroscopic variables of the system, such as the pressure, temperature, and volume of a gas.

To better understand these ideas on a quantitative basis, suppose that a system undergoes a change from an initial state to a final state. During this change, energy transfer by heat Q to the system occurs, and work W is done *by* the system. As an example, suppose that the system is a gas in which the pressure and volume change from P_i and V_i to P_f and V_f. If the quantity $Q - W$ is measured for various paths connecting the initial and final equilibrium states, we find that it is the same for all paths connecting the two states. We conclude that the quantity $Q - W$ is determined completely by the initial and final states of the system, and we call this quantity the **change in the internal energy** of the system. Although Q and W both depend on the path, **the quantity $Q - W$ is independent of the path.** If we use the symbol E_{int} to represent the internal energy, then the *change* in internal energy ΔE_{int} can be expressed as[5]

$$\Delta E_{\text{int}} = Q - W \qquad \textbf{(20.9)}$$

$Q - W$ is the change in internal energy

First-law equation

[5] It is an unfortunate accident of history that the traditional symbol for internal energy is U, which is also the traditional symbol for potential energy, as introduced in Chapter 8. To avoid confusion between potential energy and internal energy, we use the symbol E_{int} for internal energy in this book. If you take an advanced course in thermodynamics, however, be prepared to see U used as the symbol for internal energy.

where all quantities must have the same units of measure for energy.[6] Equation 20.9 is known as the **first-law equation** and is a key concept in many applications. As a reminder, we use the convention that Q is positive when energy enters the system and negative when energy leaves the system, and that W is positive when the system does work on the surroundings and negative when work is done on the system.

When a system undergoes an infinitesimal change in state in which a small amount of energy dQ is transferred by heat and a small amount of work dW is done, the internal energy changes by a small amount dE_{int}. Thus, for infinitesimal processes we can express the first-law equation as[7]

First-law equation for infinitesimal changes

$$dE_{int} = dQ - dW$$

The first-law equation is an energy conservation equation specifying that the only type of energy that changes in the system is the internal energy E_{int}. Let us look at some special cases in which this condition exists.

Isolated system

First, let us consider an *isolated system*—that is, one that does not interact with its surroundings. In this case, no energy transfer by heat takes place and the value of the work done by the system is zero; hence, the internal energy remains constant. That is, because $Q = W = 0$, it follows that $\Delta E_{int} = 0$, and thus $E_{int, i} = E_{int, f}$. We conclude that **the internal energy E_{int} of an isolated system remains constant.**

Next, we consider the case of a system (one not isolated from its surroundings) that is taken through a **cyclic process**—that is, a process that starts and ends at the same state. In this case, the change in the internal energy must again be zero, and therefore the energy Q added to the system must equal the work W done by the system during the cycle. That is, in a cyclic process,

Cyclic process

$$\Delta E_{int} = 0 \qquad \text{and} \qquad Q = W$$

On a *PV* diagram, a cyclic process appears as a closed curve. (The processes described in Figure 20.5 are represented by open curves because the initial and final states differ.) It can be shown that **in a cyclic process, the net work done by the system per cycle equals the area enclosed by the path representing the process on a *PV* diagram.**

If the value of the work done by the system during some process is zero, then the change in internal energy ΔE_{int} equals the energy transfer Q into or out of the system:

$$\Delta E_{int} = Q$$

If energy enters the system, then Q is positive and the internal energy increases. For a gas, we can associate this increase in internal energy with an increase in the kinetic energy of the molecules. Conversely, if no energy transfer occurs during some process but work is done by the system, then the change in internal energy equals the negative value of the work done by the system:

$$\Delta E_{int} = -W$$

---

[6] For the definition of work from our mechanics studies, the first law would be written as $\Delta E_{int} = Q + W$ because energy transfer into the system by either work or heat would increase the internal energy of the system. Because of the reversal of the definition of positive work discussed in Section 20.4, the first law appears as in Equation 20.9, with a minus sign.

[7] Note that dQ and dW are not true differential quantities; however, dE_{int} is. Because dQ and dW are *inexact differentials*, they are often represented by the symbols dQ and dW. For further details on this point, see an advanced text on thermodynamics, such as R. P. Bauman, *Modern Thermodynamics and Statistical Mechanics*, New York, Macmillan Publishing Co., 1992.

For example, if a gas is compressed by a moving piston in an insulated cylinder, no energy is transferred by heat and the work done by the gas is negative; thus, the internal energy increases because kinetic energy is transferred from the moving piston to the gas molecules.

On a microscopic scale, no distinction exists between the result of heat and that of work. Both heat and work can produce a change in the internal energy of a system. Although the macroscopic quantities Q and W are *not* properties of a system, they are related to the change of the internal energy of a system through the first-law equation. Once we define a process, or path, we can either calculate or measure Q and W, and we can find the change in the system's internal energy using the first-law equation.

One of the important consequences of the first law of thermodynamics is that there exists a quantity known as internal energy whose value is determined by the state of the system. The internal energy function is therefore called a *state function.*

20.6 ▷ SOME APPLICATIONS OF THE FIRST LAW OF THERMODYNAMICS

Before we apply the first law of thermodynamics to specific systems, it is useful for us to first define some common thermodynamic processes. An **adiabatic process** is one during which no energy enters or leaves the system by heat—that is, $Q = 0$. An adiabatic process can be achieved either by thermally insulating the system from its surroundings (as shown in Fig. 20.6b) or by performing the process rapidly, so that there is little time for energy to transfer by heat. Applying the first law of thermodynamics to an adiabatic process, we see that

$$\Delta E_{\text{int}} = - W \qquad \text{(adiabatic process)} \qquad \textbf{(20.10)}$$

From this result, we see that if a gas expands adiabatically such that W is positive, then ΔE_{int} is negative and the temperature of the gas decreases. Conversely, the temperature of a gas increases when the gas is compressed adiabatically.

Adiabatic processes are very important in engineering practice. Some common examples are the expansion of hot gases in an internal combustion engine, the liquefaction of gases in a cooling system, and the compression stroke in a diesel engine.

The process described in Figure 20.6b, called an **adiabatic free expansion,** is unique. The process is adiabatic because it takes place in an insulated container. Because the gas expands into a vacuum, it does not apply a force on a piston as was depicted in Figure 20.6a, so no work is done on or by the gas. Thus, in this adiabatic process, both $Q = 0$ and $W = 0$. As a result, $\Delta E_{\text{int}} = 0$ for this process, as we can see from the first law. That is, **the initial and final internal energies of a gas are equal in an adiabatic free expansion.** As we shall see in the next chapter, the internal energy of an ideal gas depends only on its temperature. Thus, we expect no change in temperature during an adiabatic free expansion. This prediction is in accord with the results of experiments performed at low pressures. (Experiments performed at high pressures for real gases show a slight decrease or increase in temperature after the expansion. This change is due to intermolecular interactions, which represent a deviation from the model of an ideal gas.)

A process that occurs at constant pressure is called an **isobaric process.** In such a process, the values of the heat and the work are both usually nonzero. The

In an adiabatic process, $Q = 0$.

First-law equation for an adiabatic process

In an adiabatic free expansion, $\Delta E_{\text{int}} = 0$.

In an isobaric process, P remains constant.

work done by the gas is simply

$$W = P(V_f - V_i) \qquad \text{(isobaric process)} \qquad \textbf{(20.11)}$$

where P is the constant pressure.

A process that takes place at constant volume is called an **isovolumetric process.** In such a process, the value of the work done is clearly zero because the volume does not change. Hence, from the first law we see that in an isovolumetric process, because $W = 0$,

$$\Delta E_{int} = Q \qquad \text{(isovolumetric process)} \qquad \textbf{(20.12)}$$

This expression specifies that **if energy is added by heat to a system kept at constant volume, then all of the transferred energy remains in the system as an increase of the internal energy of the system.** For example, when a can of spray paint is thrown into a fire, energy enters the system (the gas in the can) by heat through the metal walls of the can. Consequently, the temperature, and thus the pressure, in the can increases until the can possibly explodes.

A process that occurs at constant temperature is called an **isothermal process.** A plot of P versus V at constant temperature for an ideal gas yields a hyperbolic curve called an *isotherm*. The internal energy of an ideal gas is a function of temperature only. Hence, in an isothermal process involving an ideal gas, $\Delta E_{int} = 0$. For an isothermal process, then, we conclude from the first law that the energy transfer Q must be equal to the work done by the gas—that is, $Q = W$. Any energy that enters the system by heat is transferred out of the system by work; as a result, no change of the internal energy of the system occurs.

Quick Quiz 20.4

In the last three columns of the following table, fill in the boxes with −, +, or 0. For each situation, the system to be considered is identified.

Situation	System	Q	W	ΔE_{int}
(a) Rapidly pumping up a bicycle tire	Air in the pump			
(b) Pan of room-temperature water sitting on a hot stove	Water in the pan			
(c) Air quickly leaking out of a balloon	Air originally in balloon			

Isothermal Expansion of an Ideal Gas

Suppose that an ideal gas is allowed to expand quasi-statically at constant temperature, as described by the PV diagram shown in Figure 20.7. The curve is a hyperbola (see Appendix B, Eq. B.23), and the equation of state of an ideal gas with T constant indicates that the equation of this curve is PV = constant. The isothermal expansion of the gas can be achieved by placing the gas in thermal contact with an energy reservoir at the same temperature, as shown in Figure 20.6a.

Let us calculate the work done by the gas in the expansion from state i to state f. The work done by the gas is given by Equation 20.8. Because the gas is ideal and the process is quasi-static, we can use the expression $PV = nRT$ for each point on

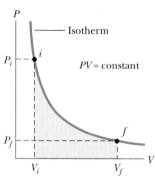

Figure 20.7 The PV diagram for an isothermal expansion of an ideal gas from an initial state to a final state. The curve is a hyperbola.

the path. Therefore, we have

$$W = \int_{V_i}^{V_f} P\, dV = \int_{V_i}^{V_f} \frac{nRT}{V}\, dV$$

Because T is constant in this case, it can be removed from the integral along with n and R:

$$W = nRT \int_{V_i}^{V_f} \frac{dV}{V} = nRT \ln V \Big|_{V_i}^{V_f}$$

To evaluate the integral, we used $\int (dx/x) = \ln x$. Evaluating this at the initial and final volumes, we have

$$W = nRT \ln\left(\frac{V_f}{V_i}\right) \qquad\qquad (20.13)$$

Work done by an ideal gas in an isothermal process

Numerically, this work W equals the shaded area under the PV curve shown in Figure 20.7. Because the gas expands, $V_f > V_i$, and the value for the work done by the gas is positive, as we expect. If the gas is compressed, then $V_f < V_i$, and the work done by the gas is negative.

Quick Quiz 20.5

Characterize the paths in Figure 20.8 as isobaric, isovolumetric, isothermal, or adiabatic. Note that $Q = 0$ for path B.

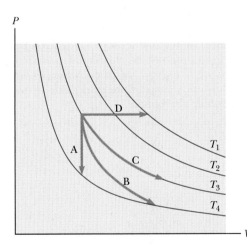

Figure 20.8 Identify the nature of paths A, B, C, and D.

EXAMPLE 20.6   An Isothermal Expansion

A 1.0-mol sample of an ideal gas is kept at 0.0°C during an expansion from 3.0 L to 10.0 L. (a) How much work is done by the gas during the expansion?

Solution Substituting the values into Equation 20.13, we have

$$W = nRT \ln\left(\frac{V_f}{V_i}\right)$$

$$W = (1.0\ \text{mol})(8.31\ \text{J/mol}\cdot\text{K})(273\ \text{K}) \ln\left(\frac{10.0}{3.0}\right)$$

$$= 2.7 \times 10^3\ \text{J}$$

(b) How much energy transfer by heat occurs with the surroundings in this process?

Solution   From the first law, we find that

$$\Delta E_{\text{int}} = Q - W$$
$$0 = Q - W$$

$$Q = W = \boxed{2.7 \times 10^3 \, \text{J}}$$

(c) If the gas is returned to the original volume by means of an isobaric process, how much work is done by the gas?

Solution   The work done in an isobaric process is given by Equation 20.11. We are not given the pressure, so we need to incorporate the ideal gas law:

$$W = P(V_f - V_i) = \frac{nRT_i}{V_i}(V_f - V_i)$$

$$= \frac{(1.0 \, \text{mol})(8.31 \, \text{J/mol} \cdot \text{K})(273 \, \text{K})}{10.0 \times 10^{-3} \, \text{m}^3}$$

$$\times (3.0 \times 10^{-3} \, \text{m}^3 - 10.0 \times 10^{-3} \, \text{m}^3)$$

$$= \boxed{-1.6 \times 10^3 \, \text{J}}$$

Notice that we use the initial temperature and volume to determine the value of the constant pressure because we do not know the final temperature. The work done by the gas is negative because the gas is being compressed.

EXAMPLE 20.7   Boiling Water

Suppose 1.00 g of water vaporizes isobarically at atmospheric pressure (1.013×10^5 Pa). Its volume in the liquid state is $V_i = V_{\text{liquid}} = 1.00 \, \text{cm}^3$, and its volume in the vapor state is $V_f = V_{\text{vapor}} = 1\,671 \, \text{cm}^3$. Find the work done in the expansion and the change in internal energy of the system. Ignore any mixing of the steam and the surrounding air—imagine that the steam simply pushes the surrounding air out of the way.

Solution   Because the expansion takes place at constant pressure, the work done by the system in pushing away the surrounding air is, from Equation 20.11,

$$W = P(V_f - V_i)$$
$$= (1.013 \times 10^5 \, \text{Pa})(1\,671 \times 10^{-6} \, \text{m}^3 - 1.00 \times 10^{-6} \, \text{m}^3)$$

$$= \boxed{169 \, \text{J}}$$

To determine the change in internal energy, we must know the energy transfer Q needed to vaporize the water. Using Equation 20.6 and the latent heat of vaporization for water, we have

$$Q = mL_v = (1.00 \times 10^{-3} \, \text{kg})(2.26 \times 10^6 \, \text{J/kg}) = 2\,260 \, \text{J}$$

Hence, from the first law, the change in internal energy is

$$\Delta E_{\text{int}} = Q - W = 2\,260 \, \text{J} - 169 \, \text{J} = \boxed{2.09 \, \text{kJ}}$$

The positive value for ΔE_{int} indicates that the internal energy of the system increases. We see that most ($2\,090 \, \text{J}/2\,260 \, \text{J} = 93\%$) of the energy transferred to the liquid goes into increasing the internal energy of the system. Only $169 \, \text{J}/2\,260 \, \text{J} = 7\%$ leaves the system by work done by the steam on the surrounding atmosphere.

EXAMPLE 20.8   Heating a Solid

A 1.0-kg bar of copper is heated at atmospheric pressure. If its temperature increases from 20°C to 50°C, (a) what is the work done by the copper on the surrounding atmosphere?

Solution   Because the process is isobaric, we can find the work done by the copper using Equation 20.11, $W = P(V_f - V_i)$. We can calculate the change in volume of the copper using Equation 19.6. Using the average linear expansion coefficient for copper given in Table 19.2, and remembering that $\beta = 3\alpha$, we obtain

$$\Delta V = \beta V_i \Delta T$$
$$= [5.1 \times 10^{-5} (°\text{C})^{-1}](50°\text{C} - 20°\text{C}) V_i = 1.5 \times 10^{-3} \, V_i$$

The volume V_i is equal to m/ρ, and Table 15.1 indicates that the density of copper is $8.92 \times 10^3 \, \text{kg/m}^3$. Hence,

$$\Delta V = (1.5 \times 10^{-3}) \left(\frac{1.0 \, \text{kg}}{8.92 \times 10^3 \, \text{kg/m}^3} \right) = 1.7 \times 10^{-7} \, \text{m}^3$$

The work done is

$$W = P \Delta V = (1.013 \times 10^5 \, \text{N/m}^2)(1.7 \times 10^{-7} \, \text{m}^3)$$

$$= \boxed{1.7 \times 10^{-2} \, \text{J}}$$

(b) What quantity of energy is transferred to the copper by heat?

Solution   Taking the specific heat of copper from Table 20.1 and using Equation 20.4, we find that the energy transferred by heat is

$$Q = mc\Delta T = (1.0\ \text{kg})(387\ \text{J/kg}\cdot{}^\circ\text{C})(30^\circ\text{C}) = \boxed{1.2 \times 10^4\ \text{J}}$$

(c) What is the increase in internal energy of the copper?

Solution From the first law of thermodynamics, we have

$$\Delta E_{\text{int}} = Q - W = 1.2 \times 10^4\ \text{J} - 1.7 \times 10^{-2}\ \text{J} = \boxed{1.2 \times 10^4\ \text{J}}$$

Note that almost all of the energy transferred into the system by heat goes into increasing the internal energy. The fraction of energy used to do work on the surrounding atmosphere is only about 10^{-6}! Hence, when analyzing the thermal expansion of a solid or a liquid, the small amount of work done by the system is usually ignored.

20.7 ENERGY TRANSFER MECHANISMS

It is important to understand the rate at which energy is transferred between a system and its surroundings and the mechanisms responsible for the transfer. Therefore, let us now look at three common energy transfer mechanisms that can result in a change in internal energy of a system.

Thermal Conduction

The energy transfer process that is most clearly associated with a temperature difference is **thermal conduction.** In this process, the transfer can be represented on an atomic scale as an exchange of kinetic energy between microscopic particles—molecules, atoms, and electrons—in which less energetic particles gain energy in collisions with more energetic particles. For example, if you hold one end of a long metal bar and insert the other end into a flame, you will find that the temperature of the metal in your hand soon increases. The energy reaches your hand by means of conduction. We can understand the process of conduction by examining what is happening to the microscopic particles in the metal. Initially, before the rod is inserted into the flame, the microscopic particles are vibrating about their equilibrium positions. As the flame heats the rod, those particles near the flame begin to vibrate with greater and greater amplitudes. These particles, in turn, collide with their neighbors and transfer some of their energy in the collisions. Slowly, the amplitudes of vibration of metal atoms and electrons farther and farther from the flame increase until, eventually, those in the metal near your hand are affected. This increased vibration represents an increase in the temperature of the metal and of your potentially burned hand.

The rate of thermal conduction depends on the properties of the substance being heated. For example, it is possible to hold a piece of asbestos in a flame indefinitely. This implies that very little energy is conducted through the asbestos. In general, metals are good thermal conductors, and materials such as asbestos, cork, paper, and fiberglass are poor conductors. Gases also are poor conductors because the separation distance between the particles is so great. Metals are good thermal conductors because they contain large numbers of electrons that are relatively free to move through the metal and so can transport energy over large distances. Thus, in a good conductor, such as copper, conduction takes place both by means of the vibration of atoms and by means of the motion of free electrons.

Conduction occurs only if there is a difference in temperature between two parts of the conducting medium. Consider a slab of material of thickness Δx and cross-sectional area A. One face of the slab is at a temperature T_1, and the other face is at a temperature $T_2 > T_1$ (Fig. 20.9). Experimentally, it is found that the

Melted snow pattern on a parking lot surface indicates the presence of underground hot water pipes used to aid snow removal. Energy from the water is conducted from the pipes to the pavement, where it causes the snow to melt. *(Courtesy of Dr. Albert A. Bartlett, University of Colorado, Boulder, CO)*

energy Q transferred in a time Δt flows from the hotter face to the colder one. The rate $Q/\Delta t$ at which this energy flows is found to be proportional to the cross-sectional area and the temperature difference $\Delta T = T_2 - T_1$, and inversely proportional to the thickness:

$$\frac{Q}{\Delta t} \propto A \frac{\Delta T}{\Delta x}$$

It is convenient to use the symbol for power $\mathcal{P}$ to represent the **rate of energy transfer:** $\mathcal{P} = Q/\Delta t$. Note that $\mathcal{P}$ has units of watts when Q is in joules and Δt is in seconds. For a slab of infinitesimal thickness dx and temperature difference dT, we can write the **law of thermal conduction** as

Law of thermal conduction

$$\mathcal{P} = kA \left| \frac{dT}{dx} \right| \tag{20.14}$$

where the proportionality constant k is the **thermal conductivity** of the material and $|dT/dx|$ is the **temperature gradient** (the variation of temperature with position).

Suppose that a long, uniform rod of length L is thermally insulated so that energy cannot escape by heat from its surface except at the ends, as shown in Figure 20.10. One end is in thermal contact with an energy reservoir at temperature T_1, and the other end is in thermal contact with a reservoir at temperature $T_2 > T_1$. When a steady state has been reached, the temperature at each point along the rod is constant in time. In this case if we assume that k is not a function of temperature, the temperature gradient is the same everywhere along the rod and is

$$\left| \frac{dT}{dx} \right| = \frac{T_2 - T_1}{L}$$

Thus the rate of energy transfer by conduction through the rod is

$$\mathcal{P} = kA \frac{(T_2 - T_1)}{L} \tag{20.15}$$

Substances that are good thermal conductors have large thermal conductivity values, whereas good thermal insulators have low thermal conductivity values. Table 20.3 lists thermal conductivities for various substances. Note that metals are generally better thermal conductors than nonmetals are.

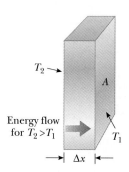

Figure 20.9 Energy transfer through a conducting slab with a cross-sectional area A and a thickness Δx. The opposite faces are at different temperatures T_1 and T_2.

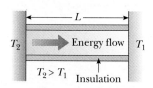

Figure 20.10 Conduction of energy through a uniform, insulated rod of length L. The opposite ends are in thermal contact with energy reservoirs at different temperatures.

Quick Quiz 20.6

Will an ice cube wrapped in a wool blanket remain frozen for (a) a shorter length of time, (b) the same length of time, or (c) a longer length of time than an identical ice cube exposed to air at room temperature?

For a compound slab containing several materials of thicknesses L_1, L_2, . . . and thermal conductivities k_1, k_2, . . . , the rate of energy transfer through the slab at steady state is

$$\mathcal{P} = \frac{A(T_2 - T_1)}{\sum_i (L_i/k_i)} \tag{20.16}$$

TABLE 20.3	Thermal Conductivities
Substance	**Thermal Conductivity (W/m·°C)**
Metals (at 25°C)	
Aluminum	238
Copper	397
Gold	314
Iron	79.5
Lead	34.7
Silver	427
Nonmetals (approximate values)	
Asbestos	0.08
Concrete	0.8
Diamond	2 300
Glass	0.8
Ice	2
Rubber	0.2
Water	0.6
Wood	0.08
Gases (at 20°C)	
Air	0.023 4
Helium	0.138
Hydrogen	0.172
Nitrogen	0.023 4
Oxygen	0.023 8

where T_1 and T_2 are the temperatures of the outer surfaces (which are held constant) and the summation is over all slabs. The following example shows how this equation results from a consideration of two thicknesses of materials.

EXAMPLE 20.9   Energy Transfer Through Two Slabs

Two slabs of thickness L_1 and L_2 and thermal conductivities k_1 and k_2 are in thermal contact with each other, as shown in Figure 20.11. The temperatures of their outer surfaces are T_1 and T_2, respectively, and $T_2 > T_1$. Determine the temperature at the interface and the rate of energy transfer by conduction through the slabs in the steady-state condition.

Solution   If T is the temperature at the interface, then the rate at which energy is transferred through slab 1 is

$$(1) \qquad \mathscr{P}_1 = \frac{k_1 A (T - T_1)}{L_1}$$

The rate at which energy is transferred through slab 2 is

$$(2) \qquad \mathscr{P}_2 = \frac{k_2 A (T_2 - T)}{L_2}$$

When a steady state is reached, these two rates must be equal; hence,

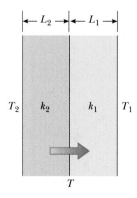

Figure 20.11   Energy transfer by conduction through two slabs in thermal contact with each other. At steady state, the rate of energy transfer through slab 1 equals the rate of energy transfer through slab 2.

$$\frac{k_1 A(T - T_1)}{L_1} = \frac{k_2 A(T_2 - T)}{L_2}$$

Solving for T gives

$$(3) \qquad T = \frac{k_1 L_2 T_1 + k_2 L_1 T_2}{k_1 L_2 + k_2 L_1}$$

Substituting (3) into either (1) or (2), we obtain

$$\mathcal{P} = \frac{A(T_2 - T_1)}{(L_1/k_1) + (L_2/k_2)}$$

Extension of this model to several slabs of materials leads to Equation 20.16.

Home Insulation

In engineering practice, the term L/k for a particular substance is referred to as the **R value** of the material. Thus, Equation 20.16 reduces to

$$\mathcal{P} = \frac{A(T_2 - T_1)}{\sum\limits_i R_i} \qquad \textbf{(20.17)}$$

where $R_i = L_i/k_i$. The R values for a few common building materials are given in Table 20.4. In the United States, the insulating properties of materials used in buildings are usually expressed in engineering units, not SI units. Thus, in Table 20.4, measurements of R values are given as a combination of British thermal units, feet, hours, and degrees Fahrenheit.

At any vertical surface open to the air, a very thin stagnant layer of air adheres to the surface. One must consider this layer when determining the R value for a wall. The thickness of this stagnant layer on an outside wall depends on the speed of the wind. Energy loss from a house on a windy day is greater than the loss on a day when the air is calm. A representative R value for this stagnant layer of air is given in Table 20.4.

Energy is conducted from the inside to the exterior more rapidly on the part of the roof where the snow has melted. The dormer appears to have been added and insulated. The main roof does not appear to be well insulated. *(Courtesy of Dr. Albert A. Bartlett, University of Colorado, Boulder, CO)*

TABLE 20.4 **R Values for Some Common Building Materials**

Material	R value (ft$^2 \cdot$ °F $\cdot$ h/Btu)
Hardwood siding (1 in. thick)	0.91
Wood shingles (lapped)	0.87
Brick (4 in. thick)	4.00
Concrete block (filled cores)	1.93
Fiberglass batting (3.5 in. thick)	10.90
Fiberglass batting (6 in. thick)	18.80
Fiberglass board (1 in. thick)	4.35
Cellulose fiber (1 in. thick)	3.70
Flat glass (0.125 in. thick)	0.89
Insulating glass (0.25-in. space)	1.54
Air space (3.5 in. thick)	1.01
Stagnant air layer	0.17
Drywall (0.5 in. thick)	0.45
Sheathing (0.5 in. thick)	1.32

This thermogram of a home, made during cold weather, shows colors ranging from white and orange (areas of greatest energy loss) to blue and purple (areas of least energy loss). *(Daedalus Enterprises, Inc./Peter Arnold, Inc.)*

EXAMPLE 20.10 The *R* Value of a Typical Wall

Calculate the total *R* value for a wall constructed as shown in Figure 20.12a. Starting outside the house (toward the front in the figure) and moving inward, the wall consists of 4-in. brick, 0.5-in. sheathing, an air space 3.5 in. thick, and 0.5-in. drywall. Do not forget the stagnant air layers inside and outside the house.

Solution Referring to Table 20.4, we find that

R_1 (outside stagnant air layer)	$= 0.17 \ \text{ft}^2 \cdot {}^\circ\text{F} \cdot \text{h/Btu}$
R_2 (brick)	$= 4.00 \ \text{ft}^2 \cdot {}^\circ\text{F} \cdot \text{h/Btu}$
R_3 (sheathing)	$= 1.32 \ \text{ft}^2 \cdot {}^\circ\text{F} \cdot \text{h/Btu}$
R_4 (air space)	$= 1.01 \ \text{ft}^2 \cdot {}^\circ\text{F} \cdot \text{h/Btu}$
R_5 (drywall)	$= 0.45 \ \text{ft}^2 \cdot {}^\circ\text{F} \cdot \text{h/Btu}$
R_6 (inside stagnant air layer)	$= 0.17 \ \text{ft}^2 \cdot {}^\circ\text{F} \cdot \text{h/Btu}$
R_{total}	$= \boxed{7.12 \ \text{ft}^2 \cdot {}^\circ\text{F} \cdot \text{h/Btu}}$

Exercise If a layer of fiberglass insulation 3.5 in. thick is placed inside the wall to replace the air space, as shown in Figure 20.12b, what is the new total *R* value? By what factor is the energy loss reduced?

Answer $R = 17 \ \text{ft}^2 \cdot {}^\circ\text{F} \cdot \text{h/Btu}$; 2.4.

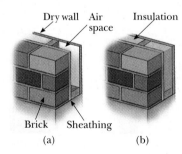

Figure 20.12 An exterior house wall containing (a) an air space and (b) insulation.

Convection

At one time or another, you probably have warmed your hands by holding them over an open flame. In this situation, the air directly above the flame is heated and expands. As a result, the density of this air decreases and the air rises. This warmed mass of air heats your hands as it flows by. **Energy transferred by the movement of a heated substance is said to have been transferred by convection.** When the movement results from differences in density, as with air around a fire, it is referred to as *natural convection*. Air flow at a beach is an example of natural convection, as is the mixing that occurs as surface water in a lake cools and sinks (see

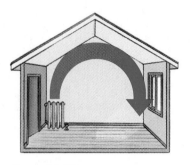

Figure 20.13   Convection currents are set up in a room heated by a radiator.

Chapter 19). When the heated substance is forced to move by a fan or pump, as in some hot-air and hot-water heating systems, the process is called *forced convection*.

If it were not for convection currents, it would be very difficult to boil water. As water is heated in a teakettle, the lower layers are warmed first. The heated water expands and rises to the top because its density is lowered. At the same time, the denser, cool water at the surface sinks to the bottom of the kettle and is heated.

The same process occurs when a room is heated by a radiator. The hot radiator warms the air in the lower regions of the room. The warm air expands and rises to the ceiling because of its lower density. The denser, cooler air from above sinks, and the continuous air current pattern shown in Figure 20.13 is established.

Radiation

The third means of energy transfer that we shall discuss is **radiation.** All objects radiate energy continuously in the form of electromagnetic waves (see Chapter 34) produced by thermal vibrations of the molecules. You are likely familiar with electromagnetic radiation in the form of the orange glow from an electric stove burner, an electric space heater, or the coils of a toaster.

The rate at which an object radiates energy is proportional to the fourth power of its absolute temperature. This is known as **Stefan's law** and is expressed in equation form as

$$\mathcal{P} = \sigma A e T^4 \tag{20.18}$$

where $\mathcal{P}$ is the power in watts radiated by the object, σ is a constant equal to $5.669\ 6 \times 10^{-8}$ W/m$^2 \cdot$K^4, A is the surface area of the object in square meters, e is the **emissivity** constant, and T is the surface temperature in kelvins. The value of e can vary between zero and unity, depending on the properties of the surface of the object. The emissivity is equal to the fraction of the incoming radiation that the surface absorbs.

Approximately 1 340 J of electromagnetic radiation from the Sun passes perpendicularly through each 1 m^2 at the top of the Earth's atmosphere every second. This radiation is primarily visible and infrared light accompanied by a significant amount of ultraviolet radiation. We shall study these types of radiation in detail in Chapter 34. Some of this energy is reflected back into space, and some is absorbed by the atmosphere. However, enough energy arrives at the surface of the Earth each day to supply all our energy needs on this planet hundreds of times over—if only it could be captured and used efficiently. The growth in the number of solar energy–powered houses built in this country reflects the increasing efforts being made to use this abundant energy. Radiant energy from the Sun affects our day-to-day existence in a number of ways. For example, it influences the Earth's average temperature, ocean currents, agriculture, and rain patterns.

What happens to the atmospheric temperature at night is another example of the effects of energy transfer by radiation. If there is a cloud cover above the Earth, the water vapor in the clouds absorbs part of the infrared radiation emitted by the Earth and re-emits it back to the surface. Consequently, temperature levels at the surface remain moderate. In the absence of this cloud cover, there is nothing to prevent this radiation from escaping into space; thus the temperature decreases more on a clear night than on a cloudy one.

As an object radiates energy at a rate given by Equation 20.18, it also absorbs electromagnetic radiation. If the latter process did not occur, an object would eventually radiate all its energy, and its temperature would reach absolute zero. The energy an object absorbs comes from its surroundings, which consist of other objects that radiate energy. If an object is at a temperature T and its surroundings are at a temperature T_0, then the net energy gained or lost each second by the object as a result of radiation is

$$\mathcal{P}_{net} = \sigma A e (T^4 - T_0^4) \qquad \text{(20.19)}$$

When an object is in equilibrium with its surroundings, it radiates and absorbs energy at the same rate, and so its temperature remains constant. When an object is hotter than its surroundings, it radiates more energy than it absorbs, and its temperature decreases. An **ideal absorber** is defined as an object that absorbs all the energy incident on it, and for such a body, $e = 1$. Such an object is often referred to as a **black body.** An ideal absorber is also an ideal radiator of energy. In contrast, an object for which $e = 0$ absorbs none of the energy incident on it. Such an object reflects all the incident energy, and thus is an **ideal reflector.**

The Dewar Flask

The *Dewar flask*[8] is a container designed to minimize energy losses by conduction, convection, and radiation. Such a container is used to store either cold or hot liquids for long periods of time. (A Thermos bottle is a common household equivalent of a Dewar flask.) The standard construction (Fig. 20.14) consists of a double-walled Pyrex glass vessel with silvered walls. The space between the walls is evacuated to minimize energy transfer by conduction and convection. The silvered surfaces minimize energy transfer by radiation because silver is a very good reflector and has very low emissivity. A further reduction in energy loss is obtained by reducing the size of the neck. Dewar flasks are commonly used to store liquid nitrogen (boiling point: 77 K) and liquid oxygen (boiling point: 90 K).

To confine liquid helium (boiling point: 4.2 K), which has a very low heat of vaporization, it is often necessary to use a double Dewar system in which the Dewar flask containing the liquid is surrounded by a second Dewar flask. The space between the two flasks is filled with liquid nitrogen.

Newer designs of storage containers use "super insulation" that consists of many layers of reflecting material separated by fiberglass. All of this is in a vacuum, and no liquid nitrogen is needed with this design.

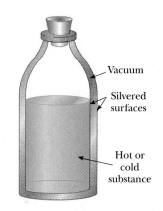

Vacuum

Silvered surfaces

Hot or cold substance

Figure 20.14 A cross-sectional view of a Dewar flask, which is used to store hot or cold substances.

EXAMPLE 20.11   Who Turned Down the Thermostat?

A student is trying to decide what to wear. The surroundings (his bedroom) are at 20.0°C. If the skin temperature of the unclothed student is 35°C, what is the net energy loss from his body in 10.0 min by radiation? Assume that the emissivity of skin is 0.900 and that the surface area of the student is 1.50 m².

Solution   Using Equation 20.19, we find that the net rate of energy loss from the skin is

$$\mathcal{P}_{net} = \sigma A e (T^4 - T_0^4)$$
$$= (5.67 \times 10^{-8} \text{ W/m}^2 \cdot \text{K}^4)(1.50 \text{ m}^2)$$
$$\times (0.900)[(308 \text{ K})^4 - (293 \text{ K})^4] = 125 \text{ W}$$

(Why is the temperature given in kelvins?) At this rate, the total energy lost by the skin in 10 min is

$$Q = \mathcal{P}_{net} \times \Delta t = (125 \text{ W})(600 \text{ s}) = \boxed{7.5 \times 10^4 \text{ J}}$$

Note that the energy radiated by the student is roughly equivalent to that produced by two 60-W light bulbs!

---

[8] Invented by Sir James Dewar (1842–1923).

SUMMARY

Internal energy is all of a system's energy that is associated with the system's microscopic components. Internal energy includes kinetic energy of translation, rotation, and vibration of molecules, potential energy within molecules, and potential energy between molecules.

Heat is the transfer of energy across the boundary of a system resulting from a temperature difference between the system and its surroundings. We use the symbol Q for the amount of energy transferred by this process.

The **calorie** is the amount of energy necessary to raise the temperature of 1 g of water from 14.5°C to 15.5°C. The **mechanical equivalent of heat** is 1 cal = 4.186 J.

The **heat capacity** C of any sample is the amount of energy needed to raise the temperature of the sample by 1°C. The energy Q required to change the temperature of a mass m of a substance by an amount ΔT is

$$Q = mc\Delta T \qquad (20.4)$$

where c is the **specific heat** of the substance.

The energy required to change the phase of a pure substance of mass m is

$$Q = mL \qquad (20.6)$$

where L is the **latent heat** of the substance and depends on the nature of the phase change and the properties of the substance.

The **work done** by a gas as its volume changes from some initial value V_i to some final value V_f is

$$W = \int_{V_i}^{V_f} P \, dV \qquad (20.8)$$

where P is the pressure, which may vary during the process. In order to evaluate W, the process must be fully specified—that is, P and V must be known during each step. In other words, the work done depends on the path taken between the initial and final states.

The **first law of thermodynamics** states that when a system undergoes a change from one state to another, the change in its internal energy is

$$\Delta E_{int} = Q - W \qquad (20.9)$$

where Q is the energy transferred into the system by heat and W is the work done by the system. Although Q and W both depend on the path taken from the initial state to the final state, the quantity ΔE_{int} is path-independent. This central equation is a statement of conservation of energy that includes changes in internal energy.

In a **cyclic process** (one that originates and terminates at the same state), $\Delta E_{int} = 0$ and, therefore, $Q = W$. That is, the energy transferred into the system by heat equals the work done by the system during the process.

In an **adiabatic process,** no energy is transferred by heat between the system and its surroundings ($Q = 0$). In this case, the first law gives $\Delta E_{int} = -W$. That is, the internal energy changes as a consequence of work being done by the system. In the **adiabatic free expansion** of a gas, $Q = 0$ and $W = 0$; thus, $\Delta E_{int} = 0$. That is, the internal energy of the gas does not change in such a process.

An **isobaric process** is one that occurs at constant pressure. The work done in such a process is $W = P(V_f - V_i)$.

An **isovolumetric process** is one that occurs at constant volume. No work is done in such a process, so $\Delta E_{int} = Q$.

An **isothermal process** is one that occurs at constant temperature. The work done by an ideal gas during an isothermal process is

$$W = nRT \ln\left(\frac{V_f}{V_i}\right) \qquad (20.13)$$

Energy may be transferred by work, which we addressed in Chapter 7, and by conduction, convection, or radiation. **Conduction** can be viewed as an exchange of kinetic energy between colliding molecules or electrons. The rate at which energy flows by conduction through a slab of area A is

$$\mathcal{P} = kA\left|\frac{dT}{dx}\right| \qquad (20.14)$$

where k is the **thermal conductivity** of the material from which the slab is made and $|dT/dx|$ is the **temperature gradient.** This equation can be used in many situations in which the rate of transfer of energy through materials is important.

In **convection,** a heated substance moves from one place to another.

All bodies emit **radiation** in the form of electromagnetic waves at the rate

$$\mathcal{P} = \sigma A e T^4 \qquad (20.18)$$

A body that is hotter than its surroundings radiates more energy than it absorbs, whereas a body that is cooler than its surroundings absorbs more energy than it radiates.

QUESTIONS

1. The specific heat of water is about two times that of ethyl alcohol. Equal masses of alcohol and water are contained in separate beakers and are supplied with the same amount of energy. Compare the temperature increases of the two liquids.
2. Give one reason why coastal regions tend to have a more moderate climate than inland regions do.
3. A small metal crucible is taken from a 200°C oven and immersed in a tub full of water at room temperature (this process is often referred to as *quenching*). What is the approximate final equilibrium temperature?
4. What is the major problem that arises in measuring specific heats if a sample with a temperature greater than 100°C is placed in water?
5. In a daring lecture demonstration, an instructor dips his wetted fingers into molten lead (327°C) and withdraws them quickly, without getting burned. How is this possible? (This is a dangerous experiment that you should *not* attempt.)
6. The pioneers found that placing a large tub of water in a storage cellar would prevent their food from freezing on really cold nights. Explain why.
7. What is wrong with the statement, "Given any two bodies, the one with the higher temperature contains more heat."
8. Why is it possible for you to hold a lighted match, even when it is burned to within a few millimeters of your fingertips?
9. Why is it more comfortable to hold a cup of hot tea by the handle than by wrapping your hands around the cup itself?

10. Figure Q20.10 shows a pattern formed by snow on the roof of a barn. What causes the alternating pattern of snowcover and exposed roof?

Figure Q20.10 Alternating pattern on a snow-covered roof. *(Courtesy of Dr. Albert A. Bartlett, University of Colorado, Boulder, CO)*

11. Why is a person able to remove a piece of dry aluminum foil from a hot oven with bare fingers but burns his or her fingers if there is moisture on the foil?
12. A tile floor in a bathroom may feel uncomfortably cold to your bare feet, but a carpeted floor in an adjoining room at the same temperature feels warm. Why?

13. Why can potatoes be baked more quickly when a metal skewer has been inserted through them?

14. Explain why a Thermos bottle has silvered walls and a vacuum jacket.

15. A piece of paper is wrapped around a rod made half of wood and half of copper. When held over a flame, the paper in contact with the wood burns but the paper in contact with the metal does not. Explain.

16. Why is it necessary to store liquid nitrogen or liquid oxygen in vessels equipped with either polystyrene insulation or a double-evacuated wall?

17. Why do heavy draperies hung over the windows help keep a home warm in the winter and cool in the summer?

18. If you wish to cook a piece of meat thoroughly on an open fire, why should you not use a high flame? (*Note:* Carbon is a good thermal insulator.)

19. When insulating a wood-frame house, is it better to place the insulation against the cooler, outside wall or against the warmer, inside wall? (In either case, an air barrier must be considered.)

20. In an experimental house, polystyrene beads were pumped into the air space between the panes of glass in double-pane windows at night in the winter, and they were pumped out to holding bins during the day. How would this procedure assist in conserving energy in the house?

21. Pioneers stored fruits and vegetables in underground cellars. Discuss the advantages of choosing this location as a storage site.

22. Concrete has a higher specific heat than soil does. Use this fact to explain (partially) why cities have a higher average night-time temperature than the surrounding countryside does. If a city is hotter than the surrounding countryside, would you expect breezes to blow from city to country or from country to city? Explain.

23. When camping in a canyon on a still night, a hiker notices that a breeze begins to stir as soon as the Sun strikes the surrounding peaks. What causes the breeze?

24. Updrafts of air are familiar to all pilots and are used to keep non-motorized gliders aloft. What causes these currents?

25. If water is a poor thermal conductor, why can it be heated quickly when placed over a flame?

26. The United States penny is now made of copper-coated zinc. Can a calorimetric experiment be devised to test for the metal content in a collection of pennies? If so, describe such a procedure.

27. If you hold water in a paper cup over a flame, you can bring the water to a boil without burning the cup. How is this possible?

28. When a sealed Thermos bottle full of hot coffee is shaken, what are the changes, if any, in (a) the temperature of the coffee and (b) the internal energy of the coffee?

29. Using the first law of thermodynamics, explain why the *total* energy of an isolated system is always constant.

30. Is it possible to convert internal energy into mechanical energy? Explain using examples.

31. Suppose that you pour hot coffee for your guests and one of them chooses to drink the coffee after it has been in the cup for several minutes. For the coffee to be warmest, should the person add the cream just after the coffee is poured or just before drinking it? Explain.

32. Suppose that you fill two identical cups both at room temperature with the same amount of hot coffee. One cup contains a metal spoon, while the other does not. If you wait for several minutes, which of the two contains the warmer coffee? Which energy transfer process accounts for this result?

33. A warning sign often seen on highways just before a bridge is "Caution—Bridge Surface Freezes Before Road Surface." Which of the three energy transfer processes is most important in causing a bridge surface to freeze before a road surface on very cold days?

PROBLEMS

1, 2, 3 = straightforward, intermediate, challenging   ☐ = full solution available in the *Student Solutions Manual and Study Guide*
WEB = solution posted at http://www.saunderscollege.com/physics/   🖥 = Computer useful in solving problem   ⚙ = Interactive Physics
☐ = paired numerical/symbolic problems

Section 20.1   Heat and Internal Energy

1. Water at the top of Niagara Falls has a temperature of $10.0°C$. It falls through a distance of 50.0 m. Assuming that all of its potential energy goes into warming of the water, calculate the temperature of the water at the bottom of the Falls.

2. Consider Joule's apparatus described in Figure 20.1. Each of the two masses is 1.50 kg, and the tank is filled with 200 g of water. What is the increase in the temperature of the water after the masses fall through a distance of 3.00 m?

Section 20.2   Heat Capacity and Specific Heat

3. The temperature of a silver bar rises by $10.0°C$ when it absorbs 1.23 kJ of energy by heat. The mass of the bar is 525 g. Determine the specific heat of silver.

4. A 50.0-g sample of copper is at $25.0°C$. If $1\,200$ J of energy is added to it by heat, what is its final temperature?

WEB 5. A 1.50-kg iron horseshoe initially at $600°C$ is dropped into a bucket containing 20.0 kg of water at $25.0°C$. What is the final temperature? (Neglect the heat capacity of the container and assume that a negligible amount of water boils away.)

6. An aluminum cup with a mass of 200 g contains 800 g of water in thermal equilibrium at 80.0°C. The combination of cup and water is cooled uniformly so that the temperature decreases at a rate of 1.50°C/min. At what rate is energy being removed by heat? Express your answer in watts.

7. An aluminum calorimeter with a mass of 100 g contains 250 g of water. The calorimeter and water are in thermal equilibrium at 10.0°C. Two metallic blocks are placed into the water. One is a 50.0-g piece of copper at 80.0°C; the other block has a mass of 70.0 g and is originally at a temperature of 100°C. The entire system stabilizes at a final temperature of 20.0°C. (a) Determine the specific heat of the unknown sample. (b) Guess the material of the unknown, using the data given in Table 20.1.

8. Lake Erie contains roughly 4.00×10^{11} m^3 of water. (a) How much energy is required to raise the temperature of this volume of water from 11.0°C to 12.0°C? (b) Approximately how many years would it take to supply this amount of energy with the use of a 1 000-MW wasted energy output of an electric power plant?

9. A 3.00-g copper penny at 25.0°C drops from a height of 50.0 m to the ground. (a) If 60.0% of the change in potential energy goes into increasing the internal energy, what is its final temperature? (b) Does the result you obtained in (a) depend on the mass of the penny? Explain.

10. If a mass m_h of water at T_h is poured into an aluminum cup of mass m_{Al} containing mass m_c of water at T_c, where $T_h > T_c$, what is the equilibrium temperature of the system?

11. A water heater is operated by solar power. If the solar collector has an area of 6.00 m^2 and the power delivered by sunlight is 550 W/m^2, how long does it take to increase the temperature of 1.00 m^3 of water from 20.0°C to 60.0°C?

Section 20.3   Latent Heat

12. How much energy is required to change a 40.0-g ice cube from ice at -10.0°C to steam at 110°C?

13. A 3.00-g lead bullet at 30.0°C is fired at a speed of 240 m/s into a large block of ice at 0°C, in which it becomes embedded. What quantity of ice melts?

14. Steam at 100°C is added to ice at 0°C. (a) Find the amount of ice melted and the final temperature when the mass of steam is 10.0 g and the mass of ice is 50.0 g. (b) Repeat this calculation, taking the mass of steam as 1.00 g and the mass of ice as 50.0 g.

15. A 1.00-kg block of copper at 20.0°C is dropped into a large vessel of liquid nitrogen at 77.3 K. How many kilograms of nitrogen boil away by the time the copper reaches 77.3 K? (The specific heat of copper is 0.092 0 cal/g·°C. The latent heat of vaporization of nitrogen is 48.0 cal/g.)

16. A 50.0-g copper calorimeter contains 250 g of water at 20.0°C. How much steam must be condensed into the water if the final temperature of the system is to reach 50.0°C?

17. **WEB** In an insulated vessel, 250 g of ice at 0°C is added to 600 g of water at 18.0°C. (a) What is the final temperature of the system? (b) How much ice remains when the system reaches equilibrium?

18. **Review Problem.** Two speeding lead bullets, each having a mass of 5.00 g, a temperature of 20.0°C, and a speed of 500 m/s, collide head-on. Assuming a perfectly inelastic collision and no loss of energy to the atmosphere, describe the final state of the two-bullet system.

19. If 90.0 g of molten lead at 327.3°C is poured into a 300-g casting form made of iron and initially at 20.0°C, what is the final temperature of the system? (Assume that no energy loss to the environment occurs.)

Section 20.4   Work and Heat in Thermodynamic Processes

20. Gas in a container is at a pressure of 1.50 atm and a volume of 4.00 m^3. What is the work done by the gas (a) if it expands at constant pressure to twice its initial volume? (b) If it is compressed at constant pressure to one quarter of its initial volume?

21. **WEB** A sample of ideal gas is expanded to twice its original volume of 1.00 m^3 in a quasi-static process for which $P = \alpha V^2$, with $\alpha = 5.00$ atm/m^6, as shown in Figure P20.21. How much work is done by the expanding gas?

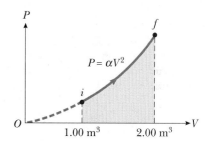

Figure P20.21

22. (a) Determine the work done by a fluid that expands from i to f as indicated in Figure P20.22. (b) How much

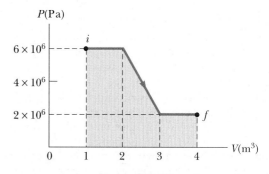

Figure P20.22

work is performed by the fluid if it is compressed from *f* to *i* along the same path?

23. One mole of an ideal gas is heated slowly so that it goes from *PV* state (P_i, V_i) to $(3P_i, 3V_i)$ in such a way that the pressure of the gas is directly proportional to the volume. (a) How much work is done in the process? (b) How is the temperature of the gas related to its volume during this process?

24. A sample of helium behaves as an ideal gas as energy is added by heat at constant pressure from 273 K to 373 K. If the gas does 20.0 J of work, what is the mass of helium present?

WEB 25. An ideal gas is enclosed in a cylinder with a movable piston on top. The piston has a mass of 8 000 g and an area of 5.00 cm² and is free to slide up and down, keeping the pressure of the gas constant. How much work is done as the temperature of 0.200 mol of the gas is raised from 20.0°C to 300°C?

26. An ideal gas is enclosed in a cylinder that has a movable piston on top. The piston has a mass *m* and an area *A* and is free to slide up and down, keeping the pressure of the gas constant. How much work is done as the temperature of *n* mol of the gas is raised from T_1 to T_2?

27. A gas expands from *I* to *F* along three possible paths, as indicated in Figure P20.27. Calculate the work in joules done by the gas along the paths *IAF*, *IF*, and *IBF*.

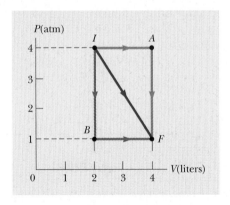

Figure P20.27

Section 20.5   The First Law of Thermodynamics

28. A gas is compressed from 9.00 L to 2.00 L at a constant pressure of 0.800 atm. In the process, 400 J of energy leaves the gas by heat. (a) What is the work done by the gas? (b) What is the change in its internal energy?

29. A thermodynamic system undergoes a process in which its internal energy decreases by 500 J. If, at the same time, 220 J of work is done on the system, what is the energy transferred to or from it by heat?

30. A gas is taken through the cyclic process described in Figure P20.30. (a) Find the net energy transferred to the system by heat during one complete cycle. (b) If the

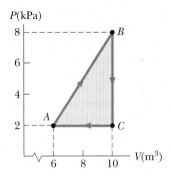

Figure P20.30   Problems 30 and 31.

cycle is reversed—that is, if the process follows the path *ACBA*—what is the net energy input per cycle by heat?

31. Consider the cyclic process depicted in Figure P20.30. If *Q* is negative for the process *BC*, and if ΔE_{int} is negative for the process *CA*, what are the signs of *Q*, *W*, and ΔE_{int} that are associated with each process?

32. A sample of an ideal gas goes through the process shown in Figure P20.32. From *A* to *B*, the process is adiabatic; from *B* to *C*, it is isobaric, with 100 kJ of energy flowing into the system by heat. From *C* to *D*, the process is isothermal; from *D* to *A*, it is isobaric, with 150 kJ of energy flowing out of the system by heat. Determine the difference in internal energy, $E_{int, B} - E_{int, A}$.

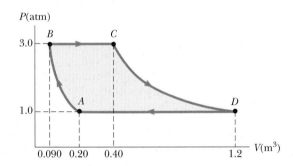

Figure P20.32

Section 20.6   Some Applications of the First Law of Thermodynamics

33. An ideal gas initially at 300 K undergoes an isobaric expansion at 2.50 kPa. If the volume increases from 1.00 m³ to 3.00 m³ and if 12.5 kJ of energy is transferred to the gas by heat, what are (a) the change in its internal energy and (b) its final temperature?

34. One mole of an ideal gas does 3 000 J of work on its surroundings as it expands isothermally to a final pressure of 1.00 atm and a volume of 25.0 L. Determine (a) the initial volume and (b) the temperature of the gas.

35. How much work is done by the steam when 1.00 mol of water at 100°C boils and becomes 1.00 mol of steam at

100°C and at 1.00 atm pressure? Assuming the steam to be an ideal gas, determine the change in internal energy of the steam as it vaporizes.

36. A 1.00-kg block of aluminum is heated at atmospheric pressure such that its temperature increases from 22.0°C to 40.0°C. Find (a) the work done by the aluminum, (b) the energy added to it by heat, and (c) the change in its internal energy.

37. A 2.00-mol sample of helium gas initially at 300 K and 0.400 atm is compressed isothermally to 1.20 atm. Assuming the behavior of helium to be that of an ideal gas, find (a) the final volume of the gas, (b) the work done by the gas, and (c) the energy transferred by heat.

38. One mole of water vapor at a temperature of 373 K cools down to 283 K. The energy given off from the cooling vapor by heat is absorbed by 10.0 mol of an ideal gas, causing it to expand at a constant temperature of 273 K. If the final volume of the ideal gas is 20.0 L, what is the initial volume of the ideal gas?

39. An ideal gas is carried through a thermodynamic cycle consisting of two isobaric and two isothermal processes, as shown in Figure P20.39. Show that the net work done in the entire cycle is given by the equation

$$W_{net} = P_1(V_2 - V_1) \ln \frac{P_2}{P_1}$$

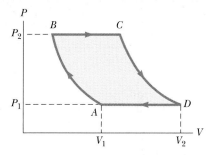

Figure P20.39

40. In Figure P20.40, the change in internal energy of a gas that is taken from A to C is + 800 J. The work done along the path ABC is + 500 J. (a) How much energy must be added to the system by heat as it goes from A through B and on to C? (b) If the pressure at point A is five times that at point C, what is the work done by the system in going from C to D? (c) What is the energy exchanged with the surroundings by heat as the gas is taken from C to A along the green path? (d) If the change in internal energy in going from point D to point A is + 500 J, how much energy must be added to the system by heat as it goes from point C to point D?

Section 20.7 Energy Transfer Mechanisms

41. A steam pipe is covered with 1.50-cm-thick insulating material with a thermal conductivity of 0.200 cal/cm · °C · s.

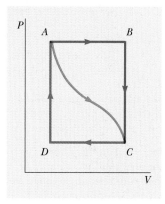

Figure P20.40

How much energy is lost every second by heat when the steam is at 200°C and the surrounding air is at 20.0°C? The pipe has a circumference of 20.0 cm and a length of 50.0 m. Neglect losses through the ends of the pipe.

42. A box with a total surface area of 1.20 m² and a wall thickness of 4.00 cm is made of an insulating material. A 10.0-W electric heater inside the box maintains the inside temperature at 15.0°C above the outside temperature. Find the thermal conductivity k of the insulating material.

43. A glass window pane has an area of 3.00 m² and a thickness of 0.600 cm. If the temperature difference between its surfaces is 25.0°C, what is the rate of energy transfer by conduction through the window?

44. A thermal window with an area of 6.00 m² is constructed of two layers of glass, each 4.00 mm thick and separated from each other by an air space of 5.00 mm. If the inside surface is at 20.0°C and the outside is at − 30.0°C, what is the rate of energy transfer by conduction through the window?

45. A bar of gold is in thermal contact with a bar of silver of the same length and area (Fig. P20.45). One end of the compound bar is maintained at 80.0°C, while the opposite end is at 30.0°C. When the rate of energy transfer by conduction reaches steady state, what is the temperature at the junction?

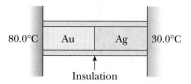

Figure P20.45

46. Two rods of the same length but made of different materials and having different cross-sectional areas are placed side by side, as shown in Figure P20.46. Deter-

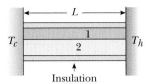

Figure P20.46

mine the rate of energy transfer by conduction in terms of the thermal conductivity and the area of each rod. Generalize your result to a system consisting of several rods.

47. Calculate the R value of (a) a window made of a single pane of flat glass $\frac{1}{8}$ in. thick; (b) a thermal window made of two single panes, each $\frac{1}{8}$ in. thick and separated by a $\frac{1}{4}$-in. air space. (c) By what factor is the thermal conduction reduced if the thermal window replaces the single-pane window?

48. The surface of the Sun has a temperature of about 5 800 K. The radius of the Sun is 6.96×10^8 m. Calculate the total energy radiated by the Sun each second. (Assume that $e = 0.965$.)

49. A large, hot pizza floats in outer space. What is the order of magnitude (a) of its rate of energy loss? (b) of its rate of temperature change? List the quantities you estimate and the value you estimate for each.

50. The tungsten filament of a certain 100-W light bulb radiates 2.00 W of light. (The other 98 W is carried away by convection and conduction.) The filament has a surface area of 0.250 mm^2 and an emissivity of 0.950. Find the filament's temperature. (The melting point of tungsten is 3 683 K.)

51. At high noon, the Sun delivers 1 000 W to each square meter of a blacktop road. If the hot asphalt loses energy only by radiation, what is its equilibrium temperature?

52. At our distance from the Sun, the intensity of solar radiation is 1 340 W/m^2. The temperature of the Earth is affected by the so-called "greenhouse effect" of the atmosphere. This effect makes our planet's emissivity for visible light higher than its emissivity for infrared light. For comparison, consider a spherical object with no atmosphere at the same distance from the Sun as the Earth. Assume that its emissivity is the same for all kinds of electromagnetic waves and that its temperature is uniform over its surface. Identify the projected area over which it absorbs sunlight and the surface area over which it radiates. Compute its equilibrium temperature. Chilly, isn't it? Your calculation applies to (a) the average temperature of the Moon, (b) astronauts in mortal danger aboard the crippled *Apollo 13* spacecraft, and (c) global catastrophe on the Earth if widespread fires caused a layer of soot to accumulate throughout the upper atmosphere so that most of the radiation from the Sun was absorbed there rather than at the surface below the atmosphere.

ADDITIONAL PROBLEMS

53. One hundred grams of liquid nitrogen at 77.3 K is stirred into a beaker containing 200 g of water at 5.00°C. If the nitrogen leaves the solution as soon as it turns to gas, how much water freezes? (The latent heat of vaporization of nitrogen is 48.0 cal/g, and the latent heat of fusion of water is 79.6 cal/g.)

54. A 75.0-kg cross-country skier moves across the snow (Fig. P20.54). The coefficient of friction between the skis and the snow is 0.200. Assume that all the snow beneath his skis is at 0°C and that all the internal energy generated by friction is added to the snow, which sticks to his skis until it melts. How far would he have to ski to melt 1.00 kg of snow?

Figure P20.54   A cross-country skier. *(Nathan Bilow/Leo de Wys, Inc.)*

55. An aluminum rod 0.500 m in length and with a cross-sectional area 2.50 cm^2 is inserted into a thermally insulated vessel containing liquid helium at 4.20 K. The rod is initially at 300 K. (a) If one half of the rod is inserted into the helium, how many liters of helium boil off by the time the inserted half cools to 4.20 K? (Assume that the upper half does not yet cool.) (b) If the upper end of the rod is maintained at 300 K, what is the approximate boil-off rate of liquid helium after the lower half has reached 4.20 K? (Aluminum has thermal conductivity of 31.0 J/s·cm·K at 4.2 K; ignore its temperature variation. Aluminum has a specific heat of 0.210 cal/g·°C and density of 2.70 g/cm^3. The density of liquid helium is 0.125 g/cm^3.)

56. On a cold winter day, you buy a hot dog from a street vendor. Into the pocket of your down parka you put the change he gives you: coins consisting of 9.00 g of copper at −12.0°C. Your pocket already contains 14.0 g of silver coins at 30.0°C. A short time later, the temperature of the copper coins is 4.00°C and is increasing at a rate of 0.500°C/s. At this time (a) what is the temperature of the silver coins, and (b) at what rate is it changing? (Neglect energy transferred to the surroundings.)

57. A *flow calorimeter* is an apparatus used to measure the specific heat of a liquid. The technique of flow calorimetry involves measuring the temperature difference between the input and output points of a flowing stream of the liquid while energy is added by heat at a known rate. In one particular experiment, a liquid with a density of 0.780 g/cm^3 flows through the calorimeter at the rate of $4.00 \text{ cm}^3/\text{s}$. At steady state, a temperature difference of $4.80°C$ is established between the input and output points when energy is supplied by heat at the rate of 30.0 J/s. What is the specific heat of the liquid?

58. A *flow calorimeter* is an apparatus used to measure the specific heat of a liquid. The technique of flow calorimetry involves measuring the temperature difference between the input and output points of a flowing stream of the liquid while energy is added by heat at a known rate. In one particular experiment, a liquid of density ρ flows through the calorimeter with volume flow rate R. At steady state, a temperature difference ΔT is established between the input and output points when energy is supplied at the rate $\mathscr{P}$. What is the specific heat of the liquid?

59. One mole of an ideal gas, initially at 300 K, is cooled at constant volume so that the final pressure is one-fourth the initial pressure. The gas then expands at constant pressure until it reaches the initial temperature. Determine the work done by the gas.

60. One mole of an ideal gas is contained in a cylinder with a movable piston. The initial pressure, volume, and temperature are P_i, V_i, and T_i, respectively. Find the work done by the gas for the following processes and show each process on a PV diagram: (a) An isobaric compression in which the final volume is one-half the initial volume. (b) An isothermal compression in which the final pressure is four times the initial pressure. (c) An isovolumetric process in which the final pressure is triple the initial pressure.

61. An ideal gas initially at P_i, V_i, and T_i is taken through a cycle as shown in Figure P20.61. (a) Find the net work done by the gas per cycle. (b) What is the net energy added by heat to the system per cycle? (c) Obtain a nu-

merical value for the net work done per cycle for 1.00 mol of gas initially at 0°C.

62. **Review Problem.** An iron plate is held against an iron wheel so that a sliding frictional force of 50.0 N acts between the two pieces of metal. The relative speed at which the two surfaces slide over each other is 40.0 m/s. (a) Calculate the rate at which mechanical energy is converted to internal energy. (b) The plate and the wheel each have a mass of 5.00 kg, and each receives 50.0% of the internal energy. If the system is run as described for 10.0 s and each object is then allowed to reach a uniform internal temperature, what is the resultant temperature increase?

WEB 63. A "solar cooker" consists of a curved reflecting mirror that focuses sunlight onto the object to be warmed (Fig. P20.63). The solar power per unit area reaching the Earth at the location is 600 W/m^2, and the cooker has a diameter of 0.600 m. Assuming that 40.0% of the incident energy is transferred to the water, how long does it take to completely boil off 0.500 L of water initially at 20.0°C? (Neglect the heat capacity of the container.)

Figure P20.63

64. Water in an electric teakettle is boiling. The power absorbed by the water is 1.00 kW. Assuming that the pressure of the vapor in the kettle equals atmospheric pressure, determine the speed of effusion of vapor from the kettle's spout if the spout has a cross-sectional area of 2.00 cm^2.

65. Liquid water evaporates and even boils at temperatures other than 100°C, depending on the ambient pressure. Suppose that the latent heat of vaporization in Table 20.2 describes the liquid–vapor transition at all temperatures. A chamber contains 1.00 kg of water at 0°C under a piston, which just touches the water's surface. The piston is then raised quickly so that part of the water is vaporized and the other part is frozen (no liquid remains). Assuming that the temperature remains con-

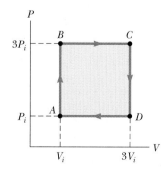

Figure P20.61

stant at 0°C, determine the mass of the ice that forms in the chamber.

66. A cooking vessel on a slow burner contains 10.0 kg of water and an unknown mass of ice in equilibrium at 0°C at time $t = 0$. The temperature of the mixture is measured at various times, and the result is plotted in Figure P20.66. During the first 50.0 min, the mixture remains at 0°C. From 50.0 min to 60.0 min, the temperature increases to 2.00°C. Neglecting the heat capacity of the vessel, determine the initial mass of the ice.

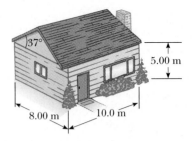

Figure P20.68

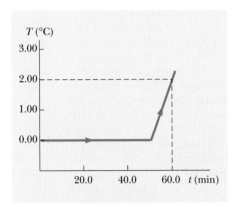

Figure P20.66

67. **Review Problem.** (a) In air at 0°C, a 1.60-kg copper block at 0°C is set sliding at 2.50 m/s over a sheet of ice at 0°C. Friction brings the block to rest. Find the mass of the ice that melts. To describe the process of slowing down, identify the energy input Q, the work output W, the change in internal energy ΔE_{int}, and the change in mechanical energy ΔK for both the block and the ice. (b) A 1.60-kg block of ice at 0°C is set sliding at 2.50 m/s over a sheet of copper at 0°C. Friction brings the block to rest. Find the mass of the ice that melts. Identify Q, W, ΔE_{int}, and ΔK for the block and for the metal sheet during the process. (c) A thin 1.60-kg slab of copper at 20°C is set sliding at 2.50 m/s over an identical stationary slab at the same temperature. Friction quickly stops the motion. If no energy is lost to the environment by heat, find the change in temperature of both objects. Identify Q, W, ΔE_{int}, and ΔK for each object during the process.

68. The average thermal conductivity of the walls (including the windows) and roof of the house depicted in Figure P20.68 is 0.480 W/m · °C, and their average thickness is 21.0 cm. The house is heated with natural gas having a heat of combustion (that is, the energy provided per cubic meter of gas burned) of 9 300 kcal/m^3. How many cubic meters of gas must be burned each day to maintain an inside temperature of 25.0°C if the outside temperature is 0.0°C? Disregard radiation and the energy lost by heat through the ground.

69. A pond of water at 0°C is covered with a layer of ice 4.00 cm thick. If the air temperature stays constant at −10.0°C, how long does it take the ice's thickness to increase to 8.00 cm? (*Hint:* To solve this problem, use Equation 20.14 in the form

$$\frac{dQ}{dt} = kA \frac{\Delta T}{x}$$

and note that the incremental energy dQ extracted from the water through the thickness x of ice is the amount required to freeze a thickness dx of ice. That is, $dQ = L\rho A \, dx$, where ρ is the density of the ice, A is the area, and L is the latent heat of fusion.)

70. The inside of a hollow cylinder is maintained at a temperature T_a while the outside is at a lower temperature T_b (Fig. P20.70). The wall of the cylinder has a thermal conductivity k. Neglecting end effects, show that the rate of energy conduction from the inner to the outer wall in the radial direction is

$$\frac{dQ}{dt} = 2\pi Lk \left[\frac{T_a - T_b}{\ln(b/a)} \right]$$

(*Hint:* The temperature gradient is dT/dr. Note that a radial flow of energy occurs through a concentric cylinder of area $2\pi rL$.)

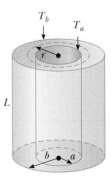

Figure P20.70

Answers to Quick Quizzes

71. The passenger section of a jet airliner has the shape of a cylindrical tube with a length of 35.0 m and an inner radius of 2.50 m. Its walls are lined with an insulating material 6.00 cm in thickness and having a thermal conductivity of 4.00×10^{-5} cal/s·cm·°C. A heater must maintain the interior temperature at 25.0°C while the outside temperature is at -35.0°C. What power must be supplied to the heater if this temperature difference is to be maintained? (Use the result you obtained in Problem 70.)

72. A student obtains the following data in a calorimetry experiment designed to measure the specific heat of aluminum:

Initial temperature of water and calorimeter	70°C
Mass of water	0.400 kg
Mass of calorimeter	0.040 kg
Specific heat of calorimeter	0.63 kJ/kg·°C
Initial temperature of aluminum	27°C
Mass of aluminum	0.200 kg
Final temperature of mixture	66.3°C

Use these data to determine the specific heat of aluminum. Your result should be within 15% of the value listed in Table 20.1.

Answers to Quick Quizzes

20.1 (a) Water, glass, iron. Because water has the highest specific heat (4 186 J/kg·°C), it has the smallest change in temperature. Glass is next (837 J/kg·°C), and iron is last (448 J/kg·°C). (b) Iron, glass, water. For a given temperature increase, the energy transfer by heat is proportional to the specific heat.

20.2 Steam. According to Table 20.2, a kilogram of 100°C steam releases 2.26×10^6 J of energy as it condenses to 100°C water. After it releases this much energy into your skin, it is identical to 100°C water and will continue to burn you.

20.3 C, A, E. The slope is the ratio of the temperature change to the amount of energy input. Thus, the slope is proportional to the reciprocal of the specific heat. Water, which has the highest specific heat, has the least slope.

20.4

Situation	System	Q	W	ΔE_{int}
(a) Rapidly pumping up a bicycle tire	Air in the pump	0	−	+
(b) Pan of room-temperature water sitting on a hot stove	Water in the pan	+	0	+
(c) Air quickly leaking out of a balloon	Air originally in the balloon	0	+	−

(a) Because the pumping is rapid, no energy enters or leaves the system by heat; thus, $Q = 0$. Because work is done *on* the system, this work is negative. Thus, $\Delta E_{int} = Q - W$ must be positive. The air in the pump is warmer. (b) No work is done either by or on the system, but energy flows into the water by heat from the hot burner, making both Q and ΔE_{int} positive. (c) Because the leak is rapid, no energy flows into or out of the system by heat; hence, $Q = 0$. The air molecules escaping from the balloon do work on the surrounding air molecules as they push them out of the way. Thus, W is positive and ΔE_{int} is negative. The decrease in internal energy is evidenced by the fact that the escaping air becomes cooler.

20.5 A is isovolumetric, B is adiabatic, C is isothermal, and D is isobaric.

20.6 c. The blanket acts as a thermal insulator, slowing the transfer of energy by heat from the air into the cube.

c h a p t e r

The Kinetic Theory of Gases

Chapter Outline

In Chapter 19 we discussed the properties of an ideal gas, using such macroscopic variables as pressure, volume, and temperature. We shall now show that such large-scale properties can be described on a microscopic scale, where matter is treated as a collection of molecules. Newton's laws of motion applied in a statistical manner to a collection of particles provide a reasonable description of thermodynamic processes. To keep the mathematics relatively simple, we shall consider molecular behavior of gases only, because in gases the interactions between molecules are much weaker than they are in liquids or solids. In the current view of gas behavior, called the *kinetic theory*, gas molecules move about in a random fashion, colliding with the walls of their container and with each other. Perhaps the most important feature of this theory is that it demonstrates that the kinetic energy of molecular motion and the internal energy of a gas system are equivalent. Furthermore, the kinetic theory provides us with a physical basis for our understanding of the concept of temperature.

In the simplest model of a gas, each molecule is considered to be a hard sphere that collides elastically with other molecules and with the container's walls. The hard-sphere model assumes that the molecules do not interact with each other except during collisions and that they are not deformed by collisions. This description is adequate only for monatomic gases, for which the energy is entirely translational kinetic energy. One must modify the theory for more complex molecules, such as oxygen (O_2) and carbon dioxide (CO_2), to include the internal energy associated with rotations and vibrations of the molecules.

21.1 MOLECULAR MODEL OF AN IDEAL GAS

10.5 We begin this chapter by developing a microscopic model of an ideal gas. The model shows that the pressure that a gas exerts on the walls of its container is a consequence of the collisions of the gas molecules with the walls. As we shall see, the model is consistent with the macroscopic description of Chapter 19. In developing this model, we make the following assumptions:

- The number of molecules is large, and the average separation between molecules is great compared with their dimensions. This means that the volume of the molecules is negligible when compared with the volume of the container.
- The molecules obey Newton's laws of motion, but as a whole they move randomly. By "randomly" we mean that any molecule can move in any direction with equal probability. We also assume that the distribution of speeds does not change in time, despite the collisions between molecules. That is, at any given moment, a certain percentage of molecules move at high speeds, a certain percentage move at low speeds, and a certain percentage move at speeds intermediate between high and low.
- The molecules undergo elastic collisions with each other and with the walls of the container. Thus, in the collisions, both kinetic energy and momentum are constant.
- The forces between molecules are negligible except during a collision. The forces between molecules are short-range, so the molecules interact with each other only during collisions.
- The gas under consideration is a pure substance. That is, all of its molecules are identical.

Although we often picture an ideal gas as consisting of single atoms, we can assume that the behavior of molecular gases approximates that of ideal gases rather

Assumptions of the molecular model of an ideal gas

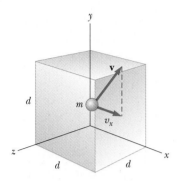

Figure 21.1 A cubical box with sides of length d containing an ideal gas. The molecule shown moves with velocity **v**.

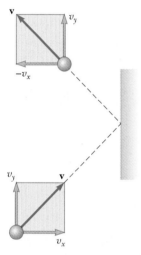

Figure 21.2 A molecule makes an elastic collision with the wall of the container. Its x component of momentum is reversed, while its y component remains unchanged. In this construction, we assume that the molecule moves in the xy plane.

well at low pressures. Molecular rotations or vibrations have no effect, on the average, on the motions that we considered here.

Now let us derive an expression for the pressure of an ideal gas consisting of N molecules in a container of volume V. The container is a cube with edges of length d (Fig. 21.1). Consider the collision of one molecule moving with a velocity **v** toward the right-hand face of the box. The molecule has velocity components v_x, v_y, and v_z. Previously, we used m to represent the mass of a sample, but throughout this chapter we shall use m to represent the mass of one molecule. As the molecule collides with the wall elastically, its x component of velocity is reversed, while its y and z components of velocity remain unaltered (Fig. 21.2). Because the x component of the momentum of the molecule is mv_x before the collision and $-mv_x$ after the collision, the change in momentum of the molecule is

$$\Delta p_x = -mv_x - (mv_x) = -2mv_x$$

Applying the impulse–momentum theorem (Eq. 9.9) to the molecule gives

$$F_1\,\Delta t = \Delta p_x = -2mv_x$$

where F_1 is the magnitude of the average force exerted by the wall on the molecule in the time Δt. The subscript 1 indicates that we are currently considering only *one* molecule. For the molecule to collide twice with the same wall, it must travel a distance $2d$ in the x direction. Therefore, the time interval between two collisions with the same wall is $\Delta t = 2d/v_x$. Over a time interval that is long compared with Δt, the average force exerted on the molecule for each collision is

$$F_1 = \frac{-2mv_x}{\Delta t} = \frac{-2mv_x}{2d/v_x} = \frac{-mv_x^2}{d} \qquad \textbf{(21.1)}$$

According to Newton's third law, the average force exerted by the molecule on the wall is equal in magnitude and opposite in direction to the force in Equation 21.1:

$$F_{1,\,\text{on wall}} = -F_1 = -\left(\frac{-mv_x^2}{d}\right) = \frac{mv_x^2}{d}$$

Each molecule of the gas exerts a force F_1 on the wall. We find the total force F exerted by all the molecules on the wall by adding the forces exerted by the individual molecules:

$$F = \frac{m}{d}\,(v_{x1}^2 + v_{x2}^2 + \cdots)$$

In this equation, v_{x1} is the x component of velocity of molecule 1, v_{x2} is the x component of velocity of molecule 2, and so on. The summation terminates when we reach N molecules because there are N molecules in the container.

To proceed further, we must note that the average value of the square of the velocity in the x direction for N molecules is

$$\overline{v_x^2} = \frac{v_{x1}^2 + v_{x2}^2 + \cdots + v_{xN}^2}{N}$$

Thus, the total force exerted on the wall can be written

$$F = \frac{Nm}{d}\,\overline{v_x^2}$$

Now let us focus on one molecule in the container whose velocity components are v_x, v_y, and v_z. The Pythagorean theorem relates the square of the speed of this molecule to the squares of these components:

$$v^2 = v_x{}^2 + v_y{}^2 + v_z{}^2$$

Hence, the average value of v^2 for all the molecules in the container is related to the average values of $v_x{}^2$, $v_y{}^2$, and $v_z{}^2$ according to the expression

$$\overline{v^2} = \overline{v_x{}^2} + \overline{v_y{}^2} + \overline{v_z{}^2}$$

Because the motion is completely random, the average values $\overline{v_x{}^2}$, $\overline{v_y{}^2}$, and $\overline{v_z{}^2}$ are equal to each other. Using this fact and the previous equation, we find that

$$\overline{v^2} = 3\overline{v_x{}^2}$$

Thus, the total force exerted on the wall is

$$F = \frac{N}{3}\left(\frac{m\overline{v^2}}{d}\right)$$

Using this expression, we can find the total pressure exerted on the wall:

$$P = \frac{F}{A} = \frac{F}{d^2} = \frac{1}{3}\left(\frac{N}{d^3}\,m\overline{v^2}\right) = \frac{1}{3}\left(\frac{N}{V}\right)m\overline{v^2}$$

$$P = \frac{2}{3}\left(\frac{N}{V}\right)\left(\frac{1}{2}\,m\overline{v^2}\right) \tag{21.2}$$

Ludwig Boltzmann **Austrian theoretical physicist (1844–1906)** Boltzmann made many important contributions to the development of the kinetic theory of gases, electromagnetism, and thermodynamics. His pioneering work in the field of kinetic theory led to the branch of physics known as *statistical mechanics*. *(Courtesy of AIP Niels Bohr Library, Lande Collection)*

Relationship between pressure and molecular kinetic energy

This result indicates that **the pressure is proportional to the number of molecules per unit volume and to the average translational kinetic energy of the molecules, $\frac{1}{2}m\overline{v^2}$.** In deriving this simplified model of an ideal gas, we obtain an important result that relates the large-scale quantity of pressure to an atomic quantity—the average value of the square of the molecular speed. Thus, we have established a key link between the atomic world and the large-scale world.

You should note that Equation 21.2 verifies some features of pressure with which you are probably familiar. One way to increase the pressure inside a container is to increase the number of molecules per unit volume in the container. This is what you do when you add air to a tire. The pressure in the tire can also be increased by increasing the average translational kinetic energy of the air molecules in the tire. As we shall soon see, this can be accomplished by increasing the temperature of that air. It is for this reason that the pressure inside a tire increases as the tire warms up during long trips. The continuous flexing of the tire as it moves along the surface of a road results in work done as parts of the tire distort and in an increase in internal energy of the rubber. The increased temperature of the rubber results in the transfer of energy by heat into the air inside the tire. This transfer increases the air's temperature, and this increase in temperature in turn produces an increase in pressure.

Molecular Interpretation of Temperature

We can gain some insight into the meaning of temperature by first writing Equation 21.2 in the more familiar form

$$PV = \tfrac{2}{3}N\left(\tfrac{1}{2}m\overline{v^2}\right)$$

Let us now compare this with the equation of state for an ideal gas (Eq. 19.10):

$$PV = Nk_{\mathrm{B}}T$$

Recall that the equation of state is based on experimental facts concerning the macroscopic behavior of gases. Equating the right sides of these expressions, we find that

Temperature is proportional to average kinetic energy

$$T = \frac{2}{3k_B}\left(\frac{1}{2}m\overline{v^2}\right)$$ **(21.3)**

That is, **temperature is a direct measure of average molecular kinetic energy.**

By rearranging Equation 21.3, we can relate the translational molecular kinetic energy to the temperature:

Average kinetic energy per molecule

$$\frac{1}{2}m\overline{v^2} = \frac{3}{2}k_B T$$ **(21.4)**

That is, the average translational kinetic energy per molecule is $\frac{3}{2}k_B T$. Because $\overline{v_x^2} = \frac{1}{3}\overline{v^2}$, it follows that

$$\frac{1}{2}m\overline{v_x^2} = \frac{1}{2}k_B T$$ **(21.5)**

In a similar manner, it follows that the motions in the y and z directions give us

$$\frac{1}{2}m\overline{v_y^2} = \frac{1}{2}k_B T \qquad \text{and} \qquad \frac{1}{2}m\overline{v_z^2} = \frac{1}{2}k_B T$$

Thus, each translational degree of freedom contributes an equal amount of energy to the gas, namely, $\frac{1}{2}k_B T$. (In general, "degrees of freedom" refers to the number of independent means by which a molecule can possess energy.) A generalization of this result, known as the **theorem of equipartition of energy,** states that

Theorem of equipartition of energy

each degree of freedom contributes $\frac{1}{2}k_B T$ to the energy of a system.

The total translational kinetic energy of N molecules of gas is simply N times the average energy per molecule, which is given by Equation 21.4:

Total translational kinetic energy of N molecules

$$E_{\text{trans}} = N\left(\frac{1}{2}m\overline{v^2}\right) = \frac{3}{2}Nk_B T = \frac{3}{2}nRT$$ **(21.6)**

where we have used $k_B = R/N_A$ for Boltzmann's constant and $n = N/N_A$ for the number of moles of gas. If we consider a gas for which the only type of energy for the molecules is translational kinetic energy, we can use Equation 21.6 to express

TABLE 21.1   Some rms Speeds

Gas	Molar Mass (g/mol)	v_{rms} at 20°C (m/s)
H_2	2.02	1904
He	4.00	1352
H_2O	18.0	637
Ne	20.2	602
N_2 or CO	28.0	511
NO	30.0	494
CO_2	44.0	408
SO_2	64.1	338

the internal energy of the gas. This result implies that the internal energy of an ideal gas depends only on the temperature.

The square root of $\overline{v^2}$ is called the *root-mean-square* (rms) *speed* of the molecules. From Equation 21.4 we obtain, for the rms speed,

$$v_{rms} = \sqrt{\overline{v^2}} = \sqrt{\frac{3k_B T}{m}} = \sqrt{\frac{3RT}{M}} \qquad (21.7)$$

Root-mean-square speed

where M is the molar mass in kilograms per mole. This expression shows that, at a given temperature, lighter molecules move faster, on the average, than do heavier molecules. For example, at a given temperature, hydrogen molecules, whose molar mass is 2×10^{-3} kg/mol, have an average speed four times that of oxygen molecules, whose molar mass is 32×10^{-3} kg/mol. Table 21.1 lists the rms speeds for various molecules at 20°C.

EXAMPLE 21.1   A Tank of Helium

A tank used for filling helium balloons has a volume of 0.300 m^3 and contains 2.00 mol of helium gas at 20.0°C. Assuming that the helium behaves like an ideal gas, (a) what is the total translational kinetic energy of the molecules of the gas?

Solution Using Equation 21.6 with $n = 2.00$ mol and $T = 293$ K, we find that

$$E_{trans} = \tfrac{3}{2} nRT = \tfrac{3}{2}(2.00 \text{ mol})(8.31 \text{ J/mol} \cdot \text{K})(293 \text{ K})$$

$$= 7.30 \times 10^3 \text{ J}$$

(b) What is the average kinetic energy per molecule?

Solution Using Equation 21.4, we find that the average kinetic energy per molecule is

$$\tfrac{1}{2} m\overline{v^2} = \tfrac{3}{2} k_B T = \tfrac{3}{2}(1.38 \times 10^{-23} \text{ J/K})(293 \text{ K})$$

$$= 6.07 \times 10^{-21} \text{ J}$$

Exercise Using the fact that the molar mass of helium is 4.00×10^{-3} kg/mol, determine the rms speed of the atoms at 20.0°C.

Answer 1.35×10^3 m/s.

Quick Quiz 21.1

At room temperature, the average speed of an air molecule is several hundred meters per second. A molecule traveling at this speed should travel across a room in a small fraction of a second. In view of this, why does it take the odor of perfume (or other smells) several minutes to travel across the room?

21.2   MOLAR SPECIFIC HEAT OF AN IDEAL GAS

10.5

The energy required to raise the temperature of n moles of gas from T_i to T_f depends on the path taken between the initial and final states. To understand this, let us consider an ideal gas undergoing several processes such that the change in temperature is $\Delta T = T_f - T_i$ for all processes. The temperature change can be achieved by taking a variety of paths from one isotherm to another, as shown in Figure 21.3. Because ΔT is the same for each path, the change in internal energy ΔE_{int} is the same for all paths. However, we know from the first law, $Q = \Delta E_{int} + W$, that the heat Q is different for each path because W (the area under the curves) is different for each path. Thus, the heat associated with a given change in temperature does not have a unique value.

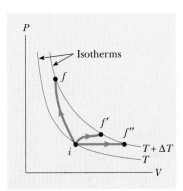

Figure 21.3 An ideal gas is taken from one isotherm at temperature T to another at temperature $T + \Delta T$ along three different paths.

We can address this difficulty by defining specific heats for two processes that frequently occur: changes at constant volume and changes at constant pressure. Because the number of moles is a convenient measure of the amount of gas, we define the **molar specific heats** associated with these processes with the following equations:

$$Q = nC_V\Delta T \qquad \text{(constant volume)} \tag{21.8}$$

$$Q = nC_P\Delta T \qquad \text{(constant pressure)} \tag{21.9}$$

where C_V is the **molar specific heat at constant volume** and C_P is the **molar specific heat at constant pressure.** When we heat a gas at constant pressure, not only does the internal energy of the gas increase, but the gas also does work because of the change in volume. Therefore, the heat $Q_{\text{constant } P}$ must account for both the increase in internal energy and the transfer of energy out of the system by work, and so $Q_{\text{constant } P}$ is greater than $Q_{\text{constant } V}$. Thus, C_P is greater than C_V.

In the previous section, we found that the temperature of a gas is a measure of the average translational kinetic energy of the gas molecules. This kinetic energy is associated with the motion of the center of mass of each molecule. It does not include the energy associated with the internal motion of the molecule—namely, vibrations and rotations about the center of mass. This should not be surprising because the simple kinetic theory model assumes a structureless molecule.

In view of this, let us first consider the simplest case of an ideal monatomic gas, that is, a gas containing one atom per molecule, such as helium, neon, or argon. When energy is added to a monatomic gas in a container of fixed volume (by heating, for example), all of the added energy goes into increasing the translational kinetic energy of the atoms. There is no other way to store the energy in a monatomic gas. Therefore, from Equation 21.6, we see that the total internal energy E_{int} of N molecules (or n mol) of an ideal monatomic gas is

> Internal energy of an ideal monatomic gas is proportional to its temperature

$$E_{\text{int}} = \tfrac{3}{2}Nk_B T = \tfrac{3}{2}nRT \tag{21.10}$$

Note that for a monatomic ideal gas, E_{int} is a function of T only, and the functional relationship is given by Equation 21.10. In general, the internal energy of an ideal gas is a function of T only, and the exact relationship depends on the type of gas, as we shall soon explore.

Quick Quiz 21.2

How does the internal energy of a gas change as its pressure is decreased while its volume is increased in such a way that the process follows the isotherm labeled T in Figure 21.4? (a) E_{int} increases. (b) E_{int} decreases. (c) E_{int} stays the same. (d) There is not enough information to determine ΔE_{int}.

If energy is transferred by heat to a system at *constant volume*, then no work is done by the system. That is, $W = \int P\,dV = 0$ for a constant-volume process. Hence, from the first law of thermodynamics, we see that

$$Q = \Delta E_{\text{int}} \tag{21.11}$$

In other words, all of the energy transferred by heat goes into increasing the internal energy (and temperature) of the system. A constant-volume process from i to f is described in Figure 21.4, where ΔT is the temperature difference between the two isotherms. Substituting the expression for Q given by Equation 21.8 into

Equation 21.11, we obtain

$$\Delta E_{\text{int}} = nC_V \Delta T \tag{21.12}$$

If the molar specific heat is constant, we can express the internal energy of a gas as

$$E_{\text{int}} = nC_V T$$

This equation applies to all ideal gases—to gases having more than one atom per molecule, as well as to monatomic ideal gases.

In the limit of infinitesimal changes, we can use Equation 21.12 to express the molar specific heat at constant volume as

$$C_V = \frac{1}{n} \frac{dE_{\text{int}}}{dT} \tag{21.13}$$

Let us now apply the results of this discussion to the monatomic gas that we have been studying. Substituting the internal energy from Equation 21.10 into Equation 21.13, we find that

$$C_V = \tfrac{3}{2}R \tag{21.14}$$

This expression predicts a value of $C_V = \tfrac{3}{2}R = 12.5 \text{ J/mol·K}$ for all monatomic gases. This is in excellent agreement with measured values of molar specific heats for such gases as helium, neon, argon, and xenon over a wide range of temperatures (Table 21.2).

Now suppose that the gas is taken along the constant-pressure path $i \rightarrow f'$ shown in Figure 21.4. Along this path, the temperature again increases by ΔT. The energy that must be transferred by heat to the gas in this process is $Q = nC_P \Delta T$. Because the volume increases in this process, the work done by the gas is $W = P\Delta V$, where P is the constant pressure at which the process occurs. Applying

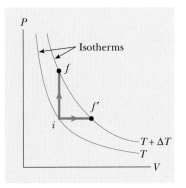

Figure 21.4 Energy is transferred by heat to an ideal gas in two ways. For the constant-volume path $i \rightarrow f$, all the energy goes into increasing the internal energy of the gas because no work is done. Along the constant-pressure path $i \rightarrow f'$, part of the energy transferred in by heat is transferred out by work done by the gas.

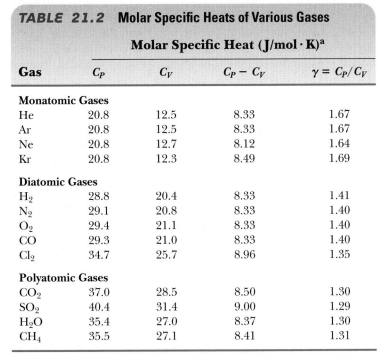

TABLE 21.2	Molar Specific Heats of Various Gases			
	Molar Specific Heat (J/mol · K)[a]			
Gas	C_P	C_V	$C_P - C_V$	$\gamma = C_P/C_V$
Monatomic Gases				
He	20.8	12.5	8.33	1.67
Ar	20.8	12.5	8.33	1.67
Ne	20.8	12.7	8.12	1.64
Kr	20.8	12.3	8.49	1.69
Diatomic Gases				
H_2	28.8	20.4	8.33	1.41
N_2	29.1	20.8	8.33	1.40
O_2	29.4	21.1	8.33	1.40
CO	29.3	21.0	8.33	1.40
Cl_2	34.7	25.7	8.96	1.35
Polyatomic Gases				
CO_2	37.0	28.5	8.50	1.30
SO_2	40.4	31.4	9.00	1.29
H_2O	35.4	27.0	8.37	1.30
CH_4	35.5	27.1	8.41	1.31

[a] All values except that for water were obtained at 300 K.

the first law to this process, we have

$$\Delta E_{\text{int}} = Q - W = nC_P \Delta T - P \Delta V \qquad \textbf{(21.15)}$$

In this case, the energy added to the gas by heat is channeled as follows: Part of it does external work (that is, it goes into moving a piston), and the remainder increases the internal energy of the gas. But the change in internal energy for the process $i \rightarrow f'$ is equal to that for the process $i \rightarrow f$ because E_{int} depends only on temperature for an ideal gas and because ΔT is the same for both processes. In addition, because $PV = nRT$, we note that for a constant-pressure process, $P \Delta V = nR \Delta T$. Substituting this value for $P \Delta V$ into Equation 21.15 with $\Delta E_{\text{int}} = nC_V \Delta T$ (Eq. 21.12) gives

$$nC_V \Delta T = nC_P \Delta T - nR \Delta T$$

$$C_P - C_V = R \qquad \textbf{(21.16)}$$

This expression applies to *any* ideal gas. It predicts that the molar specific heat of an ideal gas at constant pressure is greater than the molar specific heat at constant volume by an amount R, the universal gas constant (which has the value $8.31 \text{ J/mol} \cdot \text{K}$). This expression is applicable to real gases, as the data in Table 21.2 show.

Because $C_V = \frac{3}{2}R$ for a monatomic ideal gas, Equation 21.16 predicts a value $C_P = \frac{5}{2}R = 20.8 \text{ J/mol} \cdot \text{K}$ for the molar specific heat of a monatomic gas at constant pressure. The ratio of these heat capacities is a dimensionless quantity γ (Greek letter gamma):

Ratio of molar specific heats for a monatomic ideal gas

$$\gamma = \frac{C_P}{C_V} = \frac{(5/2)R}{(3/2)R} = \frac{5}{3} = 1.67 \qquad \textbf{(21.17)}$$

Theoretical values of C_P and γ are in excellent agreement with experimental values obtained for monatomic gases, but they are in serious disagreement with the values for the more complex gases (see Table 21.2). This is not surprising because the value $C_V = \frac{3}{2}R$ was derived for a monatomic ideal gas, and we expect some additional contribution to the molar specific heat from the internal structure of the more complex molecules. In Section 21.4, we describe the effect of molecular structure on the molar specific heat of a gas. We shall find that the internal energy—and, hence, the molar specific heat—of a complex gas must include contributions from the rotational and the vibrational motions of the molecule.

We have seen that the molar specific heats of gases at constant pressure are greater than the molar specific heats at constant volume. This difference is a consequence of the fact that in a constant-volume process, no work is done and all of the energy transferred by heat goes into increasing the internal energy (and temperature) of the gas, whereas in a constant-pressure process, some of the energy transferred by heat is transferred out as work done by the gas as it expands. In the case of solids and liquids heated at constant pressure, very little work is done because the thermal expansion is small. Consequently, C_P and C_V are approximately equal for solids and liquids.

EXAMPLE 21.2 ▸ Heating a Cylinder of Helium

A cylinder contains 3.00 mol of helium gas at a temperature of 300 K. (a) If the gas is heated at constant volume, how much energy must be transferred by heat to the gas for its temperature to increase to 500 K?

Solution   For the constant-volume process, we have

$$Q_1 = nC_V \Delta T$$

Because $C_V = 12.5 \text{ J/mol} \cdot \text{K}$ for helium and $\Delta T = 200 \text{ K}$, we obtain

$$Q_1 = (3.00 \text{ mol}) (12.5 \text{ J/mol} \cdot \text{K}) (200 \text{ K}) = \boxed{7.50 \times 10^3 \text{ J}}$$

(b) How much energy must be transferred by heat to the gas at constant pressure to raise the temperature to 500 K?

Solution   Making use of Table 21.2, we obtain

$$Q_2 = nC_P \Delta T = (3.00 \text{ mol}) (20.8 \text{ J/mol} \cdot \text{K}) (200 \text{ K})$$

$$= \boxed{12.5 \times 10^3 \text{ J}}$$

Exercise   What is the work done by the gas in this isobaric process?

Answer $W = Q_2 - Q_1 = 5.00 \times 10^3 \text{ J}$.

21.3 ADIABATIC PROCESSES FOR AN IDEAL GAS

As we noted in Section 20.6, an adiabatic process is one in which no energy is transferred by heat between a system and its surroundings. For example, if a gas is compressed (or expanded) very rapidly, very little energy is transferred out of (or into) the system by heat, and so the process is nearly adiabatic. (We must remember that the temperature of a system changes in an adiabatic process even though no energy is transferred by heat.) Such processes occur in the cycle of a gasoline engine, which we discuss in detail in the next chapter.

Another example of an adiabatic process is the very slow expansion of a gas that is thermally insulated from its surroundings. In general,

an **adiabatic process** is one in which no energy is exchanged by heat between a system and its surroundings.

Definition of an adiabatic process

Let us suppose that an ideal gas undergoes an adiabatic expansion. At any time during the process, we assume that the gas is in an equilibrium state, so that the equation of state $PV = nRT$ is valid. As we shall soon see, the pressure and volume at any time during an adiabatic process are related by the expression

$$PV^\gamma = \text{constant} \qquad \textbf{(21.18)}$$

Relationship between P and V for an adiabatic process involving an ideal gas

where $\gamma = C_P/C_V$ is assumed to be constant during the process. Thus, we see that all three variables in the ideal gas law—P, V, and T—change during an adiabatic process.

Proof That PV^γ = constant for an Adiabatic Process

When a gas expands adiabatically in a thermally insulated cylinder, no energy is transferred by heat between the gas and its surroundings; thus, $Q = 0$. Let us take the infinitesimal change in volume to be dV and the infinitesimal change in temperature to be dT. The work done by the gas is $P \, dV$. Because the internal energy of an ideal gas depends only on temperature, the change in the internal energy in an adiabatic expansion is the same as that for an isovolumetric process between the same temperatures, $dE_{\text{int}} = nC_V \, dT$ (Eq. 21.12). Hence, the first law of thermodynamics, $\Delta E_{\text{int}} = Q - W$, with $Q = 0$, becomes

$$dE_{\text{int}} = nC_V \, dT = -P \, dV$$

Taking the total differential of the equation of state of an ideal gas, $PV = nRT$, we

QuickLab

Rapidly pump up a bicycle tire and then feel the coupling at the end of the hose. Why is the coupling warm?

see that

$$P\,dV + V\,dP = nR\,dT$$

Eliminating dT from these two equations, we find that

$$P\,dV + V\,dP = -\frac{R}{C_V}\,P\,dV$$

Substituting $R = C_P - C_V$ and dividing by PV, we obtain

$$\frac{dV}{V} + \frac{dP}{P} = -\left(\frac{C_P - C_V}{C_V}\right)\frac{dV}{V} = (1 - \gamma)\frac{dV}{V}$$

$$\frac{dP}{P} + \gamma\frac{dV}{V} = 0$$

Integrating this expression, we have

$$\ln P + \gamma \ln V = \text{constant}$$

which is equivalent to Equation 21.18:

$$PV^\gamma = \text{constant}$$

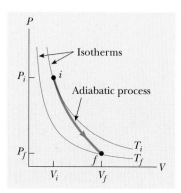

Figure 21.5 The *PV* diagram for an adiabatic expansion. Note that $T_f < T_i$ in this process.

The *PV* diagram for an adiabatic expansion is shown in Figure 21.5. Because $\gamma > 1$, the *PV* curve is steeper than it would be for an isothermal expansion. By the definition of an adiabatic process, no energy is transferred by heat into or out of the system. Hence, from the first law, we see that ΔE_{int} is negative (the gas does work, so its internal energy decreases) and so ΔT also is negative. Thus, we see that the gas cools $(T_f < T_i)$ during an adiabatic expansion. Conversely, the temperature increases if the gas is compressed adiabatically. Applying Equation 21.18 to the initial and final states, we see that

$$P_i V_i^\gamma = P_f V_f^\gamma \qquad \textbf{(21.19)}$$

Adiabatic process

Using the ideal gas law, we can express Equation 21.19 as

$$T_i V_i^{\gamma-1} = T_f V_f^{\gamma-1} \qquad \textbf{(21.20)}$$

EXAMPLE 21.3 ▶ A Diesel Engine Cylinder

Air at 20.0°C in the cylinder of a diesel engine is compressed from an initial pressure of 1.00 atm and volume of 800.0 cm³ to a volume of 60.0 cm³. Assume that air behaves as an ideal gas with $\gamma = 1.40$ and that the compression is adiabatic. Find the final pressure and temperature of the air.

Solution   Using Equation 21.19, we find that

$$P_f = P_i\left(\frac{V_i}{V_f}\right)^\gamma = (1.00\ \text{atm})\left(\frac{800.0\ \text{cm}^3}{60.0\ \text{cm}^3}\right)^{1.40}$$

$$= \boxed{37.6\ \text{atm}}$$

Because $PV = nRT$ is valid during any process and because

no gas escapes from the cylinder,

$$\frac{P_i V_i}{T_i} = \frac{P_f V_f}{T_f}$$

$$T_f = \frac{P_f V_f}{P_i V_i}\,T_i = \frac{(37.6\ \text{atm})(60.0\ \text{cm}^3)}{(1.00\ \text{atm})(800.0\ \text{cm}^3)}\,(293\ \text{K})$$

$$= \boxed{826\ \text{K} = 553°\text{C}}$$

The high compression in a diesel engine raises the temperature of the fuel enough to cause its combustion without the use of spark plugs.

THE EQUIPARTITION OF ENERGY

We have found that model predictions based on molar specific heat agree quite well with the behavior of monatomic gases but not with the behavior of complex gases (see Table 21.2). Furthermore, the value predicted by the model for the quantity $C_P - C_V = R$ is the same for all gases. This is not surprising because this difference is the result of the work done by the gas, which is independent of its molecular structure.

To clarify the variations in C_V and C_P in gases more complex than monatomic gases, let us first explain the origin of molar specific heat. So far, we have assumed that the sole contribution to the internal energy of a gas is the translational kinetic energy of the molecules. However, the internal energy of a gas actually includes contributions from the translational, vibrational, and rotational motion of the molecules. The rotational and vibrational motions of molecules can be activated by collisions and therefore are "coupled" to the translational motion of the molecules. The branch of physics known as *statistical mechanics* has shown that, for a large number of particles obeying the laws of Newtonian mechanics, the available energy is, on the average, shared equally by each independent degree of freedom. Recall from Section 21.1 that the equipartition theorem states that, at equilibrium, each degree of freedom contributes $\frac{1}{2}k_B T$ of energy per molecule.

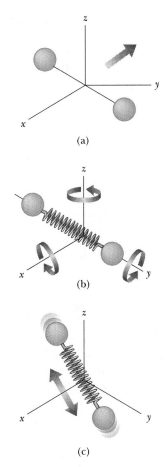

Let us consider a diatomic gas whose molecules have the shape of a dumbbell (Fig. 21.6). In this model, the center of mass of the molecule can translate in the x, y, and z directions (Fig. 21.6a). In addition, the molecule can rotate about three mutually perpendicular axes (Fig. 21.6b). We can neglect the rotation about the y axis because the moment of inertia I_y and the rotational energy $\frac{1}{2}I_y\omega^2$ about this axis are negligible compared with those associated with the x and z axes. (If the two atoms are taken to be point masses, then I_y is identically zero.) Thus, there are five degrees of freedom: three associated with the translational motion and two associated with the rotational motion. Because each degree of freedom contributes, on the average, $\frac{1}{2}k_B T$ of energy per molecule, the total internal energy for a system of N molecules is

$$E_{\text{int}} = 3N(\tfrac{1}{2}k_B T) + 2N(\tfrac{1}{2}k_B T) = \tfrac{5}{2}Nk_B T = \tfrac{5}{2}nRT$$

Figure 21.6 Possible motions of a diatomic molecule: (a) translational motion of the center of mass, (b) rotational motion about the various axes, and (c) vibrational motion along the molecular axis.

We can use this result and Equation 21.13 to find the molar specific heat at constant volume:

$$C_V = \frac{1}{n}\frac{dE_{\text{int}}}{dT} = \frac{1}{n}\frac{d}{dT}\left(\frac{5}{2}nRT\right) = \frac{5}{2}R$$

From Equations 21.16 and 21.17, we find that

$$C_P = C_V + R = \tfrac{7}{2}R$$

$$\gamma = \frac{C_P}{C_V} = \frac{\frac{7}{2}R}{\frac{5}{2}R} = \frac{7}{5} = 1.40$$

These results agree quite well with most of the data for diatomic molecules given in Table 21.2. This is rather surprising because we have not yet accounted for the possible vibrations of the molecule. In the vibratory model, the two atoms are joined by an imaginary spring (see Fig. 21.6c). The vibrational motion adds two more degrees of freedom, which correspond to the kinetic energy and the potential energy associated with vibrations along the length of the molecule. Hence, classical physics and the equipartition theorem predict an internal energy of

$$E_{\text{int}} = 3N(\tfrac{1}{2}k_B T) + 2N(\tfrac{1}{2}k_B T) + 2N(\tfrac{1}{2}k_B T) = \tfrac{7}{2}Nk_B T = \tfrac{7}{2}nRT$$

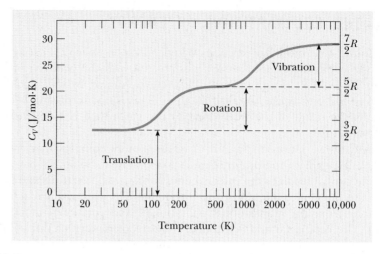

Figure 21.7 The molar specific heat of hydrogen as a function of temperature. The horizontal scale is logarithmic. Note that hydrogen liquefies at 20 K.

and a molar specific heat at constant volume of

$$C_V = \frac{1}{n}\frac{dE_{int}}{dT} = \frac{1}{n}\frac{d}{dT}\left(\frac{7}{2}\,nRT\right) = \frac{7}{2}\,R$$

This value is inconsistent with experimental data for molecules such as H_2 and N_2 (see Table 21.2) and suggests a breakdown of our model based on classical physics.

For molecules consisting of more than two atoms, the number of degrees of freedom is even larger and the vibrations are more complex. This results in an even higher predicted molar specific heat, which is in qualitative agreement with experiment. The more degrees of freedom available to a molecule, the more "ways" it can store internal energy; this results in a higher molar specific heat.

We have seen that the equipartition theorem is successful in explaining some features of the molar specific heat of gas molecules with structure. However, the theorem does not account for the observed temperature variation in molar specific heats. As an example of such a temperature variation, C_V for H_2 is $\frac{5}{2}R$ from about 250 K to 750 K and then increases steadily to about $\frac{7}{2}R$ well above 750 K (Fig. 21.7). This suggests that much more significant vibrations occur at very high temperatures. At temperatures well below 250 K, C_V has a value of about $\frac{3}{2}R$, suggesting that the molecule has only translational energy at low temperatures.

A Hint of Energy Quantization

The failure of the equipartition theorem to explain such phenomena is due to the inadequacy of classical mechanics applied to molecular systems. For a more satisfactory description, it is necessary to use a quantum-mechanical model, in which the energy of an individual molecule is quantized. The energy separation between adjacent vibrational energy levels for a molecule such as H_2 is about ten times greater than the average kinetic energy of the molecule at room temperature. Consequently, collisions between molecules at low temperatures do not provide enough energy to change the vibrational state of the molecule. It is often stated that such degrees of freedom are "frozen out." This explains why the vibrational energy does not contribute to the molar specific heats of molecules at low temperatures.

The rotational energy levels also are quantized, but their spacing at ordinary temperatures is small compared with $k_B T$. Because the spacing between quantized energy levels is small compared with the available energy, the system behaves in accordance with classical mechanics. However, at sufficiently low temperatures (typically less than 50 K), where $k_B T$ is small compared with the spacing between rotational levels, intermolecular collisions may not be sufficiently energetic to alter the rotational states. This explains why C_V reduces to $\frac{3}{2}R$ for H_2 in the range from 20 K to approximately 100 K.

The Molar Specific Heat of Solids

The molar specific heats of solids also demonstrate a marked temperature dependence. Solids have molar specific heats that generally decrease in a nonlinear manner with decreasing temperature and approach zero as the temperature approaches absolute zero. At high temperatures (usually above 300 K), the molar specific heats approach the value of $3R \approx 25$ J/mol·K, a result known as the *DuLong–Petit law.* The typical data shown in Figure 21.8 demonstrate the temperature dependence of the molar specific heats for two semiconducting solids, silicon and germanium.

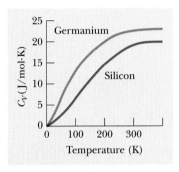

We can explain the molar specific heat of a solid at high temperatures using the equipartition theorem. For small displacements of an atom from its equilibrium position, each atom executes simple harmonic motion in the x, y, and z directions. The energy associated with vibrational motion in the x direction is

$$E = \tfrac{1}{2}mv_x^2 + \tfrac{1}{2}kx^2$$

The expressions for vibrational motions in the y and z directions are analogous. Therefore, each atom of the solid has six degrees of freedom. According to the equipartition theorem, this corresponds to an average vibrational energy of $6(\frac{1}{2}k_B T) = 3k_B T$ per atom. Therefore, the total internal energy of a solid consisting of N atoms is

$$E_{\text{int}} = 3Nk_B T = 3nRT \qquad \textbf{(21.21)}$$

From this result, we find that the molar specific heat of a solid at constant volume is

$$C_V = \frac{1}{n}\frac{dE_{\text{int}}}{dT} = 3R \qquad \textbf{(21.22)}$$

Figure 21.8 Molar specific heat of silicon and germanium. As T approaches zero, the molar specific heat also approaches zero. *(From C. Kittel,* Introduction to Solid State Physics, *New York, Wiley, 1971.)*

Total internal energy of a solid

Molar specific heat of a solid at constant volume

This result is in agreement with the empirical DuLong–Petit law. The discrepancies between this model and the experimental data at low temperatures are again due to the inadequacy of classical physics in describing the microscopic world.

21.5 THE BOLTZMANN DISTRIBUTION LAW

Thus far we have neglected the fact that not all molecules in a gas have the same speed and energy. In reality, their motion is extremely chaotic. Any individual molecule is colliding with others at an enormous rate—typically, a billion times per second. Each collision results in a change in the speed and direction of motion of each of the participant molecules. From Equation 21.7, we see that average molecular speeds increase with increasing temperature. What we would like to know now is the relative number of molecules that possess some characteristic, such as a certain percentage of the total energy or speed. The ratio of the number of molecules that have the desired characteristic to the total number of molecules is the probability that a particular molecule has that characteristic.

The Exponential Atmosphere

We begin by considering the distribution of molecules in our atmosphere. Let us determine how the number of molecules per unit volume varies with altitude. Our model assumes that the atmosphere is at a constant temperature T. (This assumption is not entirely correct because the temperature of our atmosphere decreases by about 2°C for every 300-m increase in altitude. However, the model does illustrate the basic features of the distribution.)

According to the ideal gas law, a gas containing N molecules in thermal equilibrium obeys the relationship $PV = Nk_BT$. It is convenient to rewrite this equation in terms of the **number density** $n_V = N/V$, which represents the number of molecules per unit volume of gas. This quantity is important because it can vary from one point to another. In fact, our goal is to determine how n_V changes in our atmosphere. We can express the ideal gas law in terms of n_V as $P = n_V k_B T$. Thus, if the number density n_V is known, we can find the pressure, and vice versa. The pressure in the atmosphere decreases with increasing altitude because a given layer of air must support the weight of all the atmosphere above it—that is, the greater the altitude, the less the weight of the air above that layer, and the lower the pressure.

To determine the variation in pressure with altitude, let us consider an atmospheric layer of thickness dy and cross-sectional area A, as shown in Figure 21.9. Because the air is in static equilibrium, the magnitude PA of the upward force exerted on the bottom of this layer must exceed the magnitude of the downward force on the top of the layer, $(P + dP)A$, by an amount equal to the weight of gas in this thin layer. If the mass of a gas molecule in the layer is m, and if a total of N molecules are in the layer, then the weight of the layer is given by $mgN = mgn_V V = mgn_V A \, dy$. Thus, we see that

$$PA - (P + dP)A = mgn_V A \, dy$$

This expression reduces to

$$dP = -mgn_V \, dy$$

Because $P = n_V k_B T$ and T is assumed to remain constant, we see that $dP = k_B T \, dn_V$. Substituting this result into the previous expression for dP and rearranging terms, we have

$$\frac{dn_V}{n_V} = -\frac{mg}{k_B T} \, dy$$

Integrating this expression, we find that

$$n_V(y) = n_0 e^{-mgy/k_B T} \tag{21.23}$$

where the constant n_0 is the number density at $y = 0$. This result is known as the **law of atmospheres.**

According to Equation 21.23, the number density decreases exponentially with increasing altitude when the temperature is constant. The number density of our atmosphere at sea level is about $n_0 = 2.69 \times 10^{25}$ molecules/m^3. Because the pressure is $P = n_V k_B T$, we see from Equation 21.23 that the pressure of our atmosphere varies with altitude according to the expression

$$P = P_0 e^{-mgy/k_B T} \tag{21.24}$$

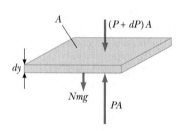

Figure 21.9 An atmospheric layer of gas in equilibrium.

where $P_0 = n_0 k_B T$. A comparison of this model with the actual atmospheric pressure as a function of altitude shows that the exponential form is a reasonable approximation to the Earth's atmosphere.

EXAMPLE 21.4 ▸ High-Flying Molecules

What is the number density of air at an altitude of 11.0 km (the cruising altitude of a commercial jetliner) compared with its number density at sea level? Assume that the air temperature at this height is the same as that at the ground, 20°C.

Solution The number density of our atmosphere decreases exponentially with altitude according to the law of atmospheres, Equation 21.23. We assume an average molecular mass of 28.9 u = 4.80×10^{-26} kg. Taking $y = 11.0$ km, we calculate the power of the exponential in Equation 21.23 to be

$$\frac{mgy}{k_B T} = \frac{(4.80 \times 10^{-26} \text{ kg})(9.80 \text{ m/s}^2)(11\,000 \text{ m})}{(1.38 \times 10^{-23} \text{ J/K})(293 \text{ K})} = 1.28$$

Thus, Equation 21.23 gives

$$n_V = n_0 e^{-mgy/k_B T} = n_0 e^{-1.28} = \boxed{0.278 n_0}$$

That is, the number density of air at an altitude of 11.0 km is only 27.8% of the number density at sea level, if we assume constant temperature. Because the temperature actually decreases with altitude, the number density of air is less than this in reality.

The pressure at this height is reduced in the same manner. For this reason, high-flying aircraft must have pressurized cabins to ensure passenger comfort and safety.

Computing Average Values

The exponential function $e^{-mgy/k_B T}$ that appears in Equation 21.23 can be interpreted as a probability distribution that gives the relative probability of finding a gas molecule at some height y. Thus, the probability distribution $p(y)$ is proportional to the number density distribution $n_V(y)$. This concept enables us to determine many properties of the atmosphere, such as the fraction of molecules below a certain height or the average potential energy of a molecule.

As an example, let us determine the average height $\bar{y}$ of a molecule in the atmosphere at temperature T. The expression for this average height is

$$\bar{y} = \frac{\int_0^\infty y n_V(y)\, dy}{\int_0^\infty n_V(y)\, dy} = \frac{\int_0^\infty y e^{-mgy/k_B T}\, dy}{\int_0^\infty e^{-mgy/k_B T}\, dy}$$

where the height of a molecule can range from 0 to ∞. The numerator in this expression represents the sum of the heights of the molecules times their number, while the denominator is the sum of the number of molecules. That is, the denominator is the total number of molecules. After performing the indicated integrations, we find that

$$\bar{y} = \frac{(k_B T/mg)^2}{k_B T/mg} = \frac{k_B T}{mg}$$

This expression states that the average height of a molecule increases as T increases, as expected.

We can use a similar procedure to determine the average potential energy of a gas molecule. Because the gravitational potential energy of a molecule at height y is $U = mgy$, the average potential energy is equal to $mg\bar{y}$. Because $\bar{y} = k_B T/mg$, we see that $\overline{U} = mg(k_B T/mg) = k_B T$. This important result indicates that the average gravitational potential energy of a molecule depends only on temperature, and not on m or g.

The Boltzmann Distribution

Because the gravitational potential energy of a molecule at height y is $U = mgy$, we can express the law of atmospheres (Eq. 21.23) as

$$n_V = n_0 e^{-U/k_B T}$$

This means that gas molecules in thermal equilibrium are distributed in space with a probability that depends on gravitational potential energy according to the exponential factor $e^{-U/k_B T}$.

This exponential expression describing the distribution of molecules in the atmosphere is powerful and applies to any type of energy. In general, the number density of molecules having energy E is

Boltzmann distribution law

$$n_V(E) = n_0 e^{-E/k_B T} \tag{21.25}$$

This equation is known as the **Boltzmann distribution law** and is important in describing the statistical mechanics of a large number of molecules. It states that **the probability of finding the molecules in a particular energy state varies exponentially as the negative of the energy divided by $k_B T$.** All the molecules would fall into the lowest energy level if the thermal agitation at a temperature T did not excite the molecules to higher energy levels.

EXAMPLE 21.5 Thermal Excitation of Atomic Energy Levels

As we discussed briefly in Section 8.10, atoms can occupy only certain discrete energy levels. Consider a gas at a temperature of 2 500 K whose atoms can occupy only two energy levels separated by 1.50 eV, where 1 eV (electron volt) is an energy unit equal to 1.6×10^{-19} J (Fig. 21.10). Determine the ratio of the number of atoms in the higher energy level to the number in the lower energy level.

Solution Equation 21.25 gives the relative number of atoms in a given energy level. In this case, the atom has two possible energies, E_1 and E_2, where E_1 is the lower energy level. Hence, the ratio of the number of atoms in the higher energy level to the number in the lower energy level is

$$\frac{n_V(E_2)}{n_V(E_1)} = \frac{n_0 e^{-E_2/k_B T}}{n_0 e^{-E_1/k_B T}} = e^{-(E_2 - E_1)/k_B T}$$

In this problem, $E_2 - E_1 = 1.50$ eV, and the denominator of the exponent is

$$k_B T = (1.38 \times 10^{-23} \text{ J/K})(2\,500 \text{ K})/1.60 \times 10^{-19} \text{ J/eV}$$
$$= 0.216 \text{ eV}$$

Therefore, the required ratio is

$$\frac{n(E_2)}{n(E_1)} = e^{-1.50 \text{ eV}/0.216 \text{ eV}} = e^{-6.94} = \boxed{9.64 \times 10^{-4}}$$

This result indicates that at $T = 2\,500$ K, only a small fraction of the atoms are in the higher energy level. In fact, for every atom in the higher energy level, there are about 1 000 atoms in the lower level. The number of atoms in the higher level increases at even higher temperatures, but the distribution law specifies that at equilibrium there are always more atoms in the lower level than in the higher level.

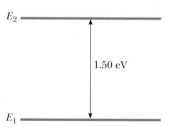

Figure 21.10 Energy level diagram for a gas whose atoms can occupy two energy levels.

21.6 DISTRIBUTION OF MOLECULAR SPEEDS

In 1860 James Clerk Maxwell (1831–1879) derived an expression that describes the distribution of molecular speeds in a very definite manner. His work and subsequent developments by other scientists were highly controversial because direct detection of molecules could not be achieved experimentally at that time. However, about 60 years later, experiments were devised that confirmed Maxwell's predictions.

Let us consider a container of gas whose molecules have some distribution of speeds. Suppose we want to determine how many gas molecules have a speed in the range from, for example, 400 to 410 m/s. Intuitively, we expect that the speed distribution depends on temperature. Furthermore, we expect that the distribution peaks in the vicinity of v_{rms}. That is, few molecules are expected to have speeds much less than or much greater than v_{rms} because these extreme speeds result only from an unlikely chain of collisions.

The observed speed distribution of gas molecules in thermal equilibrium is shown in Figure 21.11. The quantity N_v, called the **Maxwell–Boltzmann distribution function,** is defined as follows: If N is the total number of molecules, then the number of molecules with speeds between v and $v + dv$ is $dN = N_v dv$. This number is also equal to the area of the shaded rectangle in Figure 21.11. Furthermore, the fraction of molecules with speeds between v and $v + dv$ is $N_v dv/N$. This fraction is also equal to the probability that a molecule has a speed in the range v to $v + dv$.

The fundamental expression that describes the distribution of speeds of N gas molecules is

$$N_v = 4\pi N \left(\frac{m}{2\pi k_B T}\right)^{3/2} v^2 e^{-mv^2/2k_B T} \tag{21.26}$$

where m is the mass of a gas molecule, k_B is Boltzmann's constant, and T is the absolute temperature.[1] Observe the appearance of the Boltzmann factor $e^{-E/k_B T}$ with $E = \frac{1}{2}mv^2$.

As indicated in Figure 21.11, the average speed $\bar{v}$ is somewhat lower than the rms speed. The *most probable speed* v_{mp} is the speed at which the distribution curve reaches a peak. Using Equation 21.26, one finds that

$$v_{rms} = \sqrt{\overline{v^2}} = \sqrt{3k_B T/m} = 1.73\sqrt{k_B T/m} \tag{21.27}$$

$$\bar{v} = \sqrt{8k_B T/\pi m} = 1.60\sqrt{k_B T/m} \tag{21.28}$$

$$v_{mp} = \sqrt{2k_B T/m} = 1.41\sqrt{k_B T/m} \tag{21.29}$$

The details of these calculations are left for the student (see Problems 41 and 62). From these equations, we see that

$$v_{rms} > \bar{v} > v_{mp}$$

Figure 21.12 represents speed distribution curves for N_2. The curves were obtained by using Equation 21.26 to evaluate the distribution function at various speeds and at two temperatures. Note that the peak in the curve shifts to the right

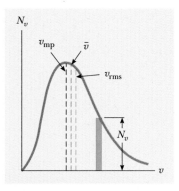

Figure 21.11 The speed distribution of gas molecules at some temperature. The number of molecules having speeds in the range dv is equal to the area of the shaded rectangle, $N_v dv$. The function N_v approaches zero as v approaches infinity.

Maxwell speed distribution function

rms speed

Average speed

Most probable speed

---

[1] For the derivation of this expression, see an advanced textbook on thermodynamics, such as that by R. P. Bauman, *Modern Thermodynamics with Statistical Mechanics,* New York, Macmillan Publishing Co., 1992.

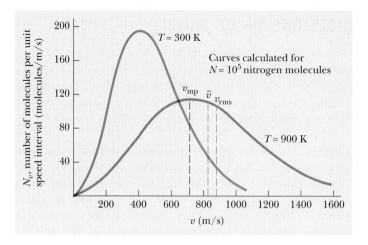

Figure 21.12   The speed distribution function for 10^5 nitrogen molecules at 300 K and 900 K. The total area under either curve is equal to the total number of molecules, which in this case equals 10^5. Note that $v_{rms} > \bar{v} > v_{mp}$.

as *T* increases, indicating that the average speed increases with increasing temperature, as expected. The asymmetric shape of the curves is due to the fact that the lowest speed possible is zero while the upper classical limit of the speed is infinity.

Quick Quiz 21.3

Consider the two curves in Figure 21.12. What is represented by the area under each of the curves between the 800-m/s and 1 000-m/s marks on the horizontal axis?

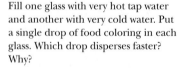

Fill one glass with very hot tap water and another with very cold water. Put a single drop of food coloring in each glass. Which drop disperses faster? Why?

Equation 21.26 shows that the distribution of molecular speeds in a gas depends both on mass and on temperature. At a given temperature, the fraction of molecules with speeds exceeding a fixed value increases as the mass decreases. This explains why lighter molecules, such as H_2 and He, escape more readily from the Earth's atmosphere than do heavier molecules, such as N_2 and O_2. (See the discussion of escape speed in Chapter 14. Gas molecules escape even more readily from the Moon's surface than from the Earth's because the escape speed on the Moon is lower than that on the Earth.)

The speed distribution curves for molecules in a liquid are similar to those shown in Figure 21.12. We can understand the phenomenon of evaporation of a liquid from this distribution in speeds, using the fact that some molecules in the liquid are more energetic than others. Some of the faster-moving molecules in the liquid penetrate the surface and leave the liquid even at temperatures well below the boiling point. The molecules that escape the liquid by evaporation are those that have sufficient energy to overcome the attractive forces of the molecules in the liquid phase. Consequently, the molecules left behind in the liquid phase have a lower average kinetic energy; as a result, the temperature of the liquid decreases. Hence, evaporation is a cooling process. For example, an alcohol-soaked cloth often is placed on a feverish head to cool and comfort a patient.

The evaporation process

EXAMPLE 21.6   A System of Nine Particles

Nine particles have speeds of 5.00, 8.00, 12.0, 12.0, 12.0, 14.0, 14.0, 17.0, and 20.0 m/s. (a) Find the particles' average speed.

Solution   The average speed is the sum of the speeds divided by the total number of particles:

$$\bar{v} = \frac{(5.00 + 8.00 + 12.0 + 12.0 + 12.0 + 14.0 + 14.0 + 17.0 + 20.0)\ \text{m/s}}{9}$$

$$= \boxed{12.7\ \text{m/s}}$$

(b) What is the rms speed?

Solution   The average value of the square of the speed is

$$\overline{v^2} = \frac{(5.00^2 + 8.00^2 + 12.0^2 + 12.0^2 + 12.0^2 + 14.0^2 + 14.0^2 + 17.0^2 + 20.0^2)\ \text{m}}{9}$$

$$= 178\ \text{m}^2/\text{s}^2$$

Hence, the rms speed is

$$v_{\text{rms}} = \sqrt{\overline{v^2}} = \sqrt{178\ \text{m}^2/\text{s}^2} = \boxed{13.3\ \text{m/s}}$$

(c) What is the most probable speed of the particles?

Solution   Three of the particles have a speed of 12 m/s, two have a speed of 14 m/s, and the remaining have different speeds. Hence, we see that the most probable speed v_{mp} is

$$\boxed{12\ \text{m/s}.}$$

Optional Section

21.7 ▶ MEAN FREE PATH

Most of us are familiar with the fact that the strong odor associated with a gas such as ammonia may take a fraction of a minute to diffuse throughout a room. However, because average molecular speeds are typically several hundred meters per second at room temperature, we might expect a diffusion time much less than 1 s. But, as we saw in Quick Quiz 21.1, molecules collide with one other because they are not geometrical points. Therefore, they do not travel from one side of a room to the other in a straight line. Between collisions, the molecules move with constant speed along straight lines. The average distance between collisions is called the **mean free path.** The path of an individual molecule is random and resembles that shown in Figure 21.13. As we would expect from this description, the mean free path is related to the diameter of the molecules and the density of the gas.

We now describe how to estimate the mean free path for a gas molecule. For this calculation, we assume that the molecules are spheres of diameter d. We see from Figure 21.14a that no two molecules collide unless their centers are less than a distance d apart as they approach each other. An equivalent way to describe the

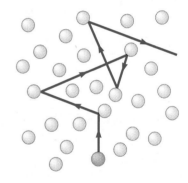

Figure 21.13   A molecule moving through a gas collides with other molecules in a random fashion. This behavior is sometimes referred to as a *random-walk process.* The mean free path increases as the number of molecules per unit volume decreases. Note that the motion is not limited to the plane of the paper.

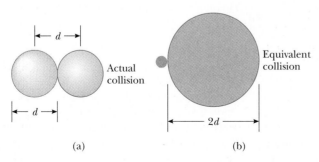

(a)   (b)

Figure 21.14   (a) Two spherical molecules, each of diameter d, collide if their centers are within a distance d of each other. (b) The collision between the two molecules is equivalent to a point molecule's colliding with a molecule having an effective diameter of $2d$.

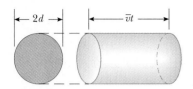

Figure 21.15   In a time t, a molecule of effective diameter $2d$ sweeps out a cylinder of length $\overline{v}t$, where $\overline{v}$ is its average speed. In this time, it collides with every point molecule within this cylinder.

collisions is to imagine that one of the molecules has a diameter $2d$ and that the rest are geometrical points (Fig. 21.14b). Let us choose the large molecule to be one moving with the average speed $\overline{v}$. In a time t, this molecule travels a distance $\overline{v}t$. In this time interval, the molecule sweeps out a cylinder having a cross-sectional area πd^2 and a length $\overline{v}t$ (Fig. 21.15). Hence, the volume of the cylinder is $\pi d^2 \overline{v}t$. If n_V is the number of molecules per unit volume, then the number of point-size molecules in the cylinder is $(\pi d^2 \overline{v}t)\,n_V$. The molecule of equivalent diameter $2d$ collides with every molecule in this cylinder in the time t. Hence, the number of collisions in the time t is equal to the number of molecules in the cylinder, $(\pi d^2 \overline{v}t)\,n_V$.

The mean free path ℓ equals the average distance $\overline{v}t$ traveled in a time t divided by the number of collisions that occur in that time:

$$\ell = \frac{\overline{v}t}{(\pi d^2 \overline{v}t)\,n_V} = \frac{1}{\pi d^2 n_V}$$

Because the number of collisions in a time t is $(\pi d^2 \overline{v}t)\,n_V$, the number of collisions per unit time, or **collision frequency f,** is

$$f = \pi d^2 \overline{v} n_V$$

The inverse of the collision frequency is the average time between collisions, known as the **mean free time.**

Our analysis has assumed that molecules in the cylinder are stationary. When the motion of these molecules is included in the calculation, the correct results are

| Mean free path |

$$\ell = \frac{1}{\sqrt{2}\,\pi d^2 n_V} \qquad \text{(21.30)}$$

| Collision frequency |

$$f = \sqrt{2}\,\pi d^2 \overline{v} n_V = \frac{\overline{v}}{\ell} \qquad \text{(21.31)}$$

EXAMPLE 21.7   Bouncing Around in the Air

Approximate the air around you as a collection of nitrogen molecules, each of which has a diameter of 2.00×10^{-10} m. (a) How far does a typical molecule move before it collides with another molecule?

Solution   Assuming that the gas is ideal, we can use the equation $PV = Nk_BT$ to obtain the number of molecules per unit volume under typical room conditions:

$$n_V = \frac{N}{V} = \frac{P}{k_BT} = \frac{1.01 \times 10^5 \text{ N/m}^2}{(1.38 \times 10^{-23} \text{ J/K})(293 \text{ K})}$$

$$= 2.50 \times 10^{25} \text{ molecules/m}^3$$

Hence, the mean free path is

$$\ell = \frac{1}{\sqrt{2}\,\pi d^2 n_V}$$

$$= \frac{1}{\sqrt{2}\,\pi (2.00 \times 10^{-10} \text{ m})^2 (2.50 \times 10^{25} \text{ molecules/m}^3)}$$

$$= 2.25 \times 10^{-7} \text{ m}$$

This value is about 10^3 times greater than the molecular diameter.

(b) On average, how frequently does one molecule collide with another?

Solution   Because the rms speed of a nitrogen molecule at 20.0°C is 511 m/s (see Table 21.1), we know from Equations 21.27 and 21.28 that $\overline{v} = (1.60/1.73)(511 \text{ m/s}) = 473$ m/s. Therefore, the collision frequency is

$$f = \frac{\overline{v}}{\ell} = \frac{473 \text{ m/s}}{2.25 \times 10^{-7} \text{ m}} = 2.10 \times 10^9\text{/s}$$

The molecule collides with other molecules at the average rate of about two billion times each second!

The mean free path ℓ is *not* the same as the average separation between particles. In fact, the average separation d between particles is approximately $n_V^{-1/3}$. In this example, the average molecular separation is

$$d = \frac{1}{n_V^{1/3}} = \frac{1}{(2.5 \times 10^{25})^{1/3}} = 3.4 \times 10^{-9} \text{ m}$$

SUMMARY

The pressure of N molecules of an ideal gas contained in a volume V is

$$P = \frac{2}{3}\frac{N}{V}\left(\frac{1}{2}m\overline{v^2}\right) \qquad \text{(21.2)}$$

The average translational kinetic energy per molecule of a gas, $\frac{1}{2}m\overline{v^2}$, is related to the temperature T of the gas through the expression

$$\frac{1}{2}m\overline{v^2} = \frac{3}{2}k_B T \qquad \text{(21.4)}$$

where k_B is Boltzmann's constant. Each translational degree of freedom (x, y, or z) has $\frac{1}{2}k_B T$ of energy associated with it.

The **theorem of equipartition of energy** states that the energy of a system in thermal equilibrium is equally divided among all degrees of freedom.

The total energy of N molecules (or n mol) of an ideal monatomic gas is

$$E_{\text{int}} = \frac{3}{2}Nk_B T = \frac{3}{2}nRT \qquad \text{(21.10)}$$

The change in internal energy for n mol of any ideal gas that undergoes a change in temperature ΔT is

$$\Delta E_{\text{int}} = nC_V\Delta T \qquad \text{(21.12)}$$

where C_V is the molar specific heat at constant volume.

The molar specific heat of an ideal monatomic gas at constant volume is $C_V = \frac{3}{2}R$; the molar specific heat at constant pressure is $C_P = \frac{5}{2}R$. The ratio of specific heats is $\gamma = C_P/C_V = \frac{5}{3}$.

If an ideal gas undergoes an adiabatic expansion or compression, the first law of thermodynamics, together with the equation of state, shows that

$$PV^{\gamma} = \text{constant} \qquad \text{(21.18)}$$

The **Boltzmann distribution law** describes the distribution of particles among available energy states. The relative number of particles having energy E is

$$n_V(E) = n_0 e^{-E/k_B T} \qquad \text{(21.25)}$$

The **Maxwell–Boltzmann distribution function** describes the distribution of speeds of molecules in a gas:

$$N_v = 4\pi N\left(\frac{m}{2\pi k_B T}\right)^{3/2} v^2 e^{-mv^2/2k_B T} \qquad \text{(21.26)}$$

This expression enables us to calculate the **root-mean-square speed,** the **average speed,** and **the most probable speed:**

$$v_{\text{rms}} = \sqrt{\overline{v^2}} = \sqrt{3k_B T/m} = 1.73\sqrt{k_B T/m} \qquad \text{(21.27)}$$

$$\overline{v} = \sqrt{8k_B T/\pi m} = 1.60\sqrt{k_B T/m} \qquad \text{(21.28)}$$

$$v_{\text{mp}} = \sqrt{2k_B T/m} = 1.41\sqrt{k_B T/m} \qquad \text{(21.29)}$$

QUESTIONS

1. Dalton's law of partial pressures states that the total pressure of a mixture of gases is equal to the sum of the partial pressures of gases making up the mixture. Give a convincing argument for this law on the basis of the kinetic theory of gases.

2. One container is filled with helium gas and another with argon gas. If both containers are at the same temperature, which gas molecules have the higher rms speed? Explain.

3. A gas consists of a mixture of He and N_2 molecules. Do the lighter He molecules travel faster than the N_2 molecules? Explain.

4. Although the average speed of gas molecules in thermal equilibrium at some temperature is greater than zero, the average velocity is zero. Explain why this statement must be true.

5. When alcohol is rubbed on your body, your body temperature decreases. Explain this effect.

6. A liquid partially fills a container. Explain why the temperature of the liquid decreases if the container is then partially evacuated. (Using this technique, one can freeze water at temperatures above 0°C.)

7. A vessel containing a fixed volume of gas is cooled. Does the mean free path of the gas molecules increase, decrease, or remain constant during the cooling process? What about the collision frequency?

8. A gas is compressed at a constant temperature. What happens to the mean free path of the molecules in the process?

9. If a helium-filled balloon initially at room temperature is placed in a freezer, will its volume increase, decrease, or remain the same?

10. What happens to a helium-filled balloon released into the air? Will it expand or contract? Will it stop rising at some height?

11. Which is heavier, dry air or air saturated with water vapor? Explain.

12. Why does a diatomic gas have a greater energy content per mole than a monatomic gas at the same temperature?

13. An ideal gas is contained in a vessel at 300 K. If the temperature is increased to 900 K, (a) by what factor does the rms speed of each molecule change? (b) By what factor does the pressure in the vessel change?

14. A vessel is filled with gas at some equilibrium pressure and temperature. Can all gas molecules in the vessel have the same speed?

15. In our model of the kinetic theory of gases, molecules were viewed as hard spheres colliding elastically with the walls of the container. Is this model realistic?

16. In view of the fact that hot air rises, why does it generally become cooler as you climb a mountain? (Note that air is a poor thermal conductor.)

PROBLEMS

1, 2, 3 = straightforward, intermediate, challenging ☐ = full solution available in the *Student Solutions Manual and Study Guide*
WEB = solution posted at **http://www.saunderscollege.com/physics/** 🖥 = Computer useful in solving problem 📱 = Interactive Physics
☐ = paired numerical/symbolic problems

Section 21.1   Molecular Model of an Ideal Gas

1. Use the definition of Avogadro's number to find the mass of a helium atom.

2. A sealed cubical container 20.0 cm on a side contains three times Avogadro's number of molecules at a temperature of 20.0°C. Find the force exerted by the gas on one of the walls of the container.

3. In a 30.0-s interval, 500 hailstones strike a glass window with an area of 0.600 m^2 at an angle of 45.0° to the window surface. Each hailstone has a mass of 5.00 g and a speed of 8.00 m/s. If the collisions are elastic, what are the average force and pressure on the window?

4. In a time t, N hailstones strike a glass window of area A at an angle θ to the window surface. Each hailstone has a mass m and a speed v. If the collisions are elastic, what are the average force and pressure on the window?

5. In a period of 1.00 s, 5.00×10^{23} nitrogen molecules strike a wall with an area of 8.00 cm^2. If the molecules

move with a speed of 300 m/s and strike the wall head-on in perfectly elastic collisions, what is the pressure exerted on the wall? (The mass of one N_2 molecule is 4.68×10^{-26} kg.)

6. A 5.00-L vessel contains 2 mol of oxygen gas at a pressure of 8.00 atm. Find the average translational kinetic energy of an oxygen molecule under these conditions.

7. A spherical balloon with a volume of 4 000 cm^3 contains helium at an (inside) pressure of 1.20×10^5 Pa. How many moles of helium are in the balloon if each helium atom has an average kinetic energy of 3.60×10^{-22} J?

8. The rms speed of a helium atom at a certain temperature is 1 350 m/s. Find by proportion the rms speed of an oxygen molecule at this temperature. (The molar mass of O_2 is 32.0 g/mol, and the molar mass of He is 4.00 g/mol.)

9. (a) How many atoms of helium gas fill a balloon of diameter 30.0 cm at 20.0°C and 1.00 atm? (b) What is the average kinetic energy of the helium atoms? (c) What is the root-mean-square speed of each helium atom?

10. A 5.00-liter vessel contains nitrogen gas at 27.0°C and 3.00 atm. Find (a) the total translational kinetic energy of the gas molecules and (b) the average kinetic energy per molecule.

WEB **11.** A cylinder contains a mixture of helium and argon gas in equilibrium at 150°C. (a) What is the average kinetic energy for each type of gas molecule? (b) What is the root-mean-square speed for each type of molecule?

12. (a) Show that $1 \text{ Pa} = 1 \text{ J/m}^3$. (b) Show that the density in space of the translational kinetic energy of an ideal gas is $3P/2$.

Section 21.2   Molar Specific Heat of an Ideal Gas

Note: You may use the data given in Table 21.2.

13. Calculate the change in internal energy of 3.00 mol of helium gas when its temperature is increased by 2.00 K.

14. One mole of air ($C_V = 5R/2$) at 300 K and confined in a cylinder under a heavy piston occupies a volume of 5.00 L. Determine the new volume of the gas if 4.40 kJ of energy is transferred to the air by heat.

WEB **15.** One mole of hydrogen gas is heated at constant pressure from 300 K to 420 K. Calculate (a) the energy transferred by heat to the gas, (b) the increase in its internal energy, and (c) the work done by the gas.

16. In a constant-volume process, 209 J of energy is transferred by heat to 1.00 mol of an ideal monatomic gas initially at 300 K. Find (a) the increase in internal energy of the gas, (b) the work it does, and (c) its final temperature.

17. A house has well-insulated walls. It contains a volume of 100 m³ of air at 300 K. (a) Calculate the energy required to increase the temperature of this air by 1.00°C. (b) If this energy could be used to lift an object of mass *m* through a height of 2.00 m, what is the value of *m*?

18. A vertical cylinder with a heavy piston contains air at 300 K. The initial pressure is 200 kPa, and the initial volume is 0.350 m³. Take the molar mass of air as 28.9 g/mol and assume that $C_V = 5R/2$. (a) Find the specific heat of air at constant volume in units of J/kg · °C. (b) Calculate the mass of the air in the cylinder. (c) Suppose the piston is held fixed. Find the energy input required to raise the temperature of the air to 700 K. (d) Assume again the conditions of the initial state and that the heavy piston is free to move. Find the energy input required to raise the temperature to 700 K.

19. A 1-L Thermos bottle is full of tea at 90°C. You pour out one cup and immediately screw the stopper back on. Make an order-of-magnitude estimate of the change in temperature of the tea remaining in the flask that results from the admission of air at room temperature. State the quantities you take as data and the values you measure or estimate for them.

20. For a diatomic ideal gas, $C_V = 5R/2$. One mole of this gas has pressure *P* and volume *V*. When the gas is heated, its pressure triples and its volume doubles. If this heating process includes two steps, the first at constant pressure and the second at constant volume, determine the amount of energy transferred to the gas by heat.

21. One mole of an ideal monatomic gas is at an initial temperature of 300 K. The gas undergoes an isovolumetric process, acquiring 500 J of energy by heat. It then undergoes an isobaric process, losing this same amount of energy by heat. Determine (a) the new temperature of the gas and (b) the work done on the gas.

22. A container has a mixture of two gases: n_1 moles of gas 1, which has a molar specific heat C_1; and n_2 moles of gas 2, which has a molar specific heat C_2. (a) Find the molar specific heat of the mixture. (b) What is the molar specific heat if the mixture has *m* gases in the amounts $n_1, n_2, n_3, \ldots, n_m$, and molar specific heats $C_1, C_2, C_3, \ldots, C_m$, respectively?

23. One mole of an ideal diatomic gas with $C_V = 5R/2$ occupies a volume V_i at a pressure P_i. The gas undergoes a process in which the pressure is proportional to the volume. At the end of the process, it is found that the rms speed of the gas molecules has doubled from its initial value. Determine the amount of energy transferred to the gas by heat.

Section 21.3   Adiabatic Processes for an Ideal Gas

24. During the compression stroke of a certain gasoline engine, the pressure increases from 1.00 atm to 20.0 atm. Assuming that the process is adiabatic and that the gas is ideal, with $\gamma = 1.40$, (a) by what factor does the volume change and (b) by what factor does the temperature change? (c) If the compression starts with 0.016 0 mol of gas at 27.0°C, find the values of Q, W, and ΔE_{int} that characterize the process.

25. Two moles of an ideal gas ($\gamma = 1.40$) expands slowly and adiabatically from a pressure of 5.00 atm and a volume of 12.0 L to a final volume of 30.0 L. (a) What is the final pressure of the gas? (b) What are the initial and final temperatures? (c) Find Q, W, and ΔE_{int}.

26. Air ($\gamma = 1.40$) at 27.0°C and at atmospheric pressure is drawn into a bicycle pump that has a cylinder with an inner diameter of 2.50 cm and a length of 50.0 cm. The down stroke adiabatically compresses the air, which reaches a gauge pressure of 800 kPa before entering the tire. Determine (a) the volume of the compressed air and (b) the temperature of the compressed air. (c) The pump is made of steel and has an inner wall that is 2.00 mm thick. Assume that 4.00 cm of the cylinder's length is allowed to come to thermal equilibrium with the air. What will be the increase in wall temperature?

27. Air in a thundercloud expands as it rises. If its initial temperature was 300 K, and if no energy is lost by thermal conduction on expansion, what is its temperature when the initial volume has doubled?

28. How much work is required to compress 5.00 mol of air at 20.0°C and 1.00 atm to one tenth of the original volume by (a) an isothermal process and (b) an adiabatic process? (c) What is the final pressure in each of these two cases?

29. Four liters of a diatomic ideal gas ($\gamma = 1.40$) confined to a cylinder is subject to a closed cycle. Initially, the gas is at 1.00 atm and at 300 K. First, its pressure is tripled under constant volume. Then, it expands adiabatically to its original pressure. Finally, the gas is compressed isobarically to its original volume. (a) Draw a *PV* diagram of this cycle. (b) Determine the volume of the gas at the end of the adiabatic expansion. (c) Find the temperature of the gas at the start of the adiabatic expansion. (d) Find the temperature at the end of the cycle. (e) What was the net work done for this cycle?

30. A diatomic ideal gas ($\gamma = 1.40$) confined to a cylinder is subjected to a closed cycle. Initially, the gas is at P_i, V_i, and T_i. First, its pressure is tripled under constant volume. Then, it expands adiabatically to its original pressure. Finally, the gas is compressed isobarically to its original volume. (a) Draw a *PV* diagram of this cycle. (b) Determine the volume of the gas at the end of the adiabatic expansion. (c) Find the temperature of the gas at the start of the adiabatic expansion. (d) Find the temperature at the end of the cycle. (e) What was the net work done for this cycle?

31. During the power stroke in a four-stroke automobile engine, the piston is forced down as the mixture of gas and air undergoes an adiabatic expansion. Assume that (1) the engine is running at 2 500 rpm, (2) the gauge pressure right before the expansion is 20.0 atm, (3) the volumes of the mixture right before and after the expansion are 50.0 and 400 cm^3, respectively (Fig.

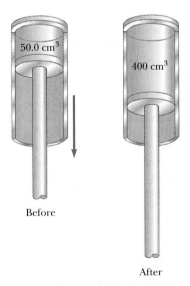

50.0 cm^3

400 cm^3

Before

After

Figure P21.31

P21.31), (4) the time involved in the expansion is one-fourth that of the total cycle, and (5) the mixture behaves like an ideal gas, with $\gamma = 1.40$. Find the average power generated during the expansion.

Section 21.4 The Equipartition of Energy

32. A certain molecule has *f* degrees of freedom. Show that a gas consisting of such molecules has the following properties: (1) its total internal energy is $fnRT/2$; (2) its molar specific heat at constant volume is $fR/2$; (3) its molar specific heat at constant pressure is $(f + 2)R/2$; (4) the ratio $\gamma = C_P/C_V = (f + 2)/f$.

WEB 33. Consider 2.00 mol of an ideal diatomic gas. Find the total heat capacity at constant volume and at constant pressure (a) if the molecules rotate but do not vibrate and (b) if the molecules both rotate and vibrate.

34. Inspecting the magnitudes of C_V and C_P for the diatomic and polyatomic gases in Table 21.2, we find that the values increase with increasing molecular mass. Give a qualitative explanation of this observation.

35. In a crude model (Fig. P21.35) of a rotating diatomic molecule of chlorine (Cl_2), the two Cl atoms are 2.00×10^{-10} m apart and rotate about their center of mass with angular speed $\omega = 2.00 \times 10^{12}$ rad/s. What is the rotational kinetic energy of one molecule of Cl_2, which has a molar mass of 70.0 g/mol?

Cl

Cl

Figure P21.35

Section 21.5 The Boltzmann Distribution Law
Section 21.6 Distribution of Molecular Speeds

36. One cubic meter of atomic hydrogen at 0°C contains approximately 2.70×10^{25} atoms at atmospheric pressure. The first excited state of the hydrogen atom has an energy of 10.2 eV above the lowest energy level, which is called the *ground state*. Use the Boltzmann factor to find the number of atoms in the first excited state at 0°C and at 10 000°C.

37. If convection currents (weather) did not keep the Earth's lower atmosphere stirred up, its chemical composition would change somewhat with altitude because the various molecules have different masses. Use the law of atmospheres to determine how the equilibrium ratio of oxygen to nitrogen molecules changes between sea level and 10.0 km. Assume a uniform temperature of 300 K and take the masses to be 32.0 u for oxygen (O_2) and 28.0 u for nitrogen (N_2).

38. A mixture of two gases diffuses through a filter at rates proportional to the gases' rms speeds. (a) Find the ratio of speeds for the two isotopes of chlorine, ^{35}Cl and ^{37}Cl, as they diffuse through the air. (b) Which isotope moves faster?

39. Fifteen identical particles have various speeds: one has a speed of 2.00 m/s; two have a speed of 3.00 m/s; three have a speed of 5.00 m/s; four have a speed of 7.00 m/s; three have a speed of 9.00 m/s; and two have a speed of 12.0 m/s. Find (a) the average speed, (b) the rms speed, and (c) the most probable speed of these particles.

40. Gaseous helium is in thermal equilibrium with liquid helium at 4.20 K. Even though it is on the point of condensation, model the gas as ideal and determine the most probable speed of a helium atom (mass = 6.64×10^{-27} kg) in it.

41. From the Maxwell–Boltzmann speed distribution, show that the most probable speed of a gas molecule is given by Equation 21.29. Note that the most probable speed corresponds to the point at which the slope of the speed distribution curve, dN_v/dv, is zero.

42. **Review Problem.** At what temperature would the average speed of helium atoms equal (a) the escape speed from Earth, 1.12×10^4 m/s, and (b) the escape speed from the Moon, 2.37×10^3 m/s ? (See Chapter 14 for a discussion of escape speed, and note that the mass of a helium atom is 6.64×10^{-27} kg.)

43. A gas is at 0°C. If we wish to double the rms speed of the gas's molecules, by how much must we raise its temperature?

44. The latent heat of vaporization for water at room temperature is 2 430 J/g. (a) How much kinetic energy does each water molecule that evaporates possess before it evaporates? (b) Find the pre-evaporation rms speed of a water molecule that is evaporating. (c) What is the effective temperature of these molecules (modeled as if they were already in a thin gas)? Why do these molecules not burn you?

(Optional)
Section 21.7   Mean Free Path

45. In an ultrahigh vacuum system, the pressure is measured to be 1.00×10^{-10} torr (where 1 torr = 133 Pa). Assume that the gas molecules have a molecular diameter of 3.00×10^{-10} m and that the temperature is 300 K. Find (a) the number of molecules in a volume of 1.00 m^3, (b) the mean free path of the molecules, and (c) the collision frequency, assuming an average speed of 500 m/s.

46. In deep space it is reported that there is only one particle per cubic meter. Using the average temperature of 3.00 K and assuming that the particle is H$_2$ (with a diameter of 0.200 nm), (a) determine the mean free path of the particle and the average time between collisions.

(b) Repeat part (a), assuming that there is only one particle per cubic centimeter.

47. Show that the mean free path for the molecules of an ideal gas at temperature T and pressure P is

$$\ell = \frac{k_B T}{\sqrt{2}\pi d^2 P}$$

where d is the molecular diameter.

48. In a tank full of oxygen, how many molecular diameters d (on average) does an oxygen molecule travel (at 1.00 atm and 20.0°C) before colliding with another O$_2$ molecule? (The diameter of the O$_2$ molecule is approximately 3.60×10^{-10} m.)

49. Argon gas at atmospheric pressure and 20.0°C is confined in a 1.00-m^3 vessel. The effective hard-sphere diameter of the argon atom is 3.10×10^{-10} m. (a) Determine the mean free path ℓ. (b) Find the pressure when the mean free path is $\ell = 1.00$ m. (c) Find the pressure when $\ell = 3.10 \times 10^{-10}$ m.

ADDITIONAL PROBLEMS

50. The dimensions of a room are 4.20 m × 3.00 m × 2.50 m. (a) Find the number of molecules of air in it at atmospheric pressure and 20.0°C. (b) Find the mass of this air, assuming that the air consists of diatomic molecules with a molar mass of 28.9 g/mol. (c) Find the average kinetic energy of a molecule. (d) Find the root-mean-square molecular speed. (e) On the assumption that the specific heat is a constant independent of temperature, we have $E_{int} = 5nRT/2$. Find the internal energy in the air. (f) Find the internal energy of the air in the room at 25.0°C.

51. The function $E_{int} = 3.50nRT$ describes the internal energy of a certain ideal gas. A sample comprising 2.00 mol of the gas always starts at pressure 100 kPa and temperature 300 K. For each one of the following processes, determine the final pressure, volume, and temperature; the change in internal energy of the gas; the energy added to the gas by heat; and the work done by the gas: (a) The gas is heated at constant pressure to 400 K. (b) The gas is heated at constant volume to 400 K. (c) The gas is compressed at constant temperature to 120 kPa. (d) The gas is compressed adiabatically to 120 kPa.

52. Twenty particles, each of mass m and confined to a volume V, have various speeds: two have speed v; three have speed $2v$; five have speed $3v$; four have speed $4v$; three have speed $5v$; two have speed $6v$; one has speed $7v$. Find (a) the average speed, (b) the rms speed, (c) the most probable speed, (d) the pressure that the particles exert on the walls of the vessel, and (e) the average kinetic energy per particle.

WEB 53. A cylinder contains n mol of an ideal gas that undergoes an adiabatic process. (a) Starting with the expression

$W = \int P\, dV$ and using the expression $PV^{\gamma} = $ constant, show that the work done is

$$W = \left(\frac{1}{\gamma - 1}\right)(P_i V_i - P_f V_f)$$

(b) Starting with the first law equation in differential form, prove that the work done also is equal to $nC_V(T_i - T_f)$. Show that this result is consistent with the equation given in part (a).

54. A vessel contains 1.00×10^4 oxygen molecules at 500 K. (a) Make an accurate graph of the Maxwell speed distribution function versus speed with points at speed intervals of 100 m/s. (b) Determine the most probable speed from this graph. (c) Calculate the average and rms speeds for the molecules and label these points on your graph. (d) From the graph, estimate the fraction of molecules having speeds in the range of 300 m/s to 600 m/s.

55. **Review Problem.**   Oxygen at pressures much greater than 1 atm is toxic to lung cells. By weight, what ratio of helium gas (He) to oxygen gas (O_2) must be used by a scuba diver who is to descend to an ocean depth of 50.0 m?

56. A cylinder with a piston contains 1.20 kg of air at 25.0°C and 200 kPa. Energy is transferred into the system by heat as it is allowed to expand, with the pressure rising to 400 kPa. Throughout the expansion, the relationship between pressure and volume is given by

$$P = CV^{1/2}$$

where C is a constant. (a) Find the initial volume. (b) Find the final volume. (c) Find the final temperature. (d) Find the work that the air does. (e) Find the energy transferred by heat. Take $M = 28.9$ g/mol.

WEB 57. The compressibility κ of a substance is defined as the fractional change in volume of that substance for a given change in pressure:

$$\kappa = -\frac{1}{V}\frac{dV}{dP}$$

(a) Explain why the negative sign in this expression ensures that κ is always positive. (b) Show that if an ideal gas is compressed isothermally, its compressibility is given by $\kappa_1 = 1/P$. (c) Show that if an ideal gas is compressed adiabatically, its compressibility is given by $\kappa_2 = 1/\gamma P$. (d) Determine values for κ_1 and κ_2 for a monatomic ideal gas at a pressure of 2.00 atm.

58. **Review Problem.**   (a) Show that the speed of sound in an ideal gas is

$$v = \sqrt{\frac{\gamma R T}{M}}$$

where M is the molar mass. Use the general expression for the speed of sound in a fluid from Section 17.1; the definition of the bulk modulus from Section 12.4; and the result of Problem 57 in this chapter. As a sound

wave passes through a gas, the compressions are either so rapid or so far apart that energy flow by heat is prevented by lack of time or by effective thickness of insulation. The compressions and rarefactions are adiabatic. (b) Compute the theoretical speed of sound in air at 20°C and compare it with the value given in Table 17.1. Take $M = 28.9$ g/mol. (c) Show that the speed of sound in an ideal gas is

$$v = \sqrt{\frac{\gamma k_B T}{m}}$$

where m is the mass of one molecule. Compare your result with the most probable, the average, and the rms molecular speeds.

59. For a Maxwellian gas, use a computer or programmable calculator to find the numerical value of the ratio $N_v(v)/N_v(v_{mp})$ for the following values of v: $v = (v_{mp}/50)$, $(v_{mp}/10)$, $(v_{mp}/2)$, v_{mp}, $2v_{mp}$, $10v_{mp}$, $50v_{mp}$. Give your results to three significant figures.

60. A pitcher throws a 0.142-kg baseball at 47.2 m/s (Fig. P21.60). As it travels 19.4 m, the ball slows to 42.5 m/s because of air resistance. Find the change in temperature of the air through which it passes. To find the greatest possible temperature change, you may make the following assumptions: Air has a molar heat capacity of $C_P = 7R/2$ and an equivalent molar mass of 28.9 g/mol. The process is so rapid that the cover of the baseball acts as thermal insulation, and the temperature of the ball itself does not change. A change in temperature happens initially only for the air in a cylinder 19.4 m in length and 3.70 cm in radius. This air is initially at 20.0°C.

Figure P21.60   Nolan Ryan hurls the baseball for his 5 000th strikeout. (*Joe Patronite/ALLSPORT*)

61. Consider the particles in a *gas centrifuge,* a device that separates particles of different mass by whirling them in a circular path of radius r at angular speed ω. Newton's second law applied to circular motion states that a force of magnitude equal to $m\omega^2 r$ acts on a particle. (a) Discuss how a gas centrifuge can be used to separate particles of different mass. (b) Show that the density of the particles as a function of r is

$$n(r) = n_0 e^{mr^2\omega^2/2k_B T}$$

62. Verify Equations 21.27 and 21.28 for the rms and average speeds of the molecules of a gas at a temperature T. Note that the average value of v^n is

$$\overline{v^n} = \frac{1}{N}\int_0^\infty v^n N_v\, dv$$

and make use of the definite integrals

$$\int_0^\infty x^3 e^{-ax^2}\, dx = \frac{1}{2a^2} \qquad \int_0^\infty x^4 e^{-ax^2}\, dx = \frac{3}{8a^2}\sqrt{\frac{\pi}{a}}$$

63. A sample of a monatomic ideal gas occupies 5.00 L at atmospheric pressure and 300 K (point A in Figure P21.63). It is heated at constant volume to 3.00 atm (point B). Then, it is allowed to expand isothermally to 1.00 atm (point C) and at last is compressed isobarically to its original state. (a) Find the number of moles in the sample. (b) Find the temperatures at points B and C and the volume at point C. (c) Assuming that the specific heat does not depend on temperature, so that $E_{int} = 3nRT/2$, find the internal energy at points A, B,

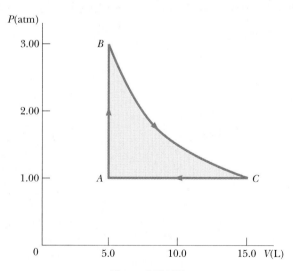

Figure P21.63

and C. (d) Tabulate P, V, T, and E_{int} at the states at points A, B, and C. (e) Now consider the processes $A \rightarrow B$, $B \rightarrow C$, and $C \rightarrow A$. Describe just how to carry out each process experimentally. (f) Find Q, W, and ΔE_{int} for each of the processes. (g) For the whole cycle $A \rightarrow B \rightarrow C \rightarrow A$, find Q, W, and ΔE_{int}.

64. *If you can't walk to outer space, can you walk at least half way?* (a) Show that the fraction of particles below an altitude h in the atmosphere is

$$f = 1 - e^{(-mgh/k_B T)}$$

(b) Use this result to show that half the particles are below the altitude $h' = k_B T \ln(2)/mg$. What is the value of h' for the Earth? (Assume a temperature of 270 K, and note that the average molar mass for air is 28.9 g/mol.)

65. This problem will help you to think about the size of molecules. In the city of Beijing, a restaurant keeps a pot of chicken broth simmering continuously. Every morning it is topped off to contain 10.0 L of water, along with a fresh chicken, vegetables, and spices. The soup is thoroughly stirred. The molar mass of water is 18.0 g/mol. (a) Find the number of molecules of water in the pot. (b) During a certain month, 90.0% of the broth was served each day to people who then emigrated immediately. Of the water molecules present in the pot on the first day of the month, when was the last one likely to have been ladled out of the pot? (c) The broth has been simmering for centuries, through wars, earthquakes, and stove repairs. Suppose that the water that was in the pot long ago has thoroughly mixed into the Earth's hydrosphere, of mass 1.32×10^{21} kg. How many of the water molecules originally in the pot are likely to be present in it again today?

66. Review Problem.   (a) If it has enough kinetic energy, a molecule at the surface of the Earth can escape the Earth's gravitation. Using the principle of conservation of energy, show that the minimum kinetic energy needed for escape is mgR, where m is the mass of the molecule, g is the free-fall acceleration at the surface of the Earth, and R is the radius of the Earth. (b) Calculate the temperature for which the minimum escape kinetic energy is ten times the average kinetic energy of an oxygen molecule.

67. Using multiple laser beams, physicists have been able to cool and trap sodium atoms in a small region. In one experiment, the temperature of the atoms was reduced to 0.240 mK. (a) Determine the rms speed of the sodium atoms at this temperature. The atoms can be trapped for about 1.00 s. The trap has a linear dimension of roughly 1.00 cm. (b) Approximately how long would it take an atom to wander out of the trap region if there were no trapping action?

ANSWERS TO QUICK QUIZZES

21.1 Although a molecule moves very rapidly, it does not travel far before it collides with another molecule. The collision deflects the molecule from its original path. Eventually, a perfume molecule will make its way from one end of the room to the other, but the path it takes is much longer than the straight-line distance from the perfume bottle to your nose.

21.2 (c) E_{int} stays the same. According to Equation 21.10, E_{int} is a function of temperature only. Along an isotherm, T is constant by definition. Therefore, the internal energy of the gas does not change.

21.3 The area under each curve represents the number of molecules in that particular velocity range. The $T = 900$ K curve has many more molecules moving between 800 m/s and 1000 m/s than does the $T = 300$ K curve.

By permission of John Hart and Field Enterprises, Inc.

P U Z Z L E R

The purpose of a refrigerator is to keep its contents cool. Beyond the attendant increase in your electricity bill, there is another good reason you should not try to cool the kitchen on a hot day by leaving the refrigerator door open. What might this reason be? *(Charles D. Winters)*

c h a p t e r

22

Heat Engines, Entropy, and the Second Law of Thermodynamics

Chapter Outline

The first law of thermodynamics, which we studied in Chapter 20, is a statement of conservation of energy, generalized to include internal energy. This law states that a change in internal energy in a system can occur as a result of energy transfer by heat or by work, or by both. As was stated in Chapter 20, the law makes no distinction between the results of heat and the results of work—either heat or work can cause a change in internal energy. However, an important distinction between the two is not evident from the first law. One manifestation of this distinction is that it is impossible to convert internal energy completely to mechanical energy by taking a substance through a thermodynamic cycle such as in a *heat engine*, a device we study in this chapter.

Although the first law of thermodynamics is very important, it makes no distinction between processes that occur spontaneously and those that do not. However, we find that only certain types of energy-conversion and energy-transfer processes actually take place. The *second law of thermodynamics*, which we study in this chapter, establishes which processes do and which do not occur in nature. The following are examples of processes that proceed in only one direction, governed by the second law:

- When two objects at different temperatures are placed in thermal contact with each other, energy always flows by heat from the warmer to the cooler, never from the cooler to the warmer.
- A rubber ball dropped to the ground bounces several times and eventually comes to rest, but a ball lying on the ground never begins bouncing on its own.
- An oscillating pendulum eventually comes to rest because of collisions with air molecules and friction at the point of suspension. The mechanical energy of the system is converted to internal energy in the air, the pendulum, and the suspension; the reverse conversion of energy never occurs.

All these processes are *irreversible*—that is, they are processes that occur naturally in one direction only. No irreversible process has ever been observed to run backward—if it were to do so, it would violate the second law of thermodynamics.[1]

From an engineering standpoint, perhaps the most important implication of the second law is the limited efficiency of heat engines. The second law states that a machine capable of continuously converting internal energy completely to other forms of energy in a cyclic process cannot be constructed.

22.1 ▶ HEAT ENGINES AND THE SECOND LAW OF THERMODYNAMICS

A **heat engine** is a device that converts internal energy to mechanical energy. For instance, in a typical process by which a power plant produces electricity, coal or some other fuel is burned, and the high-temperature gases produced are used to convert liquid water to steam. This steam is directed at the blades of a turbine, setting it into rotation. The mechanical energy associated with this rotation is used to drive an electric generator. Another heat engine—the internal combustion engine in an automobile—uses energy from a burning fuel to perform work that results in the motion of the automobile.

---

[1] Although we have never *observed* a process occurring in the time-reversed sense, it is *possible* for it to occur. As we shall see later in the chapter, however, such a process is highly improbable. From this viewpoint, we say that processes occur with a vastly greater probability in one direction than in the opposite direction.

Figure 22.1 This steam-driven locomotive runs from Durango to Silverton, Colorado. It obtains its energy by burning wood or coal. The generated energy vaporizes water into steam, which powers the locomotive. (This locomotive must take on water from tanks located along the route to replace steam lost through the funnel.) Modern locomotives use diesel fuel instead of wood or coal. Whether old-fashioned or modern, such locomotives are heat engines, which extract energy from a burning fuel and convert a fraction of it to mechanical energy. *(© Phil Degginger/Tony Stone Images)*

A heat engine carries some working substance through a cyclic process during which (1) the working substance absorbs energy from a high-temperature energy reservoir, (2) work is done by the engine, and (3) energy is expelled by the engine to a lower-temperature reservoir. As an example, consider the operation of a steam engine (Fig. 22.1), in which the working substance is water. The water in a boiler absorbs energy from burning fuel and evaporates to steam, which then does work by expanding against a piston. After the steam cools and condenses, the liquid water produced returns to the boiler and the cycle repeats.

It is useful to represent a heat engine schematically as in Figure 22.2. The engine absorbs a quantity of energy Q_h from the hot reservoir, does work W, and then gives up a quantity of energy Q_c to the cold reservoir. Because the working substance goes through a cycle, its initial and final internal energies are equal, and so $\Delta E_{int} = 0$. Hence, from the first law of thermodynamics, $\Delta E_{int} = Q - W$, and with no change in internal energy, **the net work W done by a heat engine is equal to the net energy Q_{net} flowing through it.** As we can see from Figure 22.2, $Q_{net} = Q_h - Q_c$; therefore,

$$W = Q_h - Q_c \qquad \textbf{(22.1)}$$

In this expression and in many others throughout this chapter, to be consistent with traditional treatments of heat engines, we take both Q_h and Q_c to be positive quantities, even though Q_c represents energy leaving the engine. In discussions of heat engines, we shall describe energy leaving a system with an explicit minus sign,

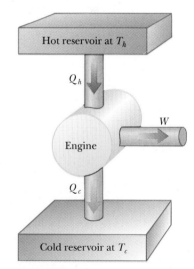

Figure 22.2 Schematic representation of a heat engine. The engine absorbs energy Q_h from the hot reservoir, expels energy Q_c to the cold reservoir, and does work W.

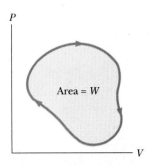

Figure 22.3 *PV* diagram for an arbitrary cyclic process. The value of the net work done equals the area enclosed by the curve.

as in Equation 22.1. Also note that we model the energy input and output for the heat engine as heat, as it often is; however, the energy transfer could occur by another mechanism.

The net work done in a cyclic process is the area enclosed by the curve representing the process on a *PV* diagram. This is shown for an arbitrary cyclic process in Figure 22.3.

The **thermal efficiency** e of a heat engine is defined as the ratio of the net work done by the engine during one cycle to the energy absorbed at the higher temperature during the cycle:

$$e = \frac{W}{Q_h} = \frac{Q_h - Q_c}{Q_h} = 1 - \frac{Q_c}{Q_h} \qquad \textbf{(22.2)}$$

We can think of the efficiency as the ratio of what you get (mechanical work) to what you give (energy transfer at the higher temperature). In practice, we find that all heat engines expel only a fraction of the absorbed energy as mechanical work and that consequently the efficiency is less than 100%. For example, a good automobile engine has an efficiency of about 20%, and diesel engines have efficiencies ranging from 35% to 40%.

Equation 22.2 shows that a heat engine has 100% efficiency ($e = 1$) only if $Q_c = 0$—that is, if no energy is expelled to the cold reservoir. In other words, a heat engine with perfect efficiency would have to expel all of the absorbed energy as mechanical work. On the basis of the fact that efficiencies of real engines are well below 100%, the **Kelvin–Planck** form of the **second law of thermodynamics** states the following:

Kelvin–Planck statement of the second law of thermodynamics

It is impossible to construct a heat engine that, operating in a cycle, produces no effect other than the absorption of energy from a reservoir and the performance of an equal amount of work.

This statement of the second law means that, during the operation of a heat engine, W can never be equal to Q_h, or, alternatively, that some energy Q_c must be

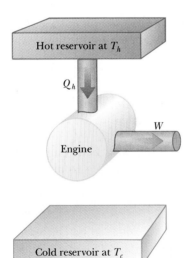

The impossible engine

Figure 22.4 Schematic diagram of a heat engine that absorbs energy Q_h from a hot reservoir and does an equivalent amount of work. It is impossible to construct such a perfect engine.

rejected to the environment. Figure 22.4 is a schematic diagram of the impossible "perfect" heat engine.

The first and second laws of thermodynamics can be summarized as follows: The first law specifies that **we cannot get more energy out of a cyclic process by work than the amount of energy we put in,** and the second law states that **we cannot break even because we must put more energy in, at the higher temperature, than the net amount of energy we get out by work.**

EXAMPLE 22.1 ▶ **The Efficiency of an Engine**

Find the efficiency of a heat engine that absorbs 2 000 J of energy from a hot reservoir and exhausts 1 500 J to a cold reservoir.

Solution To calculate the efficiency of the engine, we use

Equation 22.2:

$$e = 1 - \frac{Q_c}{Q_h} = 1 - \frac{1\ 500\ \text{J}}{2\ 000\ \text{J}} = 0.25,\ \text{or}\ \boxed{25\%}$$

Refrigerators and Heat Pumps

Refrigerators and **heat pumps** are heat engines running in reverse. Here, we introduce them briefly for the purposes of developing an alternate statement of the second law; we shall discuss them more fully in Section 22.5.

In a refrigerator or heat pump, the engine absorbs energy Q_c from a cold reservoir and expels energy Q_h to a hot reservoir (Fig. 22.5). This can be accomplished only if work is done *on* the engine. From the first law, we know that the energy given up to the hot reservoir must equal the sum of the work done and the energy absorbed from the cold reservoir. Therefore, the refrigerator or heat pump transfers energy from a colder body (for example, the contents of a kitchen refrigerator or the winter air outside a building) to a hotter body (the air in the kitchen or a room in the building). In practice, it is desirable to carry out this process with a minimum of work. If it could be accomplished without doing any work, then the refrigerator or heat pump would be "perfect" (Fig. 22.6). Again, the existence of

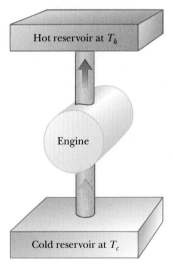

Impossible refrigerator

Figure 22.6 Schematic diagram of an impossible refrigerator or heat pump—that is, one that absorbs energy Q_c from a cold reservoir and expels an equivalent amount of energy to a hot reservoir with $W = 0$.

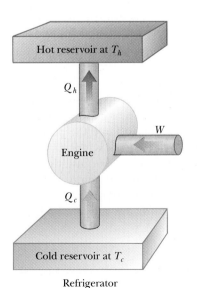

Refrigerator

Figure 22.5 Schematic diagram of a refrigerator, which absorbs energy Q_c from a cold reservoir and expels energy Q_h to a hot reservoir. Work W is done *on* the refrigerator. A heat pump, which can be used to heat or cool a building, works the same way.

such a device would be in violation of the second law of thermodynamics, which in the form of the **Clausius statement**[2] states:

> It is impossible to construct a cyclical machine whose sole effect is the continuous transfer of energy from one object to another object at a higher temperature without the input of energy by work.

In simpler terms, **energy does not flow spontaneously from a cold object to a hot object.** For example, we cool homes in summer using heat pumps called *air conditioners.* The air conditioner pumps energy from the cool room in the home to the warm air outside. This direction of energy transfer requires an input of energy to the air conditioner, which is supplied by the electric power company.

The Clausius and Kelvin–Planck statements of the second law of thermodynamics appear, at first sight, to be unrelated, but in fact they are equivalent in all respects. Although we do not prove so here, if either statement is false, then so is the other.[3]

22.2 REVERSIBLE AND IRREVERSIBLE PROCESSES

In the next section we discuss a theoretical heat engine that is the most efficient possible. To understand its nature, we must first examine the meaning of reversible and irreversible processes. In a **reversible** process, the system undergoing the process can be returned to its initial conditions along the same path shown on a *PV* diagram, and every point along this path is an equilibrium state. A process that does not satisfy these requirements is **irreversible.**

All natural processes are known to be irreversible. From the endless number of examples that could be selected, let us examine the adiabatic free expansion of a gas, which was already discussed in Section 20.6, and show that it cannot be reversible. The system that we consider is a gas in a thermally insulated container, as shown in Figure 22.7. A membrane separates the gas from a vacuum. When the membrane is punctured, the gas expands freely into the vacuum. As a result of the puncture, the system has changed because it occupies a greater volume after the expansion. Because the gas does not exert a force through a distance on the surroundings, it does no work on the surroundings as it expands. In addition, no energy is transferred to or from the gas by heat because the container is insulated from its surroundings. Thus, in this adiabatic process, the system has changed but the surroundings have not.

For this process to be reversible, we need to be able to return the gas to its original volume and temperature without changing the surroundings. Imagine that we try to reverse the process by compressing the gas to its original volume. To do so, we fit the container with a piston and use an engine to force the piston inward. During this process, the surroundings change because work is being done by an outside agent on the system. In addition, the system changes because the compression increases the temperature of the gas. We can lower the temperature of the gas by allowing it to come into contact with an external energy reservoir. Although this step returns the gas to its original conditions, the surroundings are

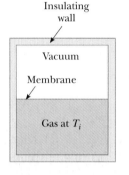

Figure 22.7 Adiabatic free expansion of a gas.

---

[2] First expressed by Rudolf Clausius (1822–1888).

[3] See, for example, R. P. Bauman, *Modern Thermodynamics and Statistical Mechanics,* New York, Macmillan Publishing Co., 1992.

again affected because energy is being added to the surroundings from the gas. If this energy could somehow be used to drive the engine that we have used to compress the gas, then the net energy transfer to the surroundings would be zero. In this way, the system and its surroundings could be returned to their initial conditions, and we could identify the process as reversible. However, the Kelvin–Planck statement of the second law specifies that the energy removed from the gas to return the temperature to its original value cannot be completely converted to mechanical energy in the form of the work done by the engine in compressing the gas. Thus, we must conclude that the process is irreversible.

We could also argue that the adiabatic free expansion is irreversible by relying on the portion of the definition of a reversible process that refers to equilibrium states. For example, during the expansion, significant variations in pressure occur throughout the gas. Thus, there is no well-defined value of the pressure for the entire system at any time between the initial and final states. In fact, the process cannot even be represented as a path on a *PV* diagram. The *PV* diagram for an adiabatic free expansion would show the initial and final conditions as points, but these points would not be connected by a path. Thus, because the intermediate conditions between the initial and final states are not equilibrium states, the process is irreversible.

Although all real processes are always irreversible, some are almost reversible. If a real process occurs very slowly such that the system is always very nearly in an equilibrium state, then the process can be approximated as reversible. For example, let us imagine that we compress a gas very slowly by dropping some grains of sand onto a frictionless piston, as shown in Figure 22.8. We make the process isothermal by placing the gas in thermal contact with an energy reservoir, and we transfer just enough energy from the gas to the reservoir during the process to keep the temperature constant. The pressure, volume, and temperature of the gas are all well defined during the isothermal compression, so each state during the process is an equilibrium state. Each time we add a grain of sand to the piston, the volume of the gas decreases slightly while the pressure increases slightly. Each grain we add represents a change to a new equilibrium state. We can reverse the process by slowly removing grains from the piston.

A general characteristic of a reversible process is that no dissipative effects (such as turbulence or friction) that convert mechanical energy to internal energy can be present. Such effects can be impossible to eliminate completely. Hence, it is not surprising that real processes in nature are irreversible.

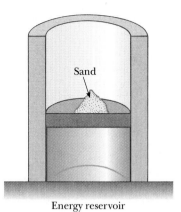

Energy reservoir

Figure 22.8 A gas in thermal contact with an energy reservoir is compressed slowly as individual grains of sand drop onto the piston. The compression is isothermal and reversible.

22.3 THE CARNOT ENGINE

In 1824 a French engineer named Sadi Carnot described a theoretical engine, now called a **Carnot engine,** that is of great importance from both practical and theoretical viewpoints. He showed that a heat engine operating in an ideal, reversible cycle—called a **Carnot cycle**—between two energy reservoirs is the most efficient engine possible. Such an ideal engine establishes an upper limit on the efficiencies of all other engines. That is, the net work done by a working substance taken through the Carnot cycle is the greatest amount of work possible for a given amount of energy supplied to the substance at the upper temperature. **Carnot's theorem** can be stated as follows:

> No real heat engine operating between two energy reservoirs can be more efficient than a Carnot engine operating between the same two reservoirs.

Sadi Carnot French physicist **(1796–1832)** Carnot was the first to show the quantitative relationship between work and heat. In 1824 he published his only work—*Reflections on the Motive Power of Heat*—which reviewed the industrial, political, and economic importance of the steam engine. In it, he defined work as "weight lifted through a height." *(FPG)*

To argue the validity of this theorem, let us imagine two heat engines operating between the *same* energy reservoirs. One is a Carnot engine with efficiency e_C, and the other is an engine with efficiency e, which is greater than e_C. We use the more efficient engine to drive the Carnot engine as a Carnot refrigerator. Thus, the output by work of the more efficient engine is matched to the input by work of the

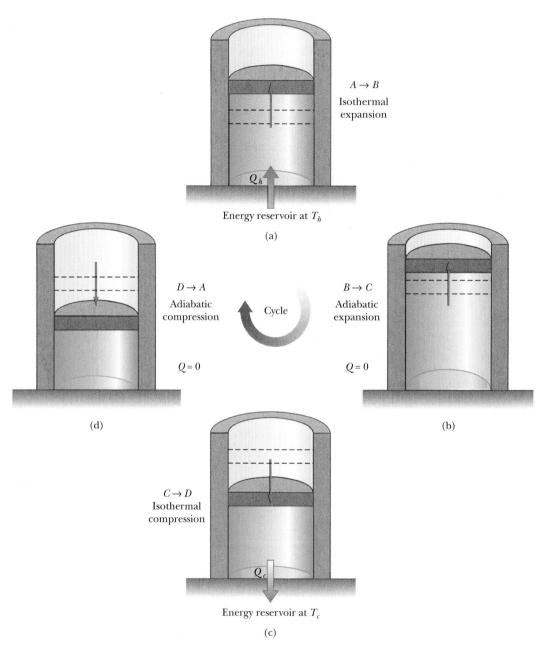

Figure 22.9 The Carnot cycle. In process $A \rightarrow B$, the gas expands isothermally while in contact with a reservoir at T_h. In process $B \rightarrow C$, the gas expands adiabatically ($Q = 0$). In process $C \rightarrow D$, the gas is compressed isothermally while in contact with a reservoir at $T_c < T_h$. In process $D \rightarrow A$, the gas is compressed adiabatically. The upward arrows on the piston indicate that weights are being removed during the expansions, and the downward arrows indicate that weights are being added during the compressions.

Carnot refrigerator. For the *combination* of the engine and refrigerator, then, no exchange by work with the surroundings occurs. Because we have assumed that the engine is more efficient than the refrigerator, the net result of the combination is a transfer of energy from the cold to the hot reservoir without work being done on the combination. According to the Clausius statement of the second law, this is impossible. Hence, the assumption that $e > e_C$ must be false. **All real engines are less efficient than the Carnot engine because they do not operate through a reversible cycle.** The efficiency of a real engine is further reduced by such practical difficulties as friction and energy losses by conduction.

To describe the Carnot cycle taking place between temperatures T_c and T_h, we assume that the working substance is an ideal gas contained in a cylinder fitted with a movable piston at one end. The cylinder's walls and the piston are thermally nonconducting. Four stages of the Carnot cycle are shown in Figure 22.9, and the *PV* diagram for the cycle is shown in Figure 22.10. The Carnot cycle consists of two adiabatic processes and two isothermal processes, all reversible:

1. Process $A \rightarrow B$ (Fig. 22.9a) is an isothermal expansion at temperature T_h. The gas is placed in thermal contact with an energy reservoir at temperature T_h. During the expansion, the gas absorbs energy Q_h from the reservoir through the base of the cylinder and does work W_{AB} in raising the piston.
2. In process $B \rightarrow C$ (Fig. 22.9b), the base of the cylinder is replaced by a thermally nonconducting wall, and the gas expands adiabatically—that is, no energy enters or leaves the system. During the expansion, the temperature of the gas decreases from T_h to T_c and the gas does work W_{BC} in raising the piston.
3. In process $C \rightarrow D$ (Fig. 22.9c), the gas is placed in thermal contact with an energy reservoir at temperature T_c and is compressed isothermally at temperature T_c. During this time, the gas expels energy Q_c to the reservoir, and the work done by the piston on the gas is W_{CD}.
4. In the final process $D \rightarrow A$ (Fig. 22.9d), the base of the cylinder is replaced by a nonconducting wall, and the gas is compressed adiabatically. The temperature of the gas increases to T_h, and the work done by the piston on the gas is W_{DA}.

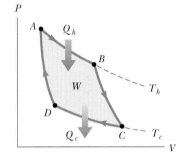

Figure 22.10 *PV* diagram for the Carnot cycle. The net work done, *W*, equals the net energy received in one cycle, $Q_h - Q_c$. Note that $\Delta E_{\text{int}} = 0$ for the cycle.

The net work done in this reversible, cyclic process is equal to the area enclosed by the path *ABCDA* in Figure 22.10. As we demonstrated in Section 22.1, because the change in internal energy is zero, the net work *W* done in one cycle equals the net energy transferred into the system, $Q_h - Q_c$. The thermal efficiency of the engine is given by Equation 22.2:

$$e = \frac{W}{Q_h} = \frac{Q_h - Q_c}{Q_h} = 1 - \frac{Q_c}{Q_h}$$

In Example 22.2, we show that for a Carnot cycle

$$\frac{Q_c}{Q_h} = \frac{T_c}{T_h} \qquad\qquad \textbf{(22.3)}$$

Ratio of energies for a Carnot cycle

Hence, the thermal efficiency of a Carnot engine is

$$e_C = 1 - \frac{T_c}{T_h} \qquad\qquad \textbf{(22.4)}$$

Efficiency of a Carnot engine

This result indicates that **all Carnot engines operating between the same two temperatures have the same efficiency.**

Equation 22.4 can be applied to any working substance operating in a Carnot cycle between two energy reservoirs. According to this equation, the efficiency is zero if $T_c = T_h$, as one would expect. The efficiency increases as T_c is lowered and as T_h is raised. However, the efficiency can be unity (100%) only if $T_c = 0$ K. Such reservoirs are not available; thus, the maximum efficiency is always less than 100%. In most practical cases, T_c is near room temperature, which is about 300 K. Therefore, one usually strives to increase the efficiency by raising T_h.

EXAMPLE 22.2   Efficiency of the Carnot Engine

Show that the efficiency of a heat engine operating in a Carnot cycle using an ideal gas is given by Equation 22.4.

Solution   During the isothermal expansion (process $A \rightarrow B$ in Figure 22.9), the temperature does not change. Thus, the internal energy remains constant. The work done by a gas during an isothermal expansion is given by Equation 20.13. According to the first law, this work is equal to Q_h, the energy absorbed, so that

$$Q_h = W_{AB} = nRT_h \ln \frac{V_B}{V_A}$$

In a similar manner, the energy transferred to the cold reservoir during the isothermal compression $C \rightarrow D$ is

$$Q_c = |W_{CD}| = nRT_c \ln \frac{V_C}{V_D}$$

We take the absolute value of the work because we are defining all values of Q for a heat engine as positive, as mentioned earlier. Dividing the second expression by the first, we find that

$$(1) \qquad \frac{Q_c}{Q_h} = \frac{T_c}{T_h} \frac{\ln(V_C/V_D)}{\ln(V_B/V_A)}$$

We now show that the ratio of the logarithmic quantities is unity by establishing a relationship between the ratio of volumes. For any quasi-static, adiabatic process, the pressure and volume are related by Equation 21.18:

$$(2) \qquad PV^\gamma = \text{constant}$$

During any reversible, quasi-static process, the ideal gas must also obey the equation of state, $PV = nRT$. Solving this ex-

pression for P and substituting into (2), we obtain

$$\frac{nRT}{V} V^\gamma = \text{constant}$$

which we can write as

$$TV^{\gamma-1} = \text{constant}$$

where we have absorbed nR into the constant right-hand side. Applying this result to the adiabatic processes $B \rightarrow C$ and $D \rightarrow A$, we obtain

$$T_h V_B{}^{\gamma-1} = T_c V_C{}^{\gamma-1}$$
$$T_h V_A{}^{\gamma-1} = T_c V_D{}^{\gamma-1}$$

Dividing the first equation by the second, we obtain

$$(V_B/V_A)^{\gamma-1} = (V_C/V_D)^{\gamma-1}$$

$$(3) \qquad \frac{V_B}{V_A} = \frac{V_C}{V_D}$$

Substituting (3) into (1), we find that the logarithmic terms cancel, and we obtain the relationship

$$\frac{Q_c}{Q_h} = \frac{T_c}{T_h}$$

Using this result and Equation 22.2, we see that the thermal efficiency of the Carnot engine is

$$e_C = 1 - \frac{Q_c}{Q_h} = 1 - \frac{T_c}{T_h}$$

which is Equation 22.4, the one we set out to prove.

EXAMPLE 22.3   The Steam Engine

A steam engine has a boiler that operates at 500 K. The energy from the burning fuel changes water to steam, and this steam then drives a piston. The cold reservoir's temperature is that of the outside air, approximately 300 K. What is the maximum thermal efficiency of this steam engine?

Solution   Using Equation 22.4, we find that the maximum thermal efficiency for any engine operating between these temperatures is

$$e_C = 1 - \frac{T_c}{T_h} = 1 - \frac{300 \text{ K}}{500 \text{ K}} = 0.4, \text{ or } \boxed{40\%}$$

You should note that this is the highest *theoretical* efficiency of the engine. In practice, the efficiency is considerably lower.

Exercise   Determine the maximum work that the engine can perform in each cycle if it absorbs 200 J of energy from the hot reservoir during each cycle.

Answer   80 J.

EXAMPLE 22.4   **The Carnot Efficiency**

The highest theoretical efficiency of a certain engine is 30%. If this engine uses the atmosphere, which has a temperature of 300 K, as its cold reservoir, what is the temperature of its hot reservoir?

Solution   We use the Carnot efficiency to find T_h:

$$e_C = 1 - \frac{T_c}{T_h}$$

$$T_h = \frac{T_c}{1 - e_C} = \frac{300 \text{ K}}{1 - 0.30} = \boxed{430 \text{ K}}$$

22.4   GASOLINE AND DIESEL ENGINES

In a gasoline engine, six processes occur in each cycle; five of these are illustrated in Figure 22.11. In this discussion, we consider the interior of the cylinder above the piston to be the system that is taken through repeated cycles in the operation of the engine. For a given cycle, the piston moves up and down twice. This represents a four-stroke cycle consisting of two upstrokes and two downstrokes. The processes in the cycle can be approximated by the **Otto cycle,** a *PV* diagram of which is illustrated in Figure 22.12:

1.   During the *intake stroke* $O \rightarrow A$ (Fig. 22.11a), the piston moves downward, and a gaseous mixture of air and fuel is drawn into the cylinder at atmospheric pressure. In this process, the volume increases from V_2 to V_1. This is the energy input part of the cycle, as energy enters the system (the interior of the cylinder) as internal energy stored in the fuel. This is energy transfer by *mass transfer*—that is, the energy is carried with a substance. It is similar to convection, which we studied in Chapter 20.

2.   During the *compression stroke* $A \rightarrow B$ (Fig. 22.11b), the piston moves upward, the air–fuel mixture is compressed adiabatically from volume V_1 to volume V_2, and the temperature increases from T_A to T_B. The work done by the gas is negative, and its value is equal to the area under the curve AB in Figure 22.12.

3.   In process $B \rightarrow C$, combustion occurs when the spark plug fires (Fig. 22.11c). This is not one of the strokes of the cycle because it occurs in a very short period of time while the piston is at its highest position. The combustion represents a rapid transformation from internal energy stored in chemical bonds in the fuel to internal energy associated with molecular motion, which is related to temperature. During this time, the pressure and temperature in the cylinder increase rapidly, with the temperature rising from T_B to T_C. The volume, however, remains approximately constant because of the short time interval. As a result, approximately no work is done by the gas. We can model this process in the *PV* diagram (Fig. 22.12) as that process in which the energy Q_h enters the system. However, in reality this process is a transformation of energy already in the cylinder (from process $O \rightarrow A$) rather than a transfer.

4.   In the *power stroke* $C \rightarrow D$ (Fig. 22.11d), the gas expands adiabatically from V_2 to

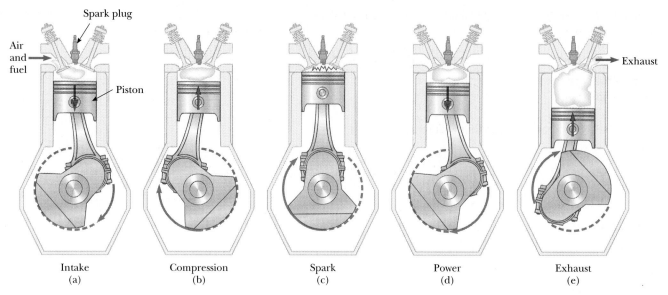

Intake   Compression   Spark   Power   Exhaust
(a)   (b)   (c)   (d)   (e)

Figure 22.11 The four-stroke cycle of a conventional gasoline engine. (a) In the intake stroke, air is mixed with fuel. (b) The intake valve is then closed, and the air–fuel mixture is compressed by the piston. (c) The mixture is ignited by the spark plug, with the result that the temperature of the mixture increases. (d) In the power stroke, the gas expands against the piston. (e) Finally, the residual gases are expelled, and the cycle repeats.

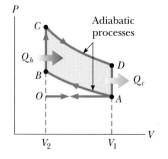

Figure 22.12 *PV* diagram for the Otto cycle, which approximately represents the processes occurring in an internal combustion engine.

Efficiency of the Otto cycle

V_1. This expansion causes the temperature to drop from T_C to T_D. Work is done by the gas in pushing the piston downward, and the value of this work is equal to the area under the curve *CD*.

5. In the process $D \rightarrow A$ (not shown in Fig. 22.11), an exhaust valve is opened as the piston reaches the bottom of its travel, and the pressure suddenly drops for a short time interval. During this interval, the piston is almost stationary and the volume is approximately constant. Energy is expelled from the interior of the cylinder and continues to be expelled during the next process.

6. In the final process, the *exhaust stroke* $A \rightarrow O$ (Fig. 22.11e), the piston moves upward while the exhaust valve remains open. Residual gases are exhausted at atmospheric pressure, and the volume decreases from V_1 to V_2. The cycle then repeats.

If the air–fuel mixture is assumed to be an ideal gas, then the efficiency of the Otto cycle is

$$e = 1 - \frac{1}{(V_1 / V_2)^{\gamma - 1}} \qquad \textbf{(22.5)}$$

where γ is the ratio of the molar specific heats C_P/C_V for the fuel–air mixture and V_1/V_2 is the **compression ratio.** Equation 22.5, which we derive in Example 22.5, shows that the efficiency increases as the compression ratio increases. For a typical compression ratio of 8 and with $\gamma = 1.4$, we predict a theoretical efficiency of 56% for an engine operating in the idealized Otto cycle. This value is much greater than that achieved in real engines (15% to 20%) because of such effects as friction, energy transfer by conduction through the cylinder walls, and incomplete combustion of the air–fuel mixture.

Diesel engines operate on a cycle similar to the Otto cycle but do not employ a spark plug. The compression ratio for a diesel engine is much greater than that for a gasoline engine. Air in the cylinder is compressed to a very small volume, and, as a consequence, the cylinder temperature at the end of the compression stroke is very high. At this point, fuel is injected into the cylinder. The temperature is high enough for the fuel–air mixture to ignite without the assistance of a spark plug. Diesel engines are more efficient than gasoline engines because of their greater compression ratios and resulting higher combustion temperatures.

EXAMPLE 22.5   Efficiency of the Otto Cycle

Show that the thermal efficiency of an engine operating in an idealized Otto cycle (see Figs. 22.11 and 22.12) is given by Equation 22.5. Treat the working substance as an ideal gas.

Solution   First, let us calculate the work done by the gas during each cycle. No work is done during processes $B \rightarrow C$ and $D \rightarrow A$. The work done by the gas during the adiabatic compression $A \rightarrow B$ is negative, and the work done by the gas during the adiabatic expansion $C \rightarrow D$ is positive. The value of the net work done equals the area of the shaded region bounded by the closed curve in Figure 22.12. Because the change in internal energy for one cycle is zero, we see from the first law that the net work done during one cycle equals the net energy flow through the system:

$$W = Q_h - Q_c$$

Because processes $B \rightarrow C$ and $D \rightarrow A$ take place at constant volume, and because the gas is ideal, we find from the definition of molar specific heat (Eq. 21.8) that

$$Q_h = nC_V(T_C - T_B) \quad \text{and} \quad Q_c = nC_V(T_D - T_A)$$

Using these expressions together with Equation 22.2, we obtain for the thermal efficiency

$$(1) \qquad e = \frac{W}{Q_h} = 1 - \frac{Q_c}{Q_h} = 1 - \frac{T_D - T_A}{T_C - T_B}$$

We can simplify this expression by noting that processes $A \rightarrow B$ and $C \rightarrow D$ are adiabatic and hence obey the relationship $TV^{\gamma-1} = $ constant, which we obtained in Example 22.2. For the two adiabatic processes, then,

$$A \rightarrow B: \qquad T_A V_A{}^{\gamma-1} = T_B V_B{}^{\gamma-1}$$

$$C \rightarrow D: \qquad T_C V_C{}^{\gamma-1} = T_D V_D{}^{\gamma-1}$$

Using these equations and relying on the fact that $V_A = V_D = V_1$ and $V_B = V_C = V_2$, we find that

$$T_A V_1{}^{\gamma-1} = T_B V_2{}^{\gamma-1}$$

$$(2) \qquad T_A = T_B \left(\frac{V_2}{V_1}\right)^{\gamma-1}$$

$$T_D V_1{}^{\gamma-1} = T_C V_2{}^{\gamma-1}$$

$$(3) \qquad T_D = T_C \left(\frac{V_2}{V_1}\right)^{\gamma-1}$$

Subtracting (2) from (3) and rearranging, we find that

$$(4) \qquad \frac{T_D - T_A}{T_C - T_B} = \left(\frac{V_2}{V_1}\right)^{\gamma-1}$$

Substituting (4) into (1), we obtain for the thermal efficiency

$$(5) \qquad e = 1 - \frac{1}{(V_1/V_2)^{\gamma-1}}$$

which is Equation 22.5.

We can also express this efficiency in terms of temperatures by noting from (2) and (3) that

$$\left(\frac{V_2}{V_1}\right)^{\gamma-1} = \frac{T_A}{T_B} = \frac{T_D}{T_C}$$

Therefore, (5) becomes

$$(6) \qquad e = 1 - \frac{T_A}{T_B} = 1 - \frac{T_D}{T_C}$$

During the Otto cycle, the lowest temperature is T_A and the highest temperature is T_C. Therefore, the efficiency of a Carnot engine operating between reservoirs at these two temperatures, which is given by the expression $e_C = 1 - (T_A/T_C)$, is *greater* than the efficiency of the Otto cycle given by (6), as expected.

APPLICATION   Models of Gasoline and Diesel Engines

We can use the thermodynamic principles discussed in this and earlier chapters to model the performance of gasoline and diesel engines. In both types of engine, a gas is first compressed in the cylinders of the engine and then the fuel–air mixture is ignited. Work is done on the gas during compression, but significantly more work is done on the piston by the mixture as the products of combustion expand in the cylinder. The power of the engine is transferred from the piston to the crankshaft by the connecting rod.

Two important quantities of either engine are the **displacement volume,** which is the volume displaced by the piston as it moves from the bottom to the top of the cylinder, and the com-

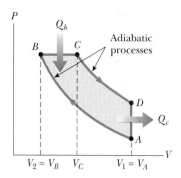

Figure 22.13   *PV* diagram for an ideal diesel engine.

pression ratio *r*, which is the ratio of the maximum and minimum volumes of the cylinder (see p. 680). In our notation, $r = V_A/V_B$, or V_1/V_2 in Eq. 22.5. Most gasoline and diesel engines operate with a four-cycle process (intake, compression, power, exhaust), in which the net work of the intake and exhaust cycles can be considered negligible. Therefore, power is developed only once for every two revolutions of the crankshaft.

In a diesel engine, only air (and no fuel) is present in the cylinder at the beginning of the compression. In the idealized diesel cycle of Figure 22.13, air in the cylinder undergoes an adiabatic compression from *A* to *B*. Starting at *B*, fuel is injected into the cylinder in such a way that the fuel–air mixture undergoes a constant-pressure expansion to an intermediate volume $V_C(B \rightarrow C)$. The high temperature of the mixture causes combustion, and the power stroke is an adiabatic expansion back to $V_D = V_A (C \rightarrow D)$. The exhaust valve is opened, and a constant-volume output of energy occurs ($D \rightarrow A$) as the cylinder empties.

To simplify our calculations, we assume that the mixture in the cylinder is air modeled as an ideal gas. We use specific heats *c* instead of molar specific heats *C* and assume constant values for air at 300 K. We express the specific heats and the universal gas constant in terms of unit masses rather than moles. Thus, $c_V = 0.718$ kJ/kg·K, $c_P = 1.005$ kJ/kg·K, $\gamma = c_P/c_V = 1.40$, and $R = c_P - c_V = 0.287$ kJ/kg·K = 0.287 kPa·m³/kg·K.

A 3.00-L Gasoline Engine

Let us calculate the power delivered by a six-cylinder gasoline engine that has a displacement volume of 3.00 L operating at 4 000 rpm and having a compression ratio of $r = 9.50$. The air–fuel mixture enters a cylinder at atmospheric pressure and an ambient temperature of 27°C. During combustion, the mixture reaches a temperature of 1 350°C.

First, let us calculate the work done by an individual cylinder. Using the initial pressure $P_A = 100$ kPa and the initial temperature $T_A = 300$ K, we calculate the initial volume and the mass of the air–fuel mixture. We know that the ratio of the initial and final volumes is the compression ratio,

$$\frac{V_A}{V_B} = r = 9.50$$

We also know that the difference in volumes is the displacement volume. The 3.00-L rating of the engine is the total displacement volume for all six cylinders. Thus, for one cylinder,

$$V_A - V_B = \frac{3.00 \text{ L}}{6} = \frac{3.00 \times 10^{-3} \text{ m}^3}{6} = 0.500 \times 10^{-3} \text{ m}^3$$

Solving these two equations simultaneously, we find the initial and final volumes:

$$V_A = 0.559 \times 10^{-3} \text{ m}^3 \qquad V_B = 0.588 \times 10^{-4} \text{ m}^3$$

Using the ideal gas law (in the form $PV = mRT$, because we are using the universal gas constant in terms of mass rather than moles), we can find the mass of the air–fuel mixture:

$$m = \frac{P_A V_A}{R T_A} = \frac{(100 \text{ kPa})(0.559 \times 10^{-3} \text{ m}^3)}{(0.287 \text{ kPa·m}^3/\text{kg·K})(300 \text{ K})}$$
$$= 6.49 \times 10^{-4} \text{ kg}$$

Process $A \rightarrow B$ (see Fig. 22.12) is an adiabatic compression, and this means that PV^γ = constant; hence,

$$P_B V_B^\gamma = P_A V_A^\gamma$$

$$P_B = P_A \left(\frac{V_A}{V_B}\right)^\gamma = P_A (r)^\gamma = (100 \text{ kPa})(9.50)^{1.40}$$

$$= 2.34 \times 10^3 \text{ kPa}$$

Using the ideal gas law, we find that the temperature after the compression is

$$T_B = \frac{P_B V_B}{mR} = \frac{(2.34 \times 10^3 \text{ kPa})(0.588 \times 10^{-4} \text{ m}^3)}{(6.49 \times 10^{-4} \text{ kg})(0.287 \text{ kPa·m}^3/\text{kg·K})}$$

$$= 739 \text{ K}$$

In process $B \rightarrow C$, the combustion that transforms the internal energy in chemical bonds into internal energy of molecular motion occurs at constant volume; thus, $V_C = V_B$. Combustion causes the temperature to increase to $T_C = 1\,350°C = 1\,623$ K. Using this value and the ideal gas law, we can calculate P_C:

$$P_C = \frac{mRT_C}{V_C}$$

$$= \frac{(6.49 \times 10^{-4} \text{ kg})(0.287 \text{ kPa·m}^3/\text{kg·K})(1\,623 \text{ K})}{(0.588 \times 10^{-4} \text{ m}^3)}$$

$$= 5.14 \times 10^3 \text{ kPa}$$

Process $C \rightarrow D$ is an adiabatic expansion; the pressure after the expansion is

$$P_D = P_C \left(\frac{V_C}{V_D}\right)^\gamma = P_C \left(\frac{V_B}{V_A}\right)^\gamma = P_C \left(\frac{1}{r}\right)^\gamma$$

$$= (5.14 \times 10^3 \text{ kPa}) \left(\frac{1}{9.50}\right)^{1.40} = 220 \text{ kPa}$$

Using the ideal gas law again, we find the final temperature:

$$T_D = \frac{P_D V_D}{mR} = \frac{(220 \text{ kPa})(0.559 \times 10^{-3} \text{ m}^3)}{(6.49 \times 10^{-4} \text{ kg})(0.287 \text{ kPa} \cdot \text{m}^3/\text{kg} \cdot \text{K})}$$

$$= 660 \text{ K}$$

Now that we have the temperatures at the beginning and end of each process of the cycle, we can calculate the net energy transfer and net work done by each cylinder every two cycles. From Equation 21.8, we can state

$$Q_h = Q_{in} = mc_V(T_C - T_B)$$

$$= (6.49 \times 10^{-4} \text{ kg})(0.718 \text{ kJ/kg} \cdot \text{K})(1\,623 \text{ K} - 739 \text{ K})$$

$$= 0.412 \text{ kJ}$$

$$Q_c = Q_{out} = mc_V(T_D - T_A)$$

$$= (6.49 \times 10^{-4} \text{ kg})(0.718 \text{ kJ/kg} \cdot \text{K})(660 \text{ K} - 300 \text{ K})$$

$$= 0.168 \text{ kJ}$$

$$W_{net} = Q_{in} - Q_{out} = 0.244 \text{ kJ}$$

From Equation 22.2, the efficiency is $e = W_{net}/Q_{in} = 59\%$. (We can also use Equation 22.5 to calculate the efficiency directly from the compression ratio.)

Recalling that power is delivered every other revolution of the crankshaft, we find that the net power for the six-cylinder engine operating at 4 000 rpm is

$$\mathcal{P}_{net} = 6\left(\frac{1}{2 \text{ rev}}\right)(4\,000 \text{ rev/min})(1 \text{ min}/60 \text{ s})(0.244 \text{ kJ})$$

$$= 49 \text{ kW} = 66 \text{ hp}$$

A 2.00-L Diesel Engine

Let us calculate the power delivered by a four-cylinder diesel engine that has a displacement volume of 2.00 L and is operating at 3 000 rpm. The compression ratio is $r = V_A/V_B = 22.0$, and the **cutoff ratio,** which is the ratio of the volume change during the constant-pressure process $B \rightarrow C$ in Figure 22.13, is $r_c = V_C/V_B = 2.00$. The air enters each cylinder at the beginning of the compression cycle at atmospheric pressure and at an ambient temperature of 27°C.

Our model of the diesel engine is similar to our model of the gasoline engine except that now the fuel is injected at point B and the mixture self-ignites near the end of the compression cycle $A \rightarrow B$, when the temperature reaches the ignition temperature. We assume that the energy input occurs in the constant-pressure process $B \rightarrow C$, and that the expansion process continues from C to D with no further energy transfer by heat.

Let us calculate the work done by an individual cylinder that has an initial volume of $V_A = (2.00 \times 10^{-3} \text{ m}^3)/4 = 0.500 \times 10^{-3} \text{ m}^3$. Because the compression ratio is quite high, we approximate the maximum cylinder volume to be the displacement volume. Using the initial pressure $P_A = 100 \text{ kPa}$ and initial temperature $T_A = 300 \text{ K}$, we can calculate the mass of the air in the cylinder using the ideal gas law:

$$m = \frac{P_A V_A}{RT_A} = \frac{(100 \text{ kPa})(0.500 \times 10^{-3} \text{ m}^3)}{(0.287 \text{ kPa} \cdot \text{m}^3/\text{kg} \cdot \text{K})(300 \text{ K})} = 5.81 \times 10^{-4} \text{ kg}$$

Process $A \rightarrow B$ is an adiabatic compression, so $PV^\gamma = $ constant; thus,

$$P_B V_B^\gamma = P_A V_A^\gamma$$

$$P_B = P_A\left(\frac{V_A}{V_B}\right)^\gamma = (100 \text{ kPa})(22.0)^{1.40} = 7.57 \times 10^3 \text{ kPa}$$

Using the ideal gas law, we find that the temperature of the air after the compression is

$$T_B = \frac{P_B V_B}{mR} = \frac{(7.57 \times 10^3 \text{ kPa})(0.500 \times 10^{-3} \text{ m}^3)\left(\frac{1}{22.0}\right)}{(5.81 \times 10^{-4} \text{ kg})(0.287 \text{ kPa} \cdot \text{m}^3/\text{kg} \cdot \text{K})}$$

$$= 1.03 \times 10^3 \text{ K}$$

Process $B \rightarrow C$ is a constant-pressure expansion; thus, $P_C = P_B$. We know from the cutoff ratio of 2.00 that the volume doubles in this process. According to the ideal gas law, a doubling of volume in an isobaric process results in a doubling of the temperature, so

$$T_C = 2T_B = 2.06 \times 10^3 \text{ K}$$

Process $C \rightarrow D$ is an adiabatic expansion; therefore,

$$P_D = P_C\left(\frac{V_C}{V_D}\right)^\gamma = P_C\left(\frac{V_C}{V_B}\frac{V_B}{V_D}\right)^\gamma = P_C\left(r_c\frac{1}{r}\right)^\gamma$$

$$= (7.57 \times 10^3 \text{ kPa})\left(\frac{2.00}{22.0}\right)^{1.40} = 264 \text{ kPa}$$

We find the temperature at D from the ideal gas law:

$$T_D = \frac{P_D V_D}{mR} = \frac{(264 \text{ kPa})(0.500 \times 10^{-3} \text{ m}^3)}{(5.81 \times 10^{-4} \text{ kg})(0.287 \text{ kPa} \cdot \text{m}^3/\text{kg} \cdot \text{K})}$$

$$= 792 \text{ K}$$

Now that we have the temperatures at the beginning and the end of each process, we can calculate the net energy transfer by heat and the net work done by each cylinder every two cycles:

$$Q_h = Q_{in} = mc_P(T_C - T_B) = 0.601 \text{ kJ}$$

$$Q_c = Q_{out} = mc_V(T_D - T_A) = 0.205 \text{ kJ}$$

$$W_{net} = Q_{in} - Q_{out} = 0.396 \text{ kJ}$$

The efficiency is $e = W_{net}/Q_{in} = 66\%$.

The net power for the four-cylinder engine operating at 3 000 rpm is

$$\mathcal{P}_{net} = 4\left(\frac{1}{2 \text{ rev}}\right)(3\,000 \text{ rev/min})(1 \text{ min}/60 \text{ s})(0.396 \text{ kJ})$$

$$= 39.6 \text{ kW} = 53 \text{ hp}$$

Of course, modern engine design goes beyond this simple thermodynamic treatment, which uses idealized cycles.

22.5 HEAT PUMPS AND REFRIGERATORS

In Section 22.1 we introduced a heat pump as a mechanical device that moves energy from a region at lower temperature to a region at higher temperature. Heat pumps have long been used for cooling homes and buildings, and they are now becoming increasingly popular for heating them as well. The heat pump contains two sets of metal coils that can exchange energy by heat with the surroundings: one set on the outside of the building, in contact with the air or buried in the ground; and the other set in the interior of the building. In the heating mode, a circulating fluid flowing through the coils absorbs energy from the outside and releases it to the interior of the building from the interior coils. The fluid is cold and at low pressure when it is in the external coils, where it absorbs energy by heat from either the air or the ground. The resulting warm fluid is then compressed and enters the interior coils as a hot, high-pressure fluid, where it releases its stored energy to the interior air.

 An air conditioner is simply a heat pump operating in the cooling mode, with its exterior and interior coils interchanged. Energy is absorbed into the circulating fluid in the interior coils; then, after the fluid is compressed, energy leaves the fluid through the external coils. The air conditioner must have a way to release energy to the outside. Otherwise, the work done on the air conditioner would represent energy added to the air inside the house, and the temperature would increase. In the same manner, a refrigerator cannot cool the kitchen if the refrigerator door is left open. The amount of energy leaving the external coils (Fig. 22.14) behind or underneath the refrigerator is greater than the amount of energy removed from the food or from the air in the kitchen if the door is left open. The difference between the energy out and the energy in is the work done by the electricity supplied to the refrigerator.

Figure 22.15 is a schematic representation of a heat pump. The cold temperature is T_c, the hot temperature is T_h, and the energy absorbed by the circulating fluid is Q_c. The heat pump does work W on the fluid, and the energy transferred from the pump to the building in the heating mode is Q_h.

The effectiveness of a heat pump is described in terms of a number called the **coefficient of performance** (COP). In the heating mode, the COP is defined as the ratio of the energy transferred to the hot reservoir to the work required to transfer that energy:

$$\text{COP (heating mode)} \equiv \frac{\text{Energy transferred at high temperature}}{\text{Work done by pump}} = \frac{Q_h}{W} \qquad \textbf{(22.6)}$$

Note that the COP is similar to the thermal efficiency for a heat engine in that it is a ratio of what you get (energy delivered to the interior of the building) to what you give (work input). Because Q_h is generally greater than W, typical values for the COP are greater than unity. It is desirable for the COP to be as high as possible, just as it is desirable for the thermal efficiency of an engine to be as high as possible.

If the outside temperature is 25°F or higher, then the COP for a heat pump is about 4. That is, the amount of energy transferred to the building is about four times greater than the work done by the motor in the heat pump. However, as the outside temperature decreases, it becomes more difficult for the heat pump to extract sufficient energy from the air, and so the COP decreases. In fact, the COP can fall below unity for temperatures below the midteens. Thus, the use of heat pumps that extract energy from the air, while satisfactory in moderate climates, is not appropriate in areas where winter temperatures are very low. It is possible to

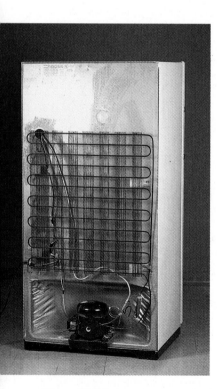

Figure 22.14 The coils on the back of a refrigerator transfer energy by heat to the air. The second law of thermodynamics states that this amount of energy must be greater than the amount of energy removed from the contents of the refrigerator (or from the air in the kitchen, if the refrigerator door is left open). *(Charles D. Winters)*

use heat pumps in colder areas by burying the external coils deep in the ground. In this case, the energy is extracted from the ground, which tends to be warmer than the air in the winter.

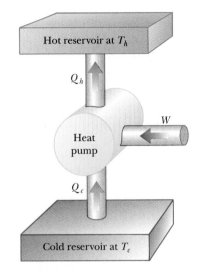

Figure 22.15 Schematic diagram of a heat pump, which absorbs energy Q_c from a cold reservoir and expels energy Q_h to a hot reservoir. Note that this diagram is the same as that for the refrigerator shown in Figure 22.5.

Quick Quiz 22.1

In an electric heater, electrical energy can be converted to internal energy with an efficiency of 100%. By what percentage does the cost of heating your home change when you replace your electric heating system with a heat pump that has a COP of 4? Assume that the motor running the heat pump is 100% efficient.

Theoretically, a Carnot-cycle heat engine run in reverse constitutes the most effective heat pump possible, and it determines the maximum COP for a given combination of hot and cold reservoir temperatures. Using Equations 22.1 and 22.3, we see that the maximum COP for a heat pump in its heating mode is

$$\text{COP}_C(\text{heating mode}) = \frac{Q_h}{W}$$

$$= \frac{Q_h}{Q_h - Q_c} = \frac{1}{1 - \dfrac{Q_c}{Q_h}} = \frac{1}{1 - \dfrac{T_c}{T_h}} = \frac{T_h}{T_h - T_c}$$

For a heat pump operating in the cooling mode, "what you get" is energy removed from the cold reservoir. The most effective refrigerator or air conditioner is one that removes the greatest amount of energy from the cold reservoir in exchange for the least amount of work. Thus, for these devices we define the COP in terms of Q_c:

$$\text{COP (cooling mode)} = \frac{Q_c}{W} \tag{22.7}$$

A good refrigerator should have a high COP, typically 5 or 6.

The greatest possible COP for a heat pump in the cooling mode is that of a heat pump whose working substance is carried through a Carnot cycle in reverse:

$$\text{COP}_C \text{ (cooling mode)} = \frac{T_c}{T_h - T_c}$$

As the difference between the temperatures of the two reservoirs approaches zero in this expression, the theoretical COP approaches infinity. In practice, the low temperature of the cooling coils and the high temperature at the compressor limit the COP to values below 10.

QuickLab

Estimate the COP of your refrigerator by making rough temperature measurements of the stored food and of the exhaust coils (found either on the back of the unit or behind a panel on the bottom). Use just your hand if no thermometer is available.

22.6 ENTROPY

10.10 & 10.11

The zeroth law of thermodynamics involves the concept of temperature, and the first law involves the concept of internal energy. Temperature and internal energy are both state functions—that is, they can be used to describe the thermodynamic state of a system. Another state function—this one related to the second law of thermodynamics—is **entropy** S. In this section we define entropy on a macroscopic scale as it was first expressed by Clausius in 1865.

Consider any infinitesimal process in which a system changes from one equilibrium state to another. If dQ_r is the amount of energy transferred by heat when the system follows a reversible path between the states, then the change in entropy dS is equal to this amount of energy for the reversible process divided by the absolute temperature of the system:

Clausius definition of change in entropy

$$dS = \frac{dQ_r}{T}$$

(22.8)

We have assumed that the temperature is constant because the process is infinitesimal. Since we have claimed that entropy is a state function, **the change in entropy during a process depends only on the end points and therefore is independent of the actual path followed.**

The subscript r on the quantity dQ_r is a reminder that the transferred energy is to be measured along a reversible path, even though the system may actually have followed some irreversible path. When energy is absorbed by the system, dQ_r is positive and the entropy of the system increases. When energy is expelled by the system, dQ_r is negative and the entropy of the system decreases. Note that Equation 22.8 defines not entropy but rather the *change* in entropy. Hence, the meaningful quantity in describing a process is the *change* in entropy.

Entropy was originally formulated as a useful concept in thermodynamics; however, its importance grew tremendously as the field of statistical mechanics developed because the analytical techniques of statistical mechanics provide an alternative means of interpreting entropy. In statistical mechanics, the behavior of a substance is described in terms of the statistical behavior of its atoms and molecules. One of the main results of this treatment is that **isolated systems tend toward disorder and that entropy is a measure of this disorder.** For example, consider the molecules of a gas in the air in your room. If half of the gas molecules had velocity vectors of equal magnitude directed toward the left and the other half had velocity vectors of the same magnitude directed toward the right, the situation would be very ordered. However, such a situation is extremely unlikely. If you could actually view the molecules, you would see that they move haphazardly in all directions, bumping into one another, changing speed upon collision, some going fast and others going slowly. This situation is highly disordered.

The cause of the tendency of an isolated system toward disorder is easily explained. To do so, we distinguish between *microstates* and *macrostates* of a system. A **microstate** is a particular description of the properties of the individual molecules of the system. For example, the description we just gave of the velocity vectors of the air molecules in your room being very ordered refers to a particular microstate, and the more likely likely haphazard motion is another microstate—one that represents disorder. A **macrostate** is a description of the conditions of the system from a macroscopic point of view and makes use of macroscopic variables such as pressure, density, and temperature. For example, in both of the microstates described for the air molecules in your room, the air molecules are distributed uniformly throughout the volume of the room; this uniform density distribution is a macrostate. We could not distinguish between our two microstates by making a macroscopic measurement—both microstates would appear to be the same macroscopically, and the two macrostates corresponding to these microstates are equivalent.

For any given macrostate of the system, a number of microstates are possible, or *accessible*. Among these microstates, it is assumed that all are equally probable. However, when all possible microstates are examined, it is found that far more of them are disordered than are ordered. Because all of the microstates are equally probable, it is highly likely that the actual macrostate is one resulting from one of the highly disordered microstates, simply because there are many more of them. Similarly, the probability of a macrostate's forming from disordered microstates is greater than the probability of a macrostate's forming from ordered microstates.

All physical processes that take place in a system tend to cause the system and its surroundings to move toward more probable macrostates. The more probable macrostate is always one of greater disorder. If we consider a system and its surroundings to include the entire Universe, then the Universe is always moving toward a macrostate corresponding to greater disorder. Because entropy is a measure of disorder, an alternative way of stating this is **the entropy of the Universe increases in all real processes.** This is yet another statement of the second law of thermodynamics that can be shown to be equivalent to the Kelvin–Planck and Clausius statements.

In real processes, the disorder of the Universe increases

To calculate the change in entropy for a finite process, we must recognize that T is generally not constant. If dQ_r is the energy transferred by heat when the system is at a temperature T, then the change in entropy in an arbitrary reversible process between an initial state and a final state is

$$\Delta S = \int_i^f dS = \int_i^f \frac{dQ_r}{T} \qquad \text{(reversible path)} \qquad \textbf{(22.9)}$$

Change in entropy for a finite process

As with an infinitesimal process, the change in entropy ΔS of a system going from one state to another has the same value for *all* paths connecting the two states. That is, the finite change in entropy ΔS of a system depends only on the properties of the initial and final equilibrium states. Thus, we are free to choose a particular reversible path over which to evaluate the entropy in place of the actual path, as long as the initial and final states are the same for both paths.

Quick Quiz 22.2

Which of the following is true for the entropy change of a system that undergoes a reversible, adiabatic process? (a) $\Delta S < 0$. (b) $\Delta S = 0$. (c) $\Delta S > 0$.

Let us consider the changes in entropy that occur in a Carnot heat engine operating between the temperatures T_c and T_h. In one cycle, the engine absorbs energy Q_h from the hot reservoir and expels energy Q_c to the cold reservoir. These energy transfers occur only during the isothermal portions of the Carnot cycle; thus, the constant temperature can be brought out in front of the integral sign in Equation 22.9. The integral then simply has the value of the total amount of energy transferred by heat. Thus, the total change in entropy for one cycle is

$$\Delta S = \frac{Q_h}{T_h} - \frac{Q_c}{T_c}$$

where the negative sign represents the fact that energy Q_c is expelled by the system, since we continue to define Q_c as a positive quantity when referring to heat engines. In Example 22.2 we showed that, for a Carnot engine,

$$\frac{Q_c}{Q_h} = \frac{T_c}{T_h}$$

Using this result in the previous expression for ΔS, we find that the total change in entropy for a Carnot engine operating in a cycle is *zero*:

$$\Delta S = 0$$

Now let us consider a system taken through an arbitrary (non-Carnot) reversible cycle. Because entropy is a state function—and hence depends only on the properties of a given equilibrium state—we conclude that $\Delta S = 0$ for *any* reversible cycle. In general, we can write this condition in the mathematical form

$$\oint \frac{dQ_r}{T} = 0 \tag{22.10}$$

where the symbol $\oint$ indicates that the integration is over a closed path.

Quasi-Static, Reversible Process for an Ideal Gas

Let us suppose that an ideal gas undergoes a quasi-static, reversible process from an initial state having temperature T_i and volume V_i to a final state described by T_f and V_f. Let us calculate the change in entropy of the gas for this process.

Writing the first law of thermodynamics in differential form and rearranging the terms, we have $dQ_r = dE_{\text{int}} + dW$, where $dW = P\,dV$. For an ideal gas, recall that $dE_{\text{int}} = nC_V\,dT$ (Eq. 21.12), and from the ideal gas law, we have $P = nRT/V$. Therefore, we can express the energy transferred by heat in the process as

$$dQ_r = dE_{\text{int}} + P\,dV = nC_V\,dT + nRT\frac{dV}{V}$$

We cannot integrate this expression as it stands because the last term contains two variables, T and V. However, if we divide all terms by T, each of the terms on the right-hand side depends on only one variable:

$$\frac{dQ_r}{T} = nC_V\frac{dT}{T} + nR\frac{dV}{V} \tag{22.11}$$

Assuming that C_V is constant over the interval in question, and integrating Equation 22.11 from the initial state to the final state, we obtain

$$\Delta S = \int_i^f \frac{dQ_r}{T} = nC_V\ln\frac{T_f}{T_i} + nR\ln\frac{V_f}{V_i} \tag{22.12}$$

This expression demonstrates mathematically what we argued earlier—that ΔS depends only on the initial and final states and is independent of the path between the states. Also, note in Equation 22.12 that ΔS can be positive or negative, depending on the values of the initial and final volumes and temperatures. Finally, for a cyclic process ($T_i = T_f$ and $V_i = V_f$), we see from Equation 22.12 that $\Delta S = 0$. This is evidence that entropy is a state function.

EXAMPLE 22.6   Change in Entropy—Melting

A solid that has a latent heat of fusion L_f melts at a temperature T_m. (a) Calculate the change in entropy of this substance when a mass m of the substance melts.

Solution Let us assume that the melting occurs so slowly that it can be considered a reversible process. In this case the temperature can be regarded as constant and equal to T_m.

Making use of Equations 22.9 and that for the latent heat of fusion $Q = mL_f$ (Eq. 20.6), we find that

$$\Delta S = \int \frac{dQ_r}{T} = \frac{1}{T_m}\int dQ = \frac{Q}{T_m} = \frac{mL_f}{T_m}$$

Note that we are able to remove T_m from the integral because the process is isothermal. Note also that ΔS is positive. This means that when a solid melts, its entropy increases because the molecules are much more disordered in the liquid state than they are in the solid state. The positive value for ΔS also means that the substance in its liquid state does not spontaneously transfer energy from itself to the surroundings and freeze because to do so would involve a spontaneous decrease in entropy.

(b) Estimate the value of the change in entropy of an ice cube when it melts.

Solution Let us assume an ice tray makes cubes that are about 3 cm on a side. The volume per cube is then (very roughly) 30 cm³. This much liquid water has a mass of 30 g. From Table 20.2 we find that the latent heat of fusion of ice is 3.33×10^5 J/kg. Substituting these values into our answer for part (a), we find that

$$\Delta S = \frac{mL_f}{T_m} = \frac{(0.03 \text{ kg})(3.33 \times 10^5 \text{ J/kg})}{273 \text{ K}} = \boxed{4 \times 10^1 \text{ J/K}}$$

We retain only one significant figure, in keeping with the nature of our estimations.

22.7 ENTROPY CHANGES IN IRREVERSIBLE PROCESSES

By definition, calculation of the change in entropy requires information about a reversible path connecting the initial and final equilibrium states. To calculate changes in entropy for real (irreversible) processes, we must remember that entropy (like internal energy) depends only on the *state* of the system. That is, entropy is a state function. Hence, the change in entropy when a system moves between any two equilibrium states depends only on the initial and final states. We can show that if this were not the case, the second law of thermodynamics would be violated.

We now calculate the entropy change in some irreversible process between two equilibrium states by devising a reversible process (or series of reversible processes) between the same two states and computing $\Delta S = \int dQ_r/T$ for the reversible process. In irreversible processes, it is critically important that we distinguish between Q, the actual energy transfer in the process, and Q_r, the energy that would have been transferred by heat along a reversible path. Only Q_r is the correct value to be used in calculating the entropy change.

As we shall see in the following examples, the change in entropy for a system and its surroundings is always positive for an irreversible process. In general, the total entropy—and therefore the disorder—always increase in an irreversible process. Keeping these considerations in mind, we can state the second law of thermodynamics as follows:

> The total entropy of an isolated system that undergoes a change can never decrease.

Furthermore, **if the process is irreversible, then the total entropy of an isolated system always increases. In a reversible process, the total entropy of an isolated system remains constant.**

When dealing with a system that is not isolated from its surroundings, remember that the increase in entropy described in the second law is that of the system *and* its surroundings. When a system and its surroundings interact in an irreversible process, the increase in entropy of one is greater than the decrease in entropy of the other. Hence, we conclude that **the change in entropy of the Universe must be greater than zero for an irreversible process and equal to zero for a reversible process.** Ultimately, the entropy of the Universe should reach a maximum value. At this value, the Universe will be in a state of uniform temperature and density. All physical, chemical, and biological processes will cease because a state of perfect disorder implies that no energy is available for doing work. This gloomy state of affairs is sometimes referred to as the heat death of the Universe.

> **Quick Quiz 22.3**
>
> In the presence of sunlight, a tree rearranges an unorganized collection of carbon dioxide and water molecules into the highly ordered collection of molecules we see as leaves and branches. True or false: This reduction of entropy in the tree is a violation of the second law of thermodynamics. Explain your response.

Entropy Change in Thermal Conduction

Let us now consider a system consisting of a hot reservoir and a cold reservoir in thermal contact with each other and isolated from the rest of the Universe. A process occurs during which energy Q is transferred by heat from the hot reservoir at temperature T_h to the cold reservoir at temperature T_c. Because the cold reservoir absorbs energy Q, its entropy increases by Q/T_c. At the same time, the hot reservoir loses energy Q, and so its entropy change is $-Q/T_h$. Because $T_h > T_c$, the increase in entropy of the cold reservoir is greater than the decrease in entropy of the hot reservoir. Therefore, the change in entropy of the system (and of the Universe) is greater than zero:

$$\Delta S_U = \frac{Q}{T_c} + \frac{-Q}{T_h} > 0$$

EXAMPLE 22.7   **Which Way Does the Energy Flow?**

A large, cold object is at 273 K, and a large, hot object is at 373 K. Show that it is impossible for a small amount of energy—for example, 8.00 J—to be transferred spontaneously from the cold object to the hot one without a decrease in the entropy of the Universe and therefore a violation of the second law.

Solution   We assume that, during the energy transfer, the two objects do not undergo a temperature change. This is not a necessary assumption; we make it only to avoid using integral calculus in our calculations. The process as described is irreversible, and so we must find an equivalent reversible process. It is sufficient to assume that the objects are connected by a poor thermal conductor whose temperature spans the range from 273 K to 373 K. This conductor transfers energy slowly, and its state does not change during the process. Under this assumption, the energy transfer to or from each object is reversible, and we may set $Q = Q_r$. The entropy change of the hot object is

$$\Delta S_h = \frac{Q_r}{T_h} = \frac{8.00 \text{ J}}{373 \text{ K}} = 0.021 \text{ 4 J/K}$$

The cold object loses energy, and its entropy change is

$$\Delta S_c = \frac{Q_r}{T_c} = \frac{-8.00 \text{ J}}{273 \text{ K}} = -0.029 \text{ 3 J/K}$$

We consider the two objects to be isolated from the rest of the Universe. Thus, the entropy change of the Universe is just that of our two-object system, which is

$$\Delta S_U = \Delta S_c + \Delta S_h = -0.007 \text{ 9 J/K}$$

This decrease in entropy of the Universe is in violation of the second law. That is, **the spontaneous transfer of energy from a cold to a hot object cannot occur.**

In terms of disorder, let us consider the violation of the second law if energy were to continue to transfer spontaneously from a cold object to a hot object. Before the transfer, a certain degree of order is associated with the different temperatures of the objects. The hot object's molecules have a higher average energy than the cold object's molecules. If energy spontaneously flows from the cold object to the hot object, then, over a period of time, the cold object will become colder and the hot object will become hotter. The difference in average molecular energy will become even greater; this would represent an increase in order for the system and a violation of the second law.

In comparison, the process that does occur naturally is the flow of energy from the hot object to the cold object. In this process, the difference in average molecular energy decreases; this represents a more random distribution of energy and an increase in disorder.

Exercise   Suppose that 8.00 J of energy is transferred from a hot object to a cold one. What is the net entropy change of the Universe?

Answer $+0.007$ 9 J/K.

Entropy Change in a Free Expansion

Let us again consider the adiabatic free expansion of a gas occupying an initial volume V_i (Fig. 22.16). A membrane separating the gas from an evacuated region is broken, and the gas expands (irreversibly) to a volume V_f. Let us find the changes in entropy of the gas and of the Universe during this process.

The process is clearly neither reversible nor quasi-static. The work done by the gas against the vacuum is zero, and because the walls are insulating, no energy is transferred by heat during the expansion. That is, $W = 0$ and $Q = 0$. Using the first law, we see that the change in internal energy is zero. Because the gas is ideal, E_{int} depends on temperature only, and we conclude that $\Delta T = 0$ or $T_i = T_f$.

To apply Equation 22.9, we cannot use $Q = 0$, the value for the irreversible process, but must instead find Q_r; that is, we must find an equivalent reversible path that shares the same initial and final states. A simple choice is an isothermal, reversible expansion in which the gas pushes slowly against a piston while energy enters the gas by heat from a reservoir to hold the temperature constant. Because T is constant in this process, Equation 22.9 gives

$$\Delta S = \int_i^f \frac{dQ_r}{T} = \frac{1}{T}\int_i^f dQ_r$$

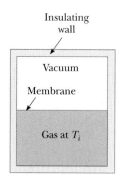

Figure 22.16 Adiabatic free expansion of a gas. When the membrane separating the gas from the evacuated region is ruptured, the gas expands freely and irreversibly. As a result, it occupies a greater final volume. The container is thermally insulated from its surroundings; thus, $Q = 0$.

For an isothermal process, the first law of thermodynamics specifies that $\int_i^f dQ_r$ is equal to the work done by the gas during the expansion from V_i to V_f, which is given by Equation 20.13. Using this result, we find that the entropy change for the gas is

$$\Delta S = nR \ln \frac{V_f}{V_i} \tag{22.13}$$

Because $V_f > V_i$, we conclude that ΔS is positive. This positive result indicates that both the entropy and the disorder of the gas increase as a result of the irreversible, adiabatic expansion.

Because the free expansion takes place in an insulated container, no energy is transferred by heat from the surroundings. (Remember that the isothermal, reversible expansion is only a *replacement* process that we use to calculate the entropy change for the gas; it is not the *actual* process.) Thus, the free expansion has no effect on the surroundings, and the entropy change of the surroundings is zero. Thus, the entropy change for the Universe is positive; this is consistent with the second law.

EXAMPLE 22.8   Free Expansion of a Gas

Calculate the change in entropy for a process in which 2.00 mol of an ideal gas undergoes a free expansion to three times its initial volume.

Solution   Using Equation 22.13 with $n = 2.00$ mol and $V_f = 3V_i$, we find that

$$\Delta S = nR \ln \frac{V_f}{V_i} = (2.00 \text{ mol})(8.31 \text{ J/mol·K})(\ln 3)$$

$$= \boxed{18.3 \text{ J/K}}$$

It is easy to see that the gas is more disordered after the expansion. Instead of being concentrated in a relatively small space, the molecules are scattered over a larger region.

Entropy Change in Calorimetric Processes

A substance of mass m_1, specific heat c_1, and initial temperature T_1 is placed in thermal contact with a second substance of mass m_2, specific heat c_2, and initial temperature $T_2 > T_1$. The two substances are contained in a calorimeter so that no energy is lost to the surroundings. The system of the two substances is allowed to reach thermal equilibrium. What is the total entropy change for the system?

First, let us calculate the final equilibrium temperature T_f. Using the techniques of Section 20.2—namely, Equation 20.5, $Q_{cold} = -Q_{hot}$, and Equation 20.4, $Q = mc\,\Delta T$, we obtain

$$m_1 c_1 \,\Delta T_1 = -m_2 c_2 \,\Delta T_2$$

$$m_1 c_1 (T_f - T_1) = -m_2 c_2 (T_f - T_2)$$

Solving for T_f, we have

$$T_f = \frac{m_1 c_1 T_1 + m_2 c_2 T_2}{m_1 c_1 + m_2 c_2} \tag{22.14}$$

The process is irreversible because the system goes through a series of non-equilibrium states. During such a transformation, the temperature of the system at any time is not well defined because different parts of the system have different temperatures. However, we can imagine that the hot substance at the initial temperature T_2 is slowly cooled to the temperature T_f as it comes into contact with a series of reservoirs differing infinitesimally in temperature, the first reservoir being at T_2 and the last being at T_f. Such a series of very small changes in temperature would approximate a reversible process. We imagine doing the same thing for the cold substance. Applying Equation 22.9 and noting that $dQ = mc\,dT$ for an infinitesimal change, we have

$$\Delta S = \int_1 \frac{dQ_{cold}}{T} + \int_2 \frac{dQ_{hot}}{T} = m_1 c_1 \int_{T_1}^{T_f} \frac{dT}{T} + m_2 c_2 \int_{T_2}^{T_f} \frac{dT}{T}$$

where we have assumed that the specific heats remain constant. Integrating, we find that

Change in entropy for a calorimetric process

$$\Delta S = m_1 c_1 \ln \frac{T_f}{T_1} + m_2 c_2 \ln \frac{T_f}{T_2} \tag{22.15}$$

where T_f is given by Equation 22.14. If Equation 22.14 is substituted into Equation 22.15, we can show that one of the terms in Equation 22.15 is always positive and the other is always negative. (You may want to verify this for yourself.) The positive term is always greater than the negative term, and this results in a positive value for ΔS. Thus, we conclude that the entropy of the Universe increases in this irreversible process.

Finally, you should note that Equation 22.15 is valid only when no mixing of different substances occurs, because a further entropy increase is associated with the increase in disorder during the mixing. If the substances are liquids or gases and mixing occurs, the result applies only if the two fluids are identical, as in the following example.

EXAMPLE 22.9   **Calculating ΔS for a Calorimetric Process**

Suppose that 1.00 kg of water at 0.00°C is mixed with an equal mass of water at 100°C. After equilibrium is reached, the mixture has a uniform temperature of 50.0°C. What is the change in entropy of the system?

Solution We can calculate the change in entropy from Equation 22.15 using the values $m_1 = m_2 = 1.00$ kg, $c_1 = c_2 = 4\,186$ J/kg·K, $T_1 = 273$ K, $T_2 = 373$ K, and $T_f = 323$ K:

$$\Delta S = m_1 c_1 \ln \frac{T_f}{T_1} + m_2 c_2 \ln \frac{T_f}{T_2}$$

$$= (1.00 \text{ kg})(4\,186 \text{ J/kg} \cdot \text{K}) \ln\left(\frac{323 \text{ K}}{273 \text{ K}}\right)$$

$$+ (1.00 \text{ kg})(4\,186 \text{ J/kg} \cdot \text{K}) \ln\left(\frac{323 \text{ K}}{373 \text{ K}}\right)$$

$$= 704 \text{ J/K} - 602 \text{ J/K} = \boxed{102 \text{ J/K}}$$

That is, as a result of this irreversible process, the increase in entropy of the cold water is greater than the decrease in entropy of the warm water. Consequently, the increase in entropy of the system is 102 J/K.

Optional Section

22.8   ENTROPY ON A MICROSCOPIC SCALE[4]

As we have seen, we can approach entropy by relying on macroscopic concepts and using parameters such as pressure and temperature. We can also treat entropy from a microscopic viewpoint through statistical analysis of molecular motions. We now use a microscopic model to investigate once again the free expansion of an ideal gas, which was discussed from a macroscopic point of view in the preceding section.

In the kinetic theory of gases, gas molecules are represented as particles moving randomly. Let us suppose that the gas is initially confined to a volume V_i, as shown in Figure 22.17a. When the partition separating V_i from a larger container is removed, the molecules eventually are distributed throughout the greater volume V_f (Fig. 22.17b). For a given uniform distribution of gas in the volume, there are a large number of equivalent microstates, and we can relate the entropy of the gas to the number of microstates corresponding to a given macrostate.

We count the number of microstates by considering the variety of molecular locations involved in the free expansion. The instant after the partition is removed (and before the molecules have had a chance to rush into the other half of the container), all the molecules are in the initial volume. We assume that each molecule occupies some microscopic volume V_m. The total number of possible locations of a single molecule in a macroscopic initial volume V_i is the ratio $w_i = V_i/V_m$, which is a huge number. We use w_i here to represent the number of *ways* that the molecule can be placed in the volume, or the number of microstates, which is equivalent to the number of available locations. We assume that the molecule's occupying each of these locations is equally probable.

As more molecules are added to the system, the number of possible ways that the molecules can be positioned in the volume multiplies. For example, in considering two molecules, for every possible placement of the first, all possible placements of the second are available. Thus, there are w_1 ways of locating the first molecule, and for each of these, there are w_2 ways of locating the second molecule. The total number of ways of locating the two molecules is $w_1 w_2$.

Neglecting the very small probability of having two molecules occupy the same location, each molecule may go into any of the V_i/V_m locations, and so the number of ways of locating N molecules in the volume becomes $W_i = w_i^N = (V_i/V_m)^N$ (W_i is not to be confused with work.) Similarly, when the volume is increased to V_f, the number of ways of locating N molecules increases to $W_f = w_f^N = (V_f/V_m)^N$ The ratio of the number of ways of placing the molecules in the volume for the

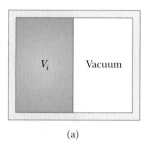

(a)

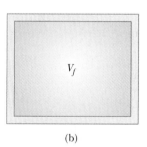

(b)

Figure 22.17 In a free expansion, the gas is allowed to expand into a region that was previously a vacuum.

[4] This section was adapted from A. Hudson and R. Nelson, *University Physics,* Philadelphia, Saunders College Publishing, 1990.

initial and final configurations is

$$\frac{W_f}{W_i} = \frac{(V_f/V_m)^N}{(V_i/V_m)^N} = \left(\frac{V_f}{V_i}\right)^N$$

If we now take the natural logarithm of this equation and multiply by Boltzmann's constant, we find that

$$k_B \ln\left(\frac{W_f}{W_i}\right) = nN_A k_B \ln\left(\frac{V_f}{V_i}\right)$$

where we have used the equality $N = nN_A$. We know from Equation 19.11 that $N_A k_B$ is the universal gas constant R; thus, we can write this equation as

$$k_B \ln W_f - k_B \ln W_i = nR \ln\left(\frac{V_f}{V_i}\right) \qquad \textbf{(22.16)}$$

From Equation 22.13 we know that when n mol of a gas undergoes a free expansion from V_i to V_f, the change in entropy is

$$S_f - S_i = nR \ln\left(\frac{V_f}{V_i}\right) \qquad \textbf{(22.17)}$$

Note that the right-hand sides of Equations 22.16 and 22.17 are identical. Thus, we make the following important connection between entropy and the number of microstates for a given macrostate:

| Entropy (microscopic definition) | $$S \equiv k_B \ln W \qquad \textbf{(22.18)}$$ |

The more microstates there are that correspond to a given macrostate, the greater is the entropy of that macrostate. As we have discussed previously, there are many more disordered microstates than ordered microstates. Thus, Equation 22.18 indicates mathematically that **entropy is a measure of microscopic disorder.** Although in our discussion we used the specific example of the free expansion of an ideal gas, a more rigorous development of the statistical interpretation of entropy would lead us to the same conclusion.

Imagine the container of gas depicted in Figure 22.18a as having all of its molecules traveling at speeds greater than the mean value on the left side and all of its molecules traveling at speeds less than the mean value on the right side (an ordered microstate). Compare this with the uniform mixture of fast- and slow-mov-

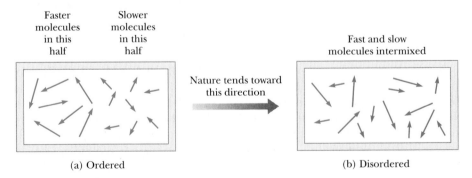

Faster molecules in this half   Slower molecules in this half

Nature tends toward this direction

Fast and slow molecules intermixed

(a) Ordered

(b) Disordered

Figure 22.18 A container of gas in two equally probable states of molecular motion. (a) An ordered arrangement, which is one of a few and therefore a collectively unlikely set. (b) A disordered arrangement, which is one of many and therefore a collectively likely set.

Figure 22.19 By tossing a coin into a jar, the carnival-goer can win the fish in the jar. It is more likely that the coin will land in a jar containing a goldfish than in the one containing the black fish.

ing molecules in Figure 22.18b (a disordered microstate). You might expect the ordered microstate to be very unlikely because random motions tend to mix the slow- and fast-moving molecules uniformly. Yet *individually* each of these microstates is equally probable. However, there are far more disordered microstates than ordered microstates, and so a macrostate corresponding to a large number of equivalent disordered microstates is much more probable than a macrostate corresponding to a small number of equivalent ordered microstates.

Figure 22.19 shows a real-world example of this concept. There are two possible macrostates for the carnival game—winning a goldfish and winning a black fish. Because only one jar in the array of jars contains a black fish, only one possible microstate corresponds to the macrostate of winning a black fish. A large number of microstates are described by the coin's falling into a jar containing a goldfish. Thus, for the macrostate of winning a goldfish, there are many equivalent microstates. As a result, the probability of winning a goldfish is much greater than the probability of winning a black fish. If there are 24 goldfish and 1 black fish, the probability of winning the black fish is 1 in 25. This assumes that all microstates have the same probability, a situation that may not be quite true for the situation shown in Figure 22.19. If you are an accurate coin tosser and you are aiming for the edge of the array of jars, then the probability of the coin's landing in a jar near the edge is likely to be greater than the probability of its landing in a jar near the center.

Let us consider a similar type of probability problem for 100 molecules in a container. At any given moment, the probability of one molecule's being in the left part of the container shown in Figure 22.20a as a result of random motion is $\frac{1}{2}$. If there are two molecules, as shown in Figure 22.20b, the probability of both being in the left part is $(\frac{1}{2})^2$ or 1 in 4. If there are three molecules (Fig. 22.20c), the probability of all of them being in the left portion at the same moment is $(\frac{1}{2})^3$, or 1 in 8. For 100 independently moving molecules, the probability that the 50 fastest ones will be found in the left part at any moment is $(\frac{1}{2})^{50}$. Likewise, the probability that the remaining 50 slower molecules will be found in the right part at any moment is $(\frac{1}{2})^{50}$. Therefore, the probability of finding this fast-slow separation as a result of random motion is the product $(\frac{1}{2})^{50}(\frac{1}{2})^{50} = (\frac{1}{2})^{100}$, which corresponds to about 1 in 10^{30}. When this calculation is extrapolated from 100 molecules to the number in 1 mol of gas (6.02×10^{23}), the ordered arrangement is found to be *extremely* improbable!

Roll a pair of dice 100 times and record the total number of spots appearing on the dice for each throw. Which total comes up most frequently? Is this expected?

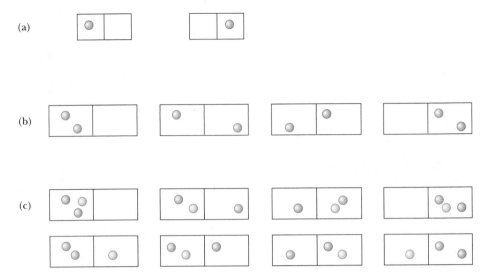

Figure 22.20   (a) One molecule in a two-sided container has a 1-in-2 chance of being on the left side. (b) Two molecules have a 1-in-4 chance of being on the left side at the same time. (c) Three molecules have a 1-in-8 chance of being on the left side at the same time.

EXAMPLE 22.10   Adiabatic Free Expansion — One Last Time

Let us verify that the macroscopic and microscopic approaches to the calculation of entropy lead to the same conclusion for the adiabatic free expansion of an ideal gas. Suppose that 1 mol of gas expands to four times its initial volume. As we have seen for this process, the initial and final temperatures are the same. (a) Using a macroscopic approach, calculate the entropy change for the gas. (b) Using statistical considerations, calculate the change in entropy for the gas and show that it agrees with the answer you obtained in part (a).

Solution   (a) Using Equation 22.13, we have

$$\Delta S = nR \ln\left(\frac{V_f}{V_i}\right) = (1)R \ln\left(\frac{4V_i}{V_i}\right) = \boxed{R \ln 4}$$

(b) The number of microstates available to a single molecule in the initial volume V_i is $w_i = V_i/V_m$. For 1 mol (N_A molecules), the number of available microstates is

$$W_i = w_i{}^{N_A} = \left(\frac{V_i}{V_m}\right)^{N_A}$$

The number of microstates for all N_A molecules in the final volume $V_f = 4V_i$ is

$$W_f = \left(\frac{V_f}{V_m}\right)^{N_A} = \left(\frac{4V_i}{V_m}\right)^{N_A}$$

Thus, the ratio of the number of final microstates to initial microstates is

$$\frac{W_f}{W_i} = 4^{N_A}$$

Using Equation 22.18, we obtain

$$\Delta S = k_B \ln W_f - k_B \ln W_i = k_B \ln\left(\frac{W_f}{W_i}\right)$$

$$= k_B \ln(4^{N_A}) = N_A k_B \ln 4 = \boxed{R \ln 4}$$

The answer is the same as that for part (a), which dealt with macroscopic parameters.

CONCEPTUAL EXAMPLE 22.11   Let's Play Marbles!

Suppose you have a bag of 100 marbles. Fifty of the marbles are red, and 50 are green. You are allowed to draw four marbles from the bag according to the following rules: Draw one marble, record its color, and return it to the bag. Then draw another marble. Continue this process until you have drawn and returned four marbles. What are the possible

macrostates for this set of events? What is the most likely macrostate? What is the least likely macrostate?

Solution   Because each marble is returned to the bag before the next one is drawn, the probability of drawing a red marble is always the same as the probability of drawing a green one. All the possible microstates and macrostates are shown in Table 22.1. As this table indicates, there is only one way to draw four red marbles, and so there is only one microstate. However, there are four possible microstates that correspond to the macrostate of one green marble and three red marbles; six microstates that correspond to two green marbles and two red marbles; four microstates that correspond to three green marbles and one red marble; and one microstate that corresponds to four green marbles. The most likely macrostate—two red marbles and two green marbles—corresponds to the most disordered microstates. The least likely macrostates—four red marbles or four green marbles—correspond to the most ordered microstates.

TABLE 22.1   Possible Results of Drawing Four Marbles from a Bag

Macrostate	Possible Microstates	Total Number of Microstates
All R	RRRR	1
1G, 3R	RRRG, RRGR, RGRR, GRRR	4
2G, 2R	RRGG, RGRG, GRRG, RGGR, GRGR, GGRR	6
3G, 1R	GGGR, GGRG, GRGG, RGGG	4
All G	GGGG	1

SUMMARY

A **heat engine** is a device that converts internal energy to other useful forms of energy. The net work done by a heat engine in carrying a working substance through a cyclic process ($\Delta E_{\text{int}} = 0$) is

$$W = Q_h - Q_c \qquad \textbf{(22.1)}$$

where Q_h is the energy absorbed from a hot reservoir and Q_c is the energy expelled to a cold reservoir.

The **thermal efficiency** e of a heat engine is

$$e = \frac{W}{Q_h} = 1 - \frac{Q_c}{Q_h} \qquad \textbf{(22.2)}$$

The **second law of thermodynamics** can be stated in the following two ways:

- It is impossible to construct a heat engine that, operating in a cycle, produces no effect other than the absorption of energy from a reservoir and the performance of an equal amount of work (the Kelvin–Planck statement).
- It is impossible to construct a cyclic machine whose sole effect is the continuous transfer of energy from one object to another object at a higher temperature without the input of energy by work (the Clausius statement).

In a **reversible** process, the system can be returned to its initial conditions along the same path shown on a PV diagram, and every point along this path is an equilibrium state. A process that does not satisfy these requirements is **irreversible. Carnot's theorem** states that no real heat engine operating (irreversibly) between the temperatures T_c and T_h can be more efficient than an engine operating reversibly in a Carnot cycle between the same two temperatures.

The **thermal efficiency** of a heat engine operating in the Carnot cycle is

$$e_{\text{C}} = 1 - \frac{T_c}{T_h} \qquad \textbf{(22.4)}$$

You should be able to use this equation (or an equivalent form involving a ratio of heats) to determine the maximum possible efficiency of any heat engine.

The second law of thermodynamics states that when real (irreversible) processes occur, the degree of disorder in the system plus the surroundings increases. When a process occurs in an isolated system, the state of the system becomes more disordered. The measure of disorder in a system is called **entropy** S. Thus, another way in which the second law can be stated is

- The entropy of the Universe increases in all real processes.

The **change in entropy** dS of a system during a process between two infinitesimally separated equilibrium states is

$$dS = \frac{dQ_r}{T} \tag{22.8}$$

where dQ_r is the energy transfer by heat for a reversible process that connects the initial and final states. The change in entropy of a system during an arbitrary process between an initial state and a final state is

$$\Delta S = \int_i^f \frac{dQ_r}{T} \tag{22.9}$$

The value of ΔS for the system is the same for all paths connecting the initial and final states. The change in entropy for a system undergoing any reversible, cyclic process is zero, and when such a process occurs, the entropy of the Universe remains constant.

From a microscopic viewpoint, entropy is defined as

$$S \equiv k_B \ln W \tag{22.18}$$

where k_B is Boltzmann's constant and W is the number of microstates available to the system for the existing macrostate. Because of the statistical tendency of systems to proceed toward states of greater probability and greater disorder, all natural processes are irreversible, and entropy increases. Thus, entropy is a measure of microscopic disorder.

QUESTIONS

1. Is it possible to convert internal energy to mechanical energy? Describe a process in which such a conversion occurs.
2. What are some factors that affect the efficiency of automobile engines?
3. In practical heat engines, which are we able to control more: the temperature of the hot reservoir, or the temperature of the cold reservoir? Explain.
4. A steam-driven turbine is one major component of an electric power plant. Why is it advantageous to have the temperature of the steam as high as possible?
5. Is it possible to construct a heat engine that creates no thermal pollution? What does this tell us about environmental considerations for an industrialized society?
6. Discuss three common examples of natural processes that involve an increase in entropy. Be sure to account for all parts of each system under consideration.
7. Discuss the change in entropy of a gas that expands (a) at constant temperature and (b) adiabatically.
8. In solar ponds constructed in Israel, the Sun's energy is concentrated near the bottom of a salty pond. With the proper layering of salt in the water, convection is prevented, and temperatures of 100°C may be reached. Can you estimate the maximum efficiency with which useful energy can be extracted from the pond?
9. The vortex tube (Fig. Q22.9) is a T-shaped device that takes in compressed air at 20 atm and 20°C and gives off air at −20°C from one flared end and air at 60°C from the other flared end. Does the operation of this device vi-

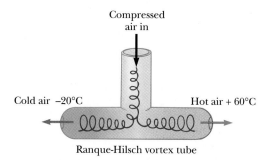

Compressed
air in

Cold air −20°C   Hot air + 60°C

Ranque-Hilsch vortex tube

Figure Q22.9

olate the second law of thermodynamics? If not, explain why not.

10. Why does your automobile burn more gas in winter than in summer?

11. Can a heat pump have a coefficient of performance (COP) less than unity? Explain.

12. Give some examples of irreversible processes that occur in nature.

13. Give an example of a process in nature that is nearly reversible.

14. A thermodynamic process occurs in which the entropy of a system changes by − 8.0 J/K. According to the second law of thermodynamics, what can you conclude about the entropy change of the environment?

15. If a supersaturated sugar solution is allowed to evaporate slowly, sugar crystals form in the container. Hence, sugar molecules go from a disordered form (in solution) to a highly ordered crystalline form. Does this process violate the second law of thermodynamics? Explain.

16. How could you increase the entropy of 1 mol of a metal that is at room temperature? How could you decrease its entropy?

17. A heat pump is to be installed in a region where the average outdoor temperature in the winter months is − 20°C. In view of this, why would it be advisable to place the outdoor compressor unit deep in the ground? Why are heat pumps not commonly used for heating in cold climates?

18. Suppose your roommate is "Mr. Clean" and tidies up your messy room after a big party. That is, your roommate is increasing order in the room. Does this represent a violation of the second law of thermodynamics?

19. Discuss the entropy changes that occur when you (a) bake a loaf of bread and (b) consume the bread.

20. The device shown in Figure Q22.20, which is called a thermoelectric converter, uses a series of semiconductor cells to convert internal energy to electrical energy. In the photograph on the left, both legs of the device are at the same temperature and no electrical energy is produced. However, when one leg is at a higher temperature than the other, as shown in the photograph on the right, electrical energy is produced as the device extracts energy from the hot reservoir and drives a small electric motor. (a) Why does the temperature differential produce electrical energy in this demonstration? (b) In what sense does this intriguing experiment demonstrate the second law of thermodynamics?

21. A classmate tells you that it is just as likely for all the air molecules in the room you are both in to be concentrated in one corner (with the rest of the room being a vacuum) as it is for the air molecules to be distributed uniformly about the room in their current state. Is this true? Why doesn't the situation he describes actually happen?

Figure Q22.20   (*Courtesy of PASCO Scientific Company*)

PROBLEMS

1, 2, 3 = straightforward, intermediate, challenging ☐ = full solution available in the *Student Solutions Manual and Study Guide*
WEB = solution posted at **http://www.saunderscollege.com/physics/** 🖥 = Computer useful in solving problem 📟 = Interactive Physics
☐ = paired numerical/symbolic problems

Section 22.1 Heat Engines and the Second Law of Thermodynamics
Section 22.2 Reversible and Irreversible Processes

1. A heat engine absorbs 360 J of energy and performs 25.0 J of work in each cycle. Find (a) the efficiency of the engine and (b) the energy expelled to the cold reservoir in each cycle.

2. The energy absorbed by an engine is three times greater than the work it performs. (a) What is its thermal efficiency? (b) What fraction of the energy absorbed is expelled to the cold reservoir?

3. A particular engine has a power output of 5.00 kW and an efficiency of 25.0%. Assuming that the engine expels 8 000 J of energy in each cycle, find (a) the energy absorbed in each cycle and (b) the time for each cycle.

4. A heat engine performs 200 J of work in each cycle and has an efficiency of 30.0%. For each cycle, how much energy is (a) absorbed and (b) expelled?

5. An ideal gas is compressed to half its original volume while its temperature is held constant. (a) If 1 000 J of energy is removed from the gas during the compression, how much work is done on the gas? (b) What is the change in the internal energy of the gas during the compression?

6. Suppose that a heat engine is connected to two energy reservoirs, one a pool of molten aluminum (660°C) and the other a block of solid mercury (−38.9°C). The engine runs by freezing 1.00 g of aluminum and melting 15.0 g of mercury during each cycle. The heat of fusion of aluminum is 3.97×10^5 J/kg; the heat of fusion of mercury is 1.18×10^4 J/kg. What is the efficiency of this engine?

Section 22.3 The Carnot Engine

7. One of the most efficient engines ever built (actual efficiency 42.0%) operates between 430°C and 1 870°C. (a) What is its maximum theoretical efficiency? (b) How much power does the engine deliver if it absorbs 1.40×10^5 J of energy each second from the hot reservoir?

8. A heat engine operating between 80.0°C and 200°C achieves 20.0% of the maximum possible efficiency. What energy input will enable the engine to perform 10.0 kJ of work?

9. A Carnot engine has a power output of 150 kW. The engine operates between two reservoirs at 20.0°C and 500°C. (a) How much energy does it absorb per hour? (b) How much energy is lost per hour in its exhaust?

10. A steam engine is operated in a cold climate where the exhaust temperature is 0°C. (a) Calculate the theoretical maximum efficiency of the engine, using an intake steam temperature of 100°C. (b) If superheated steam at 200°C were used instead, what would be the maximum possible efficiency?

11. An ideal gas is taken through a Carnot cycle. The isothermal expansion occurs at 250°C, and the isothermal compression takes place at 50.0°C. Assuming that the gas absorbs 1 200 J of energy from the hot reservoir during the isothermal expansion, find (a) the energy expelled to the cold reservoir in each cycle and (b) the net work done by the gas in each cycle.

12. The exhaust temperature of a Carnot heat engine is 300°C. What is the intake temperature if the efficiency of the engine is 30.0%?

13. A power plant operates at 32.0% efficiency during the summer when the sea water for cooling is at 20.0°C. The plant uses 350°C steam to drive turbines. Assuming that the plant's efficiency changes in the same proportion as the ideal efficiency, what would be the plant's efficiency in the winter, when the sea water is at 10.0°C?

14. Argon enters a turbine at a rate of 80.0 kg/min, a temperature of 800°C, and a pressure of 1.50 MPa. It expands adiabatically as it pushes on the turbine blades and exits at a pressure of 300 kPa. (a) Calculate its temperature at the time of exit. (b) Calculate the (maximum) power output of the turning turbine. (c) The turbine is one component of a model closed-cycle gas turbine engine. Calculate the maximum efficiency of the engine.

15. A power plant that would make use of the temperature gradient in the ocean has been proposed. The system is to operate between 5.00°C (water temperature at a depth of about 1 km) and 20.0°C (surface water temperature). (a) What is the maximum efficiency of such a system? (b) If the power output of the plant is 75.0 MW, how much energy is absorbed per hour? (c) In view of your answer to part (a), do you think such a system is worthwhile (considering that there is no charge for fuel)?

16. A 20.0%-efficient real engine is used to speed up a train from rest to 5.00 m/s. It is known that an ideal (Carnot) engine having the same cold and hot reservoirs would accelerate the same train from rest to a speed of 6.50 m/s using the same amount of fuel. Assuming that the engines use air at 300 K as a cold reservoir, find the temperature of the steam serving as the hot reservoir.

17. A firebox is at 750 K, and the ambient temperature is 300 K. The efficiency of a Carnot engine doing 150 J of work as it transports energy between these constant-temperature baths is 60.0%. The Carnot engine must absorb energy 150 J/0.600 = 250 J from the hot reser-

voir and release 100 J of energy into the environment. To follow Carnot's reasoning, suppose that some other heat engine S could have an efficiency of 70.0%. (a) Find the energy input and energy output of engine S as it does 150 J of work. (b) Let engine S operate as in part (a) and run the Carnot engine in reverse. Find the total energy the firebox puts out as both engines operate together and the total energy absorbed by the environment. Show that the Clausius statement of the second law of thermodynamics is violated. (c) Find the energy input and work output of engine S as it exhausts 100 J of energy. (d) Let engine S operate as in (c) and contribute 150 J of its work output to running the Carnot engine in reverse. Find the total energy that the firebox puts out as both engines operate together, the total work output, and the total energy absorbed by the environment. Show that the Kelvin–Planck statement of the second law is violated. Thus, our assumption about the efficiency of engine S must be false. (e) Let the engines operate together through one cycle as in part (d). Find the change in entropy of the Universe. Show that the entropy statement of the second law is violated.

18. At point A in a Carnot cycle, 2.34 mol of a monatomic ideal gas has a pressure of 1 400 kPa, a volume of 10.0 L, and a temperature of 720 K. It expands isothermally to point B, and then expands adiabatically to point C, where its volume is 24.0 L. An isothermal compression brings it to point D, where its new volume is 15.0 L. An adiabatic process returns the gas to point A. (a) Determine all the unknown pressures, volumes, and temperatures as you fill in the following table:

	P	V	T
A	1 400 kPa	10.0 L	720 K
B			
C		24.0 L	
D		15.0 L	

(b) Find the energy added by heat, the work done, and the change in internal energy for each of the following steps: $A \rightarrow B$, $B \rightarrow C$, $C \rightarrow D$, and $D \rightarrow A$. (c) Show that $W_{net}/Q_{in} = 1 - T_C/T_A$, the Carnot efficiency.

Section 22.4 Gasoline and Diesel Engines

19. In a cylinder of an automobile engine just after combustion, the gas is confined to a volume of 50.0 cm³ and has an initial pressure of 3.00×10^6 Pa. The piston moves outward to a final volume of 300 cm³, and the gas expands without energy loss by heat. (a) If $\gamma = 1.40$ for the gas, what is the final pressure? (b) How much work is done by the gas in expanding?

20. A gasoline engine has a compression ratio of 6.00 and uses a gas for which $\gamma = 1.40$. (a) What is the efficiency of the engine if it operates in an idealized Otto cycle?

(b) If the actual efficiency is 15.0%, what fraction of the fuel is wasted as a result of friction and energy losses by heat that could by avoided in a reversible engine? (Assume complete combustion of the air–fuel mixture.)

21. A 1.60-L gasoline engine with a compression ratio of 6.20 has a power output of 102 hp. Assuming that the engine operates in an idealized Otto cycle, find the energy absorbed and exhausted each second. Assume that the fuel–air mixture behaves like an ideal gas, with $\gamma = 1.40$.

22. The compression ratio of an Otto cycle, as shown in Figure 22.12, is $V_A/V_B = 8.00$. At the beginning A of the compression process, 500 cm³ of gas is at 100 kPa and 20.0°C. At the beginning of the adiabatic expansion, the temperature is $T_C = 750°C$. Model the working fluid as an ideal gas, with $E_{int} = nC_V T = 2.50nRT$ and $\gamma = 1.40$. (a) Fill in the following table to track the states of the gas:

	T (K)	P (kPa)	V (cm³)	E_{int}
A	293	100	500	
B				
C	1 023			
D				
A				

(b) Fill in the following table to track the processes:

	Q	W	ΔE_{int}
$A \rightarrow B$			
$B \rightarrow C$			
$C \rightarrow D$			
$D \rightarrow A$			
$ABCDA$			

(c) Identify the energy input Q_h, the energy exhaust Q_c, and the net output work W. (d) Calculate the thermal efficiency. (e) Find the number of revolutions per minute that the crankshaft must complete for a one-cylinder engine to have an output power of 1.00 kW = 1.34 hp. (*Hint:* The thermodynamic cycle involves four piston strokes.)

Section 22.5 Heat Pumps and Refrigerators

23. What is the coefficient of performance of a refrigerator that operates with Carnot efficiency between the temperatures − 3.00°C and + 27.0°C?

24. What is the maximum possible coefficient of performance of a heat pump that brings energy from outdoors at − 3.00°C into a 22.0°C house? (*Hint:* The heat pump does work W, which is also available to warm up the house.)

25. An ideal refrigerator or ideal heat pump is equivalent to a Carnot engine running in reverse. That is, energy Q_c is absorbed from a cold reservoir, and energy Q_h is rejected to a hot reservoir. (a) Show that the work that must be supplied to run the refrigerator or heat pump is

$$W = \frac{T_h - T_c}{T_c} Q_c$$

(b) Show that the coefficient of performance (COP) of the ideal refrigerator is

$$COP = \frac{T_c}{T_h - T_c}$$

26. A heat pump (Fig. P22.26) is essentially a heat engine run backward. It extracts energy from colder air outside and deposits it in a warmer room. Suppose that the ratio of the actual energy entering the room to the work done by the device's motor is 10.0% of the theoretical maximum ratio. Determine the energy entering the room per joule of work done by the motor when the inside temperature is 20.0°C and the outside temperature is − 5.00°C.

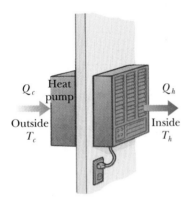

Figure P22.26

WEB 27. How much work does an ideal Carnot refrigerator require to remove 1.00 J of energy from helium at 4.00 K and reject this energy to a room-temperature (293-K) environment?

28. How much work does an ideal Carnot refrigerator require to remove energy Q from helium at T_c and reject this energy to a room-temperature environment at T_h?

29. A refrigerator has a coefficient of performance equal to 5.00. Assuming that the refrigerator absorbs 120 J of energy from a cold reservoir in each cycle, find (a) the work required in each cycle and (b) the energy expelled to the hot reservoir.

30. A refrigerator maintains a temperature of 0°C in the cold compartment with a room temperature of 25.0°C. It removes energy from the cold compartment at the rate 8 000 kJ/h. (a) What minimum power is required

to operate the refrigerator? (b) At what rate does the refrigerator exhaust energy into the room?

Section 22.6   Entropy

31. An ice tray contains 500 g of water at 0°C. Calculate the change in entropy of the water as it freezes slowly and completely at 0°C.

32. At a pressure of 1 atm, liquid helium boils at 4.20 K. The latent heat of vaporization is 20.5 kJ/kg. Determine the entropy change (per kilogram) of the helium resulting from vaporization.

33. Calculate the change in entropy of 250 g of water heated slowly from 20.0°C to 80.0°C. (*Hint:* Note that $dQ = mc \, dT$.)

34. An airtight freezer holds 2.50 mol of air at 25.0°C and 1.00 atm. The air is then cooled to − 18.0°C. (a) What is the change in entropy of the air if the volume is held constant? (b) What would the change be if the pressure were maintained at 1 atm during the cooling?

Section 22.7   Entropy Changes in Irreversible Processes

35. The temperature at the surface of the Sun is approximately 5 700 K, and the temperature at the surface of the Earth is approximately 290 K. What entropy change occurs when 1 000 J of energy is transferred by radiation from the Sun to the Earth?

36. A 1.00-kg iron horseshoe is taken from a furnace at 900°C and dropped into 4.00 kg of water at 10.0°C. Assuming that no energy is lost by heat to the surroundings, determine the total entropy change of the system (horseshoe and water).

WEB 37. A 1 500-kg car is moving at 20.0 m/s. The driver brakes to a stop. The brakes cool off to the temperature of the surrounding air, which is nearly constant at 20.0°C. What is the total entropy change?

38. How fast are you personally making the entropy of the Universe increase right now? Make an order-of-magnitude estimate, stating what quantities you take as data and the values you measure or estimate for them.

39. One mole of H_2 gas is contained in the left-hand side of the container shown in Figure P22.39, which has equal volumes left and right. The right-hand side is evacuated. When the valve is opened, the gas streams into the right-hand side. What is the final entropy change of the gas? Does the temperature of the gas change?

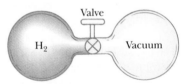

Figure P22.39

40. A rigid tank of small mass contains 40.0 g of argon, initially at 200°C and 100 kPa. The tank is placed into a reservoir at 0°C and is allowed to cool to thermal equilibrium. Calculate (a) the volume of the tank, (b) the change in internal energy of the argon, (c) the energy transferred by heat, (d) the change in entropy of the argon, and (e) the change in entropy of the constant-temperature bath.

41. A 2.00-L container has a center partition that divides it into two equal parts, as shown in Figure P22.41. The left-hand side contains H_2 gas, and the right-hand side contains O_2 gas. Both gases are at room temperature and at atmospheric pressure. The partition is removed, and the gases are allowed to mix. What is the entropy increase of the system?

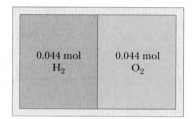

Figure P22.41

42. A 100 000-kg iceberg at $-5.00°C$ breaks away from the polar ice shelf and floats away into the ocean, at $5.00°C$. What is the final change in the entropy of the system after the iceberg has completely melted? (The specific heat of ice is 2010 J/kg · °C.)

43. One mole of an ideal monatomic gas, initially at a pressure of 1.00 atm and a volume of 0.025 0 m^3, is heated to a final state with a pressure of 2.00 atm and a volume of 0.040 0 m^3. Determine the change in entropy of the gas for this process.

44. One mole of a diatomic ideal gas, initially having pressure P and volume V, expands so as to have pressure $2P$ and volume $2V$. Determine the entropy change of the gas in the process.

(Optional)
Section 22.8   Entropy on a Microscopic Scale

45. If you toss two dice, what is the total number of ways in which you can obtain (a) a 12 and (b) a 7?

46. Prepare a table like Table 22.1 for the following occurrence. You toss four coins into the air simultaneously and then record the results of your tosses in terms of the numbers of heads and tails that result. For example, HHTH and HTHH are two possible ways in which three heads and one tail can be achieved. (a) On the basis of your table, what is the most probable result of a toss? In terms of entropy, (b) what is the most ordered state, and (c) what is the most disordered?

47. Repeat the procedure used to construct Table 22.1 (a) for the case in which you draw three marbles from your bag rather than four and (b) for the case in which you draw five rather than four.

ADDITIONAL PROBLEMS

48. Every second at Niagara Falls, some 5 000 m^3 of water falls a distance of 50.0 m (Fig. P22.48). What is the increase in entropy per second due to the falling water? (Assume that the mass of the surroundings is so great that its temperature and that of the water stay nearly constant at 20.0°C. Suppose that a negligible amount of water evaporates.)

Figure P22.48   Niagara Falls. (*Jan Kopec/Tony Stone Images*)

49. If a 35.0%-efficient Carnot heat engine is run in reverse so that it functions as a refrigerator, what would be the engine's (that is, the refrigerator's) coefficient of performance (COP)?

50. How much work does an ideal Carnot refrigerator use to change 0.500 kg of tap water at 10.0°C into ice at $-20.0°C$? Assume that the freezer compartment is held at $-20.0°C$ and that the refrigerator exhausts energy into a room at 20.0°C.

WEB 51. A house loses energy through the exterior walls and roof at a rate of 5 000 J/s = 5.00 kW when the interior temperature is 22.0°C and the outside temperature is $-5.00°C$. Calculate the electric power required to maintain the interior temperature at 22.0°C for the following two cases: (a) The electric power is used in electric resistance heaters (which convert all of the electricity supplied into internal energy). (b) The electric power is used to drive an electric motor that operates the compressor of a heat pump (which has a coefficient of performance [COP] equal to 60.0% of the Carnot-cycle value).

52. A heat engine operates between two reservoirs at $T_2 = 600$ K and $T_1 = 350$ K. It absorbs 1 000 J of energy from the higher-temperature reservoir and performs 250 J of work. Find (a) the entropy change of the Universe ΔS_U for this process and (b) the work W that could have been done by an ideal Carnot engine operating between these two reservoirs. (c) Show that the difference between the work done in parts (a) and (b) is $T_1 \Delta S_U$.

WEB 53. Figure P22.53 represents n mol of an ideal monatomic gas being taken through a cycle that consists of two isothermal processes at temperatures $3T_i$ and T_i and two constant-volume processes. For each cycle, determine,

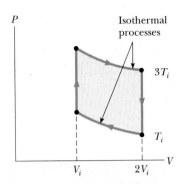

Figure P22.53

in terms of n, R, and T_i, (a) the net energy transferred by heat to the gas and (b) the efficiency of an engine operating in this cycle.

54. A refrigerator has a coefficient of performance (COP) of 3.00. The ice tray compartment is at $-20.0°C$, and the room temperature is $22.0°C$. The refrigerator can convert 30.0 g of water at $22.0°C$ to 30.0 g of ice at $-20.0°C$ each minute. What input power is required? Give your answer in watts.

55. An ideal (Carnot) freezer in a kitchen has a constant temperature of 260 K, while the air in the kitchen has a constant temperature of 300 K. Suppose that the insulation for the freezer is not perfect, such that some energy flows into the freezer at a rate of 0.150 W. Determine the average power that the freezer's motor needs to maintain the constant temperature in the freezer.

56. An electric power plant has an overall efficiency of 15.0%. The plant is to deliver 150 MW of power to a city, and its turbines use coal as the fuel. The burning coal produces steam, which drives the turbines. The steam is then condensed to water at $25.0°C$ as it passes through cooling coils in contact with river water. (a) How many metric tons of coal does the plant consume each day (1 metric ton = 10^3 kg)? (b) What is the total cost of the fuel per year if the delivered price is $8.00/metric ton? (c) If the river water is delivered at $20.0°C$, at what minimum rate must it flow over the cooling coils in order that its temperature not exceed $25.0°C$? (*Note:* The heat of combustion of coal is 33.0 kJ/g.)

57. A power plant, having a Carnot efficiency, produces 1 000 MW of electrical power from turbines that take in steam at 500 K and reject water at 300 K into a flowing river. Assuming that the water downstream is 6.00 K warmer due to the output of the power plant, determine the flow rate of the river.

58. A power plant, having a Carnot efficiency, produces electric power $\mathcal{P}$ from turbines that take in energy from steam at temperature T_h and discharge energy at temperature T_c through a heat exchanger into a flowing river. Assuming that the water downstream is warmer by ΔT due to the output of the power plant, determine the flow rate of the river.

59. An athlete whose mass is 70.0 kg drinks 16 oz (453.6 g) of refrigerated water. The water is at a temperature of $35.0°F$. (a) Neglecting the temperature change of her body that results from the water intake (that is, the body is regarded as a reservoir that is always at $98.6°F$), find the entropy increase of the entire system. (b) Assume that the entire body is cooled by the drink and that the average specific heat of a human is equal to the specific heat of liquid water. Neglecting any other energy transfers by heat and any metabolic energy release, find the athlete's temperature after she drinks the cold water, given an initial body temperature of $98.6°F$. Under *these* assumptions, what is the entropy increase of the entire system? Compare this result with the one you obtained in part (a).

60. One mole of an ideal monatomic gas is taken through the cycle shown in Figure P22.60. The process $A \rightarrow B$ is a reversible isothermal expansion. Calculate (a) the net work done by the gas, (b) the energy added to the gas, (c) the energy expelled by the gas, and (d) the efficiency of the cycle.

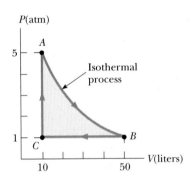

Figure P22.60

61. Calculate the increase in entropy of the Universe when you add 20.0 g of $5.00°C$ cream to 200 g of $60.0°C$ coffee. Assume that the specific heats of cream and coffee are both 4.20 J/g · °C.

62. In 1993 the federal government instituted a requirement that all room air conditioners sold in the United States must have an energy efficiency ratio (EER) of 10 or higher. The EER is defined as the ratio of the cooling capacity of the air conditioner, measured in Btu/h, to its electrical power requirement in watts. (a) Convert the EER of 10.0 to dimensionless form, using the conversion 1 Btu = 1 055 J. (b) What is the appropriate name for this dimensionless quantity? (c) In the 1970s it was common to find room air conditioners with EERs of 5 or lower. Compare the operating costs for 10 000-Btu/h air conditioners with EERs of 5.00 and 10.0 if each air conditioner were to operate for 1 500 h during the summer in a city where electricity costs 10.0¢ per kilowatt-hour.

63. One mole of a monatomic ideal gas is taken through the cycle shown in Figure P22.63. At point A, the pressure, volume, and temperature are P_i, V_i, and T_i, respectively. In terms of R and T_i, find (a) the total energy entering the system by heat per cycle, (b) the total energy leaving the system by heat per cycle, (c) the efficiency of an engine operating in this cycle, and (d) the efficiency of an engine operating in a Carnot cycle between the same temperature extremes.

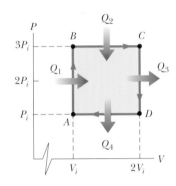

Figure P22.63

64. One mole of an ideal gas expands isothermally. (a) If the gas doubles its volume, show that the work of expansion is $W = RT \ln 2$. (b) Because the internal energy E_{int} of an ideal gas depends solely on its temperature, no change in E_{int} occurs during the expansion. It follows from the first law that the heat input to the gas during the expansion is equal to the energy output by work. Why does this conversion *not* violate the second law?

65. A system consisting of n mol of an ideal gas undergoes a reversible, *isobaric* process from a volume V_i to a volume $3V_i$. Calculate the change in entropy of the gas. (*Hint:* Imagine that the system goes from the initial state to the final state first along an isotherm and then along an adiabatic path—no change in entropy occurs along the adiabatic path.)

66. Suppose you are working in a patent office, and an inventor comes to you with the claim that her heat engine, which employs water as a working substance, has a thermodynamic efficiency of 0.61. She explains that it operates between energy reservoirs at 4°C and 0°C. It is a very complicated device, with many pistons, gears, and pulleys, and the cycle involves freezing and melting. Does her claim that $e = 0.61$ warrant serious consideration? Explain.

67. An idealized diesel engine operates in a cycle known as the *air-standard diesel cycle*, as shown in Figure 22.13. Fuel is sprayed into the cylinder at the point of maximum compression B. Combustion occurs during the expansion $B \rightarrow C$, which is approximated as an isobaric process. Show that the efficiency of an engine operating in this idealized diesel cycle is

$$e = 1 - \frac{1}{\gamma}\left(\frac{T_D - T_A}{T_C - T_B}\right)$$

68. One mole of an ideal gas ($\gamma = 1.40$) is carried through the Carnot cycle described in Figure 22.10. At point A, the pressure is 25.0 atm and the temperature is 600 K. At point C, the pressure is 1.00 atm and the temperature is 400 K. (a) Determine the pressures and volumes at points A, B, C, and D. (b) Calculate the net work done per cycle. (c) Determine the efficiency of an engine operating in this cycle.

69. A typical human has a mass of 70.0 kg and produces about 2 000 kcal (2.00×10^6 cal) of metabolic energy per day. (a) Find the rate of metabolic energy production in watts and in calories per hour. (b) If none of the metabolic energy were transferred out of the body, and the specific heat of the human body is 1.00 cal/g·°C, what is the rate at which body temperature would rise? Give your answer in degrees Celsius per hour and in degrees Fahrenheit per hour.

70. Suppose that 1.00 kg of water at 10.0°C is mixed with 1.00 kg of water at 30.0°C at constant pressure. When the mixture has reached equilibrium, (a) what is the final temperature? (b) Take $c_P = 4.19$ kJ/kg·K for water. Show that the entropy of the system increases by

$$\Delta S = 4.19 \ln\left[\left(\frac{293}{283}\right)\left(\frac{293}{303}\right)\right] \text{ kJ/K}$$

(c) Verify numerically that $\Delta S > 0$. (d) Is the mixing an irreversible process?

ANSWERS TO QUICK QUIZZES

22.1 The cost of heating your home decreases to 25% of the original cost. With electric heating, you receive the same amount of energy for heating your home as enters it by electricity. The COP of 4 for the heat pump means that you are receiving four times as much energy as the energy entering by electricity. With four times as much energy per unit of energy from electricity, you need only one-fourth as much electricity.

22.2 (b) Because the process is reversible and adiabatic, $Q_r = 0$; therefore, $\Delta S = 0$.

22.3 False. The second law states that the entropy of the *Universe* increases in real processes. Although the organization of molecules into ordered leaves and branches represents a decrease in entropy *of the tree,* this organization takes place because of a number of processes in which the tree interacts with its surroundings. If we include the entropy changes associated with all these processes, the entropy change of the Universe during the growth of a tree is still positive.

Standard Abbreviations and Symbols for Units

Symbol	Unit	Symbol	Unit
A	ampere	K	kelvin
Å	angstrom	kcal	kilocalorie
u	atomic mass unit	kg	kilogram
atm	atmosphere	kmol	kilomole
Btu	British thermal unit	L	liter
C	coulomb	lb	pound
°C	degree Celsius	m	meter
cal	calorie	min	minute
eV	electron volt	mol	mole
°F	degree Fahrenheit	N	newton
F	farad	Pa	pascal
ft	foot	rad	radian
G	gauss	rev	revolution
g	gram	s	second
H	henry	T	tesla
h	hour	V	volt
hp	horsepower	W	watt
Hz	hertz	Wb	weber
in.	inch	Ω	ohm
J	joule		

Mathematical Symbols Used in the Text and Their Meaning

Symbol	Meaning
$=$	is equal to
$\equiv$	is defined as
$\neq$	is not equal to
$\propto$	is proportional to
$\sim$	is on the order of
$>$	is greater than
$<$	is less than
$>>(<<)$	is much greater (less) than
$\approx$	is approximately equal to
Δx	the change in x
$\displaystyle\sum_{i=1}^{N} x_i$	the sum of all quantities x_i from $i = 1$ to $i = N$
$\lvert x \rvert$	the magnitude of x (always a nonnegative quantity)
$\Delta x \to 0$	Δx approaches zero
$\dfrac{dx}{dt}$	the derivative of x with respect to t
$\dfrac{\partial x}{\partial t}$	the partial derivative of x with respect to t
$\displaystyle\int$	integral